D1273000

Bridge Building in Wartime

Bridge Building in Wartime

Colonel Wesley Brainerd's Memoir of the 50th New York Volunteer Engineers

Edited by Ed Malles

Voices of the Civil War
Frank L. Byrne, Series Editor

The University of Tennessee Press
Knoxville

The Voices of the Civil War series makes available a variety of primary source materials that illuminate issues on the battlefield, the home front, and the western front, as well as other aspects of this historic era. The series contextualizes the personal accounts within the framework of the latest scholarship and expands established knowledge by offering new perspectives, new materials, and new voices.

Frontispiece. Major Wesley Brainerd. Courtesy of the William Hencken Family.

Copyright © 1997 by William E. Hencken.
All Rights Reserved. Manufactured in the United States of America.
First Edition.

The paper in this book meets the minimum requirements of the American National Standard for Permanence of Paper for Printed Library Materials.
♾ The binding materials have been chosen for strength and durability.
Printed on recycled paper.

Library of Congress Cataloging-in-Publication Data

Brainerd, Wesley, 1832–1910.
Bridge building in wartime : Colonel Wesley Brainerd's memoir of the 50th New York Volunteer Engineers / edited by Ed Malles.—1st ed.
p. cm.—(Voices of the Civil War series)
Includes bibliographical references and index.
ISBN 0-87049-977-7 (cloth: alk. paper)
1. Brainerd, Wesley, 1832–1910.
2. United States. Army. New York Infantry Regiment, 50th (1861–1865)
3. Soldiers—New York (State)—Biography.
4. Military engineers—New York (State)—Biography.
5. Bridges—United States—Design and construction—History—19th century.
6. United States—History—Civil War, 1861–1865—Personal narratives.
7. United States—History—Civil War, 1861–1865—Regimental histories.
8. United States—History—Civil War, 1861–1865—Engineering and construction.
9. New York (State)—History—Civil War, 1861–1865.
I. Malles, Ed. II. Title. III. Series.
E523.5 50th.B73 1997
973.7'447—dc21 96-45803
CIP

E
523.5
.B73
1997

Dedicated to Charlotte Phillipson Hencken,
Col. Wesley Brainerd's great-granddaughter,
the good and faithful steward of his legacy.

Contents

Illustrations

Foreword

This volume is more personal than the usual Civil War memoir. Its author expected it to be read ultimately by his son, still a very young child at the time the memoir was originally composed. Thus, Brainerd frequently addresses the boy directly, offering fatherly advice and suggestions. Besides lending a human quality to his text, this personal style gives assurance that the writer did not intend his words for publication. As a result, he is much more candid than most writers of reminiscences about his observations on fellow soldiers of all ranks. This work is by a businessman who, after his secondary education, became an engineer and a mechanic through apprenticeship. When the Civil War came, Brainerd was qualified to lead the men who did the construction essential to the movement and defense of the armies. His voice speaks for a largely overlooked element among Civil War soldiers.

The Brainerd memoir contains much of the history of the 50th New York Volunteer Engineers and some history of the 15th New York Engineers. Brainerd shows that his unit received training as infantry and sometimes served in that capacity (and his prewar militia experience gave him an advantage from the start). But the unit's main training was in such strictly engineering duties as bridge building and the construction of field fortifications in whole and part. Brainerd indicates that most of his fellow officers had some background in such work, as did many of the enlisted men. As he explains, in return for their professional skills they received such perquisites as extra pay, exemption from general picket duty, and the morbid privilege of burying their dead in wooden coffins.

Brainerd and his regiments began their serious service with bridge building during the Peninsula Campaign. Later, prior to the Battle of Fredericksburg, Brainerd conducted a reconnaissance of possible crossing points below the town, only to find himself ordered to take part in building one of the bridges directly under enemy fire. He graphically describes the horrors of the attempt; he was wounded (while, poignantly, his little daughter was almost simultaneously dying at home). As the Gettysburg Campaign began, he built a bridge across the Rappahannock in the often overlooked operation at Deep Run (Franklin's Crossing). Perhaps Brainerd's most moving writing describes Grant's Overland Campaign of 1864. At the Wilderness, when the engineers were about to be thrown in as desperately needed reinforcements, he describes quite frankly how brave men dropped out of the ranks "and stepped to the rear to attend to the demands of nature." Feeling that other descriptions of that battle did not convey its reality, he recalled with only slight exaggeration, "Two hundred thousand men, inspired with the desperation of demons, were fighting in a wilderness of fire." Of course, he also tells of the engineering work on which he and his men spent much of the campaign. Unfortunately, he did not participate in building the great bridge on which most of the Army of the Potomac crossed the James River, and his memoir ends abruptly at the approach to Petersburg.

Editor Ed Malles, having skillfully supplemented Brainerd's memoir, supplies a conclusion based on primary sources to the then colonel's military service. In his Epilogue, he also shows how Brainerd's engineering experience fit into a quite successful postwar career. The words which Brainerd originally intended only for his son should appeal to many people of subsequent generations.

Frank L. Byrne
Kent State University

Editor's Acknowledgments

I wish to thank the Hencken family for giving me the opportunity to bring this important work to print. Charlotte Hencken, Wesley Brainerd's great granddaughter, her son William Hencken and his brothers Brainerd and Jonathan and sisters Lesley and Meredith have given me every bit of support I asked for. Without their belief in this project it would never have been accomplished.

Fay Chandler has acted as transcriber, compiler, proofreader, researcher, assistant, and dauntless worker in the trenches of this project, and to her go my heartfelt thanks.

Virginia Van Dreal Herrmann in Rome, New York, helped with research from that area. Capt. Stanley Lechner of the 15th Regiment New York Volunteer Engineers gave me invaluable help in technical matters. William A. Evans of the New York State Archives was helpful in the initial stages of research. Dale Floyd, then in the Army Corps of Engineers History Office, has helped throughout the project, especially with research at the National Archives and in obtaining photographs. Carl Patin did journeyman work on geneology questions, Fred DiFilippis fearlessly piloted us through cold and storms to an important meeting at Gettysburg, and the Florida Center for Civil War Studies research library provided some important research material.

Others who helped include Richard David Brainard, genealogy research; Barton Baldwin; Fritz Updike, New York research; Mrs. Walter L. Cherry, special assistance; Dennis Ogden; Sara Dunlap; David V. Wendell, Rosehill historian; Cpl. Jerry Feinstein and the

1st Michigan Engineers & Mechanics; Kenneth M. Gambrill, biographical research; Robert and Gertrude Pasquesi, Chicago; Ralph Valentine, head of music, Choate Rosemary Hall School; and William J. Spitler.

Cyprus Amax Minerals Company of Englewood, Colorado, gave important special assistance.

I never cease to be amazed at the generosity of my colleagues, especially those on the H-Civil War list on the Internet. Special thanks to Matthew Gilmore, District of Columbia Public Library, Washingtoniana Division; Greg Urwin, University of Central Arkansas; Gus Seligman and Richard Lowe, University of North Texas; Mary L. White, Cornell University; Kerry Webb, National Library of Australia; Lynne DeMont, Portland State University Library; Mary Francis Hodges, University of Arkansas at Little Rock; Donald C. Pfanz, Staff Historian, Fredericksburg and Spotsylvania NMP; and Ron Sellers, Manuscript Division, Library of Congress.

Thanks to you all. I hope I have accomplished something that you can be proud of.

Editor's Introduction

Wesley Brainerd was born in Rome, New York, on September 27, 1832, the son of Alexander Hamilton Brainerd (1806–1879) and Mary Merritt Gouge (1808–1841). The Brainerd family can trace its American roots back to Daniel Brainerd, a settler in Haddam, Connecticut, in the early seventeenth century. Wesley's grandfather, Jeremiah, moved his family from there to Rome, where he became a contractor on the Erie Canal.[1]

Alexander Hamilton Brainerd was an industrious and successful civil engineer and railroad contractor on the Hudson River Railroad, building all the bridges on that road during the period 1848–50. He also operated railroad car manufacturing shops in Niagara, Canada, and iron mills in St. Albans, Vermont. He retired from business to Rome in the late 1870s, where he died. Mary Gouge, Wesley's mother, was the descendant of French Huguenots who settled in Trenton Falls, New York. She died at thirty-two when Wesley was nine. Wesley was the only child of his parents to survive into adulthood.[2]

Wesley was educated at the Rome Academy during the 1840s, and, at fifteen, he accompanied his father on the Hudson River Railroad contract, serving as his assistant. In 1850, Wesley Brainerd became an apprentice at the Norris Locomotive Works in Philadelphia, Pennsylvania, and finished for the trade of draughtsman. At the time, Norris was in its heyday. In business from 1833 to 1867, the company built twelve hundred locomotives for the worldwide market and was a pioneer of innovations in locomotive design and manufacture.[3]

Wesley continued with Norris as a member of locomotive starter crews throughout the United States and Canada until early 1858, when he took a position as master mechanic of the Georgia Railroad and Banking Company. This position was short-lived, however, and he returned to Rome. He married Amelia Maria Gage in Chicago on November 17, 1858, and they resided in Rome, where Wesley went into business, at first with his father and then on his own.[4]

The Civil War interrupted Wesley's plans for the future, and his memoir is the story of his service in the 50th New York Volunteer Engineers and covers the time from the beginning of the war to mid-June 1864, as the Army of the Potomac besieged Petersburg. The memoir ends abruptly, with the final sentence trailing up the margin of the page.

It is tantalizing to speculate that the rest of the narrative—that part dealing with the last year of the war—might be extant somewhere, but we have no concrete evidence of any part still undiscovered. As Wesley occasionally put the work down for long periods of time, we can only surmise that he simply put it away, meaning to return to it, but did not.

The surviving memoir was handed down in the family as a treasured heirloom. It was taken out at special times and shown to the children, who were allowed to look at but not touch it. It was protected in its original binding and wrapped in cloth. As this binding deteriorated with age, it became necessary to rebind the book, and this was done in 1976.

Charlotte Phillipson Hencken, great-granddaughter of Wesley Brainerd, handed on the stewardship of the memoir to her son, William Evans Hencken. In 1990 William Hencken brought the memoir to me for evaluation and the project to bring this unique and valuable work to publication began.

Colonel Wesley Brainerd began his memoir on December 11, 1870, writing it to his son, Irving, while the family resided in Evanston, Illinois. He wrote it on identical, ruled, sturdy business journal sheets measuring 12.5 inches tall by 7.75 inches wide. The memoir is in remarkably good shape, evidence of the care with which it has been handled by Wesley Brainerd's descendants. As the paper on which is was written was produced prior to the advent of modern papermaking techniques, there is no fading or yellowing of the pages. They seem bright and new and are not brittle or delicate.

Colonel Brainerd was a naturally curious young man and traveled throughout the eastern and southern United States and into Canada before the war. He referred to books, periodicals, and poems that an educated man would have been expected to know. He became a businessman and a mechanical engineer, versed in technical subjects as well as the humanities. He occasionally used French phrases in his narrative.

There is evidence that Colonel Brainerd wrote his memoir from a wartime diary. He mentioned such a source on text page 224 of the memoir. Since I do not have the original diary, I cannot know how much of the memoir is a literal transcription from the diary and how much is based on memory aided by hindsight. The existence of a wartime diary would explain the chronological structure of the memoir and the fresh and immediate quality of some of the passages. The text is well organized, and it is obvious that Colonel Brainerd wanted the narrative to flow freely and was conscious of his role as author. The memoir was written in an intimate, conversational tone and is full of fatherly advice to his son.

Colonel Brainerd does not mention a motivation for writing his memoir other than as instruction for his son, Irving. Because of the many sharp opinions offered in the text, it might be inferred that the Colonel did not write his memoir to be published, at least not while he was still alive. It is clear that Brainerd was a "McClellan Man" and probably a War Democrat from his comments in the text. Colonel Brainerd considered himself a patriot and the Confederates traitors.

As it was not written for publication it must be considered a rough, or first, draft. If the Colonel had been prevailed upon to publish his manuscript, I am sure that he would have carefully prepared it for submission to a publisher, going through much of the editing process I have completed.

The body text was handwritten with pen and ink and the penmanship varies widely in quality. There are times when it is obvious that Colonel Brainerd was tired or rushed, perhaps writing into the night, and other times when he was relaxed and concentrating fully on the task at hand. A few written words in the narrative are impossible to decipher, while some word meanings can be clearly deduced by usage and context.

Colonel Brainerd's handwriting is very personal in style. He did not close his a's, g's, or o's, making it difficult to separate them from u's, y's, or n's. He had two different styles of r's. He capitalized in the midst of sentences, adhering to his own rule on this, and tended to run certain different letters together in a series of identical strokes. Ascenders and descenders are sometimes not reliable when trying to decipher a word. Colonel Brainerd underlined in a puzzling fashion, both for emphasis and for seemingly no reason.

Colonel Brainerd's punctuation is even more eccentric than his handwriting. He used dashes, double-dashes, and indecipherable strokes in place of commas, periods, colons, and semicolons. When he did use commas, they are not always in the right place. He used quotation marks on a wide variety of words and phrases, sometimes not completing the pair. He did not use a single apostrophe in the entire memoir.

There are instances throughout the document where there are erratic and unintelligible marks above and below the line, sometimes three or four in a sentence. Fay Chandler, the transcriber, and I have concluded that these are not punctuation marks but evidence of Brainerd's habit of nervously tapping the nib of his pen on the paper while writing. Under magnification, these marks can be seen to "drag" in many different directions, further evidence of their accidental quality. Most of these marks occur where no punctuation is called for.

Colonel Brainerd's spelling is idiomatic. It is obvious that he applied some rules incorrectly and, when in doubt, spelled a word phonetically. He regularly used double consonants in some places where a single consonant should have been used. He regularly used "ley" instead of "ly" as a suffix. He used different spellings for the same word in different places. Sometimes his handwriting makes it impossible to figure out the exact spelling of a word whose meaning is quite clear. There are times when a word is written clearly but does not seem to fit in context. Some of the misspellings are consistent while others are not. He employed his own abbreviations and shorthand, including an abundance of this symbol, "&c."

He used nineteenth-century grammar and wrote in a free-flowing form. A few clauses in sentences were left dangling or were misplaced, forcing the reader to reread the sentence several times to plumb its meaning. He also wrote long paragraphs that might contain two, three, or more separate thoughts.

While Colonel Brainerd's handwriting, idiosyncratic punctuation, spelling, and grammar made the transcription process somewhat of a chore, it did not materially affect the overall comprehension of the meaning of the text. He wrote in a clear, straightforward manner, and the memoir was not difficult to edit from a readability standpoint.

There is some evidence that Colonel Brainerd returned to the memoir to edit the content. Although there is little marginalia, there are blanks filled in with place names or numbers or insertions written with pencil rather than the pen and ink of the rest of the copy. It appears that Colonel Brainerd returned to the manuscript and filled in facts after some research. Other than these few instances of later additions, the memoir appears to have survived as originally written. There is no evidence of later tampering by another party.

I began this project with a full understanding of the wide and diverse range of responsibilities to Colonel Brainerd, to the scholars who will use the published edition of his memoir, and to the general reader of American and Civil War history.

While I did not feel that the Brainerd memoir rated a facsimile edition or a diplomatic transcription, I did believe that the document was certainly important enough to be carefully edited. My goal was to prepare an expanded edition that was easy to read, that scrupulously preserved the meanings and desires of the author, and that was presented in a form as close as possible to the author's "voice" and would be valuable to scholars of all persuasions. This preparation was, for the most part, enjoyable, as I came to know Wesley Brainerd quite well and found him likeable.

I carefully applied modern rules of capitalization and punctuation only where such changes were deemed necessary for clarity. I corrected difficult sentences or phrases for the same reason. This was done with the strictest adherence to the principle that Colonel Brainerd's original meaning and voice would not in any way be abridged or changed.

Stray, problematic marks were considered on an individual basis with each mark's value to the meaning of the text being the deciding factor determining its retention.

Where spelling reflected the practice of the time during which the memoir was written or where placenames were spelled as they were commonly spelled then, these spellings were kept intact and were not followed with [*sic*]. Where poor handwriting obscured the exact spelling of a word but the word itself was clear, the correct spelling of the word was used. Brainerd's consistent misspellings were retained and [*sic*] was not added.

The word *ponton/pontoon* presented a problem. The latter spelling was widely used during the Civil War and is correct even now, but the single-o spelling, the French spelling, is the official form found in U.S. Army publications. Both spellings were used interchangeably by Wesley Brainerd, by newspaper writers, and by official sources. To avoid undue confusion, I have used the single-o spelling in the text, as Wesley Brainerd was a former army officer writing of his army experiences. In all other instances I have allowed the spelling to remain as it was in the original source, be it newspaper article, official document, or letter.

Where clauses were awkwardly placed in sentences, they were placed where clarity demanded. Colonel Brainerd's underlining was retained where it added emphasis and appears in italic in this work. Colonel Brainerd occasionally constructed sentences without articles and sometimes without verbs. Where it was clear what the missing word needed to be, it was inserted within brackets. All abbreviations were retained as written. Many abbreviations were written without periods and these were added silently.

There were occasional snippets of dialogue in the text and these have been placed into modern form for the sake of clarity.

Where paragraphs contained a confusing number of thoughts, they were broken up into smaller paragraphs for a more logical movement through the text. Where paragraphs needed to be combined for clarity, they were.

Where a famous general, politician or personage was mentioned, few notes were added. Where a person whom the reader would not be expected to know was mentioned, a note was made according to the information available. As records of Civil War soldiers are readily available to scholars but not always to the general reader, service records of most soldiers who were mentioned in the text were included in Appendix 1 at the end of the narrative without reference in the text or in a note.

Where a historical event was mentioned, care was taken to temper the desire to annotate. I did so only where clarification was necessary or a discrepancy existed. Brainerd mentioned many historical events without including dates. In these cases, I added a note that clarified the date of historical events that Brainerd described or sorted out discrepancies to keep the narrative in chronological context.

Where Colonel Brainerd expressed an opinion on a historical event and a note helped to explain that opinion, such an annotation was made. Otherwise, Brainerd's opinions stand on their own without undue intrusion.

Where large and well-documented campaigns or battles were mentioned an annotation of the overall strategic concerns was included for clarity and explanation. Small incidents in the war were annotated when such information helped place the incident in a larger context and therefore helped the reader's understanding of the event.

Colonel Brainerd included a large amount of supporting material—180 separate items—in his memoir. This material places Wesley Brainerd and the people and events in his memoir firmly in a time and place in our country's history. I have retained some of the supporting material in the text when I felt that it played an important role in the narrative and it should be read as an integral part of the memoir. Other supporting material was placed in appendices. Appendix 1 includes service records; Appendix 2, newspaper articles; Appendix 3, letters; Appendix 4, orders, official documents, and miscellaneous items. Only the most important examples of Brainerd's collection of supporting material was included, as much of the material proved to be redundant with Brainerd's own comments.

Throughout the editing process I took pains to insure that the finished product would be acceptable to Colonel Brainerd. He spent many hours on the writing and organizing of his memoir and I could be no less careful or committed in the editing.

Bridge Building in Wartime

Colonel Wesley Brainerd's Memoir of the 50th New York Volunteer Engineers

Chapter One

My Dear Son

Evanston Ill.
December 11th, 1870.

My Dear Son,

Once upon a time I read in some newspaper, somewhere, that the history of our great war, known as the "War of the Rebellion of 1861" had not yet been written and that it never would be until someone would write it as a Father would relate the story to his son. Now, my boy, I do not propose to write a book, but it has occured to me that as you grow older you would naturally seek to become familiar with the history of your country and should you ever arrive to man's estate, you will, I doubt not, be interested in the perusal of a memorandum written by your own Father for your especial benefit. If you find instruction and amusement in the perusal of these pages, should it serve to employ now and then a leisure hour when you might otherwise be engaged in some pastime not so profitable and above all should these pages serve to add to your love of country as a knowledge of the sacrifices made to obtain the priviledges which you now enjoy should do and make of you a good *citizen*, I will be satisfied.

Long before these pages are presented to you, you will be told that your Father served as a Soldier in the War of the Rebellion. If you possess the spirit of your family you will experience a proper degree of pride in the fact that he did so, and that for four long years, or during the entire war, he was never known to do an act unbecoming an Officer and a Gentleman. This reflection may console you for the lack of a brilliant reputation gained by many during that war and which he did not achieve. By the term "brilliant" I might better say "national," for his reputation was brilliant enough in his sphere. This without egotism he can say: that it was his constant aim to do his duty fully, not doubting but that his reputation would take care of itself.

> "Honor and fame from no condition rise
> Act full your part, there all the virtue lies."[1]

I suppose that little couplet passed through my mind at least ten thousand times a year for four years.

You will read the published histories of the war and will inform yourself of the causes which finally led to a resort to arms. You will learn how the separation of the States was brought about and by whom. You will read of the movements of great Armies from one end of the land to the other, of great battles, great victories and disastrous defeats, the gloom and despondency and the triumphant exultation which from time to time agitated the nation during all of that tremendous struggle, and yet you will be able to see but the outside, the framing to the picture. The picture itself must be *seen* to be fully appreciated. I am not a writer and cannot therefore pen pictures to dazzle your imagination. Nor would it be appropriate for me to do so. A simple statement of some of the more prominent features of events as they occured in my experience is more becoming. May our kind Father in Heaven spare you and my beloved country a repetition of those scenes.

The year 1861 found me, after a somewhat checquered life, engaged in business in Rome, New York, *settled*, as I supposed, for life. Your Mother and I were married in Chicago on the 17th of November, 1858. In order that you may have no doubt about this I commence by pasting in the marriage notice.

> MARRIED:
>
> At Chicago, on Wednesday, November 17th, by Rev. Philo Judson, Mr. WESLEY BRAINARD, of Rome, to Miss AMELIA M., daughter of Eli A. Gage, of the former place, formerly of Rome.
>
> [*Roman Citizen*, Wednesday, Nov. 24, 1858]

Mr. Sandford, the editor of the Roman Citizen, had been a personal friend to us both and was formerley my Sunday School teacher.[2] When I sent the notice of our marriage to him, I made the request that he would not mention the circumstances of the *fee* which accompanied the notice, as I disliked the custom. The result of my request appears opposite.

> Not a word.—the marriage notice of Mr. WESLEY BRAINARD, in another column, was accompanied with a handsome cash fee,—which we were requested not to mention, and shall therefore say nothing about it.—We may say, however, that the bride formerly resided in this village, and will be joyfully welcomed home by a host of friends, who parted with her so sorrowfully a few years since when she left with her parents, for their new home in the west.
>
> [*Roman Citizen*, Wednesday, Nov. 24, 1858]

My Father [Alexander Hamilton Brainerd] had closed out his business at Niagara, Canada West, had married for the third time and was anxious to have me join him in business at our native place. On his leaving Niagara, the paper there published the following note.

> Our respected fellow townsman, A.H. Brainerd Esq., has retired from his connection with the Car Factory in this town.—Mr. Brainerd has been the managing partner of the Car works since their establishment; and which he has conducted with success and satisfaction to all.
>
> We clip the above paragraph from "The Mail" published at Niagara, Canada West. It has reference to A.H. Brainerd Esq., formerly of this town, but who has for nearly ten years past been transacting business in Canada. Mr. Brainerd there, as here, has by his energy of character, generosity, and gentlemanly treatment which he bestows upon all with whom he is brought in contact, made a host of friends.
>
> We are happy to learn that he has been successful, and that having disposed of his interest at Niagara, there is some possibility of his again making Rome his permanent residence. Should he decide to do so, he will be warmly welcomed by all our citizens.
>
> [*Roman Citizen*, date unknown]

and the "Roman Citizen" [date unknown] the notice herein pasted.

> ENCOURAGING.
>
> We learned that the property commonly known as the Plaining Mill and Sash and Door Factory, formerly owned by MICHAEL BURNS, has been purchased by A.H. Brainard Esq., of this village, who in con-

nection with his son WESLEY BRAINARD, intends setting the machinery immediately in motion. We are glad that this is the case. Mr. B. is a gentleman of abundant means, and has a large business experience, and cannot fail to make the enterprise a profitable one.

The building is already well stocked with tools and machinery, and we learn that other facilities will be added, preparatory to the manufacture of Rail Road Cars on an extensive scale, which is to constitute one branch of the new business.

Manufacturing of any description, is of the highest importance to our village, and we hope this move on the part of Messrs. Brainard, is but the harbinger of better times for our village, in this respect.

We did not intend to manufacture RR Cars as mentioned above but confined its business to Plaining & Sawing, Turning &c. We also manufactured Bedsteads & Fanning Mills,[3] but after a years trial the business did not prove remunerative and I entered into an arrangement with my Father to convert the building into a Grist Mill and conduct the business alone. This was accordingly accomplished and I had the little mill with two run of stone fairly under way with a prospect of making a fair living when the war broke upon us. I will not say that it broke upon us like a thunderbolt out of a clear sky for such was not the case. We had heard its muttering afar off and for years it had been threatening. At last it came.

Your Mother and I were keeping house in a brick building not more than half a square from my place of business. We kept no servant because we could not afford to do so and our luxuries were limited I assure you. Still, where love dwells there you find happiness, and we were about as happy in our unpretending little house as the most of people.

I have found since then, my son, that money is a very desirable thing to have and that it contributes very naturally to ones happiness. But I have also observed that money of itself does not bring peace of mind or contentment of spirit. A cheerful disposition, combined with cultivated and refined tastes [and] good health are more highly to be prized than any thing else that this world affords.

Soon after we had settled in Rome a military company was organized of which I was made 1st Lieutenant. This company comprised a portion of the 46th Regiment of N.Y. State Militia and was lettered "A". We were called the Gansevoort Light Guard in honor of a distin-

Amelia Maria Gage Brainerd. Courtesy of the William Hencken Family.

guished Patriot of the Revolutionary War and the defender of old Fort Stanwix.[4]

We were a fancy dress Company of holiday Soldiers comprised chiefly of young men fond of show and with much leisure time on hand. It was customary for us to parade upon all possible occasions—4th of July, Washingtons Birthday (22d Feb), Thanksgiving Day, St. Patricks Day, *our* Annual Day, at picnics, excursions &c. For my part, I joined the company out of pure love for the science of military manoeuvers. I had always possessed a military spirit from my youth and never shall I forget my disappointment when, at the age of 14, I had placed my heart upon going to West Point [and] my Father ridiculed me out of the notion. Who can tell what I *might have been* had my wish at that time been gratified? The subsequent opportunities do not arise but once during a generation.

There was but one draw back to my happiness in connexion with this Company and that was the aversion with which your Mother regarded it. The drills one or two evenings a week kept me away from my home and she was, of course, loneley. Besides this, there was an ill defined dread of *something* in the future, a dread that she could not overcome. How well did subsequent events explain it all.

On one occasion we visited Watertown, N.Y., by invitation of the

Company there. After the drill we had a dinner and then a ball given in our honor. At the dinner sat the famous Captain Hollins, he, the *Hero of Greytown*, as he was called. He sat at the head of the table dressed in the full uniform of a Captain (as he was) of the United States Navy. I sat at his left and Capt. Skillen of our company at his right. He was called upon to respond to the toast "The Navy." His expressions of loyalty elicited the hearty applause of the entire company and there was a large crowd present besides members of the two companies. "I am a Sailor," said he, "not unaccustomed to hardships and dangers. I have many friends in the South, but *my heart beats time to the Union*," accompanying his words with violent blows upon the breast which was supposed to contain the loyal heart alluded to. In a very few months he was fighting against the glorious old flag, having joined hands with the traitors to his country.[5]

Our Company was invited to Kingston, Canada, on the occasion of the visit of the Prince of Wales. We were the invited guests of a company of Riflemen who happened to be Catholics. At that time there was much angry feeling existing between the Orangemen [Protestants] and the Catholics. We did not dream of such a state of feeling until we arrived in Kingston. Upon our arrival there we were met and escorted through the town by the Company mentioned. The Orangemen had constructed arches across several of the streets and as we approached these arches we were countermarched to the rear [by the Catholic escort] to avoid passing under them. The Orange population, supposing us to be a Catholic Company come over for the purpose of assisting the Catholics, greeted us with jeers and hisses. The excitement ran high and bloodshed would have followed but for a lucky circumstance.

It so happened that we were left alone by ourselves to march from the hotel to the armory and, with our fine band leading us, we passed directly under several of the arches. This action on our part made friends of the Orangemen and from that hour during the remainder of our visit each party vied with the other in showing us kindness.

But the excitement was so high that the Duke of Newcastle and the others of the Princes advisors decided not to land. [When] the boat touched the wharf during the night, we were notified and took our band down and serenaded him. He sent an Aide to acknowledge the compliment. The next day we gave an exhibition drill in the park and left that afternoon with the good wishes of all concerned, the crowd cheering and throwing peaches to us until we were out of reach.[6]

These, my Son, were the "piping times of peace" and such the manner in which the valiant militia fought their battles. Our broad-cloth uniforms [were] plentifully besprinkled with gold lace and tinsel. [With] burnished arms and immaculate accoutrements, gay plumes fluttering in the breeze, we stepped daintily to the delicious music of silver plated instruments. With gloved hands and full stomachs we strutted so proudly before the admiring gaze of bevies of fair ladies whose radience was reflected in the glimmer & sheen of our polished boots. Ah, how little did we then know of the *realities* that were soon to follow. [See Appendix 2, Item 1.]

When I contrast these scenes with those which met my eye in after years, how strange the contrast. Think of the real *Soldiers*, wearied with long marches on dusty roads, hungry, thirsty, foot sore, smoke, dirt & powder be-grimmed, alone, *demoralized* perhaps, perhaps bleeding and dying in some out of the way place, his comrades having abandoned him from necessity, perhaps lingering in prison or hospital, perhaps—but we will see enough of this as we progress.

In this way time passed along. The country becomeing dayly more excited and angry over the great questions at issue until the election of Abraham Lincoln for President, when the Southern people broke out in open threats of war. They no longer attempted to conceal their purposes, but frankly avowed their intention never to submit to the ruling of the majority. For the details of affairs in these days I refer you to Mr. Lossings history of the war, which I think gives as truthful and vivid [an] account of the transactions there transpiring as any history yet published.[7]

I well remember the day that the new President passed through Rome on the N. York Central R. R. on his way to Washington.[8] The adventures he met with on his way and the attempted assasination at Baltimore created the utmost feeling of excitement throughout the country. His final arrival at Washington in safety was cause of general relief and rejoicing. The war spirit was rising, but it is astonishing how both sections of the country underrated the other. The Southerners boasted that one Southern man could whip five Yankees and the people of the north responded that the 7th Regiment alone could march through the entire South.[9] Of course these recriminations tended to aggravate both sections still more bitterly.

Now my son, if ever you should be called upon to act in a military capacity, never be guilty of the blunder of underrating your enemy. I am not a prophet, or the son of a prophet, but I can truthfully say that

I never underestimated the ability or the intention of the South during all that time. It was my fortune to travel in the South during the year 1856 and I happened to be in Savannah, Georgia, at the time Senator Sumner was assaulted on the floor of the Senate by Mr. Brooks of South Carolina.[10] I saw enugh at that time to convince me that the South was desperateley in earnest.[11] When my friends talked about overrunning the South and whipping them back into the Union in three months, I excited their ridicule by saying that, in my opinion, the war would last for *three years*.

Fort Sumpter had been fired upon on the 14th and the whole nation was aroused to the terrible reality of the situation.[12]

At last came the intelligence that the Massachusetts Troops who were on their way for the protection of the Capital were fired upon in the streets of Baltimore and that several of them had been killed. This was the first blood shed during the war. I well remember that gloomy 19th day of April. The extract here shown is from the N.Y. Herald.

STARTLING FROM BALTIMORE
THE NORTHERN TROOPS MOBBED AND FIRED UPON—
THE TROOPS RETURNED THE FIRE—FOUR MASSACHUSETTS VOLUNTEERS KILLED AND SEVERAL
WOUNDED—SEVERAL OF THE
RIOTERS KILLED.

BALTIMORE, FRIDAY, APRIL 19.

There was a horrible scene on Pratt Street, today. The railroad track was taken up, and the troops attempted to march through. They were attacked by a mob with bricks and stones, and were fired upon. The fire was returned. Two of the Seventh Regiment of Pennsylvania were killed and several wounded.

It is impossible to say what portion of the troops have been attacked. They bore a white flag as they marched up Pratt Street and were greeted with showers of paving stones. The Mayor of the city went ahead of them with the police. An immense crowd blocked up the streets. The soldiers finally turned and fired on the mob. Several of the wounded have just gone up the street in carts.

At the Washington depot, an immense crowd assembled. The rioters attacked the soldiers, who fired into the mob. Several were wounded, and some fatally. It is said that four of the military and four rioters are killed. The city is in great excitement. Marshall law has been proclaimed. The military are rushing to the armories.

Civil war has commenced. The railroad track is said to be torn up outside of the city. Parties threaten to destroy the Pratt Street bridge.

As the troops passed along Pratt Street a perfect shower of paving stones rained on their heads.

The cars have left for Washington, and were stoned as they left.

It was the Seventh Regiment of Massachusetts which broke through the mob. Three of the mob are known to be dead and three soldiers. Many were wounded. Stores are closing, and the military rapidly forming. The Minute Men are turning out.

BALTIMORE, FRIDAY, APRIL 19—2:30 P.M.

Affairs are getting serious. Before all the cars got through, great crowds assembled at various points and commenced obstructing the road.

Reports are now arriving that the mob are tearing up the track.

It is understood the principal portion of the troops have got through.

BALTIMORE, FRIDAY, APRIL 19—4 P.M.

A town meeting has been called for 4 o'clock.

It is said there have been 12 lives lost.

Several are mortally wounded.

Parties of men half frantic are roaming the streets armed with guns, pistols and muskets.

The stores are closed.

Business is suspended.

A general state of dread prevails.

Parties a short time ago rushed into the telegraph, armed with hatchets, and cut the wires. Not much damage was done.

BALTIMORE, APRIL 19—5 P.M.

R.W. Davis, of the firm of Pegram, Paynter & Davis, was shot dead during the riot, near Camden Station.

It is reported that the Philadelphians are now at the outer depot.

The President of the Road has ordered the train back, at the urgent request of the Mayor and Governor. They are already off.

The citizens who were mortally wounded are, John McCan, P. Griffin, and G. Needham.

Four of the Massachusetts troops were killed and several wounded,

> but it is impossible to learn their names.
>
> As far as ascertained, only two of the Massachusetts soldiers were killed, belonging to Company C. Their bodies are now at the Police Station.
>
> At the same Station are the following wounded:
>
> Sergeant Ames, of the Lowell City Guard, wounded in the head, slightly.
>
> Private Colum, of the same place, shot in the head, not serious.
>
> Private Michael Green, of Lawrence, Mass., wounded in the head, by stones.
>
> H.W. Danforth, Company C, Sixth Regiment, of Massachusetts, slightly wounded.
>
> So far as known at present, seven citizens were killed, including Mr. Davis, before mentioned, and James Clark.
>
> Half a dozen or so are seriously wounded, though it is believed not fatally.
>
> Comparative quiet now prevails. The military are under arms, and the police are out in full force.
>
> There is a large mass meeting here to-night, addressed by the Mayor. The Governor was present.[13]

The intelligence was first communicated to me by Capt. Skillen, who I met at the Post Office. It was about 5 O'clock that afternoon. Capt. Skillen had recently been appointed Inspector of our Division of the Militia and I was then Captain of the Gansevoort Light Guard. I remember it was a raw, chilly, April day, a regular blue day. Capt. Skillen could not conceal his emotion when he informed me that in all probability the N.Y. State Militia would be called upon to march to the rescue of the Capital.

This affair at Baltimore added fuel to the flame and no one could now doubt but that *war* had actually commenced. That night I went home with feelings I will not attempt to describe for I knew the crisis had come. President Lincoln had called for 75,000 men to put down the Rebellion.[14] The call was promptly responded to. On the same day that the troops were fired upon in Baltimore, the 7th Regiment of N. York left for the Capital.[15] By the article which I enclose here you can judge somewhat of the feeling then existing. The article is from the N. Y. Times of the 20th of April. Can you imagine *my* feelings when I read that article?

THE WAR SPIRIT RISING HIGHER.

Yesterday was presented in our streets a spectacle that must continue memorable through generations yet unborn—the 19th of April, a day already famous in our history, having acquired a new and stronger claim on national recollection and gratitude.[16] Amidst unparalleled enthusiasm the volunteer soldiers of New England and New York struck hands in our streets on their march to the rescue of the National Capital. And beautiful the streets looked with bannered parapets, peopled roofs, windows thronged with sympathetic beauty, and sidewalks densely packed with multitudes of excited and applauding citizens.

But it required only a single glance at the faces of this great multitude, to become convinced that no mere gala or festive purpose had called out this magnificent demonstration. In every eye burned the unquenchable fire of patriotic ardor, and in every heart was the aspiration to join in defence of one common country. Old men, who must have seen the earlier struggles of our history, came forth to bless the young soldiers on their march to take share in a grande and more noble struggle than any the American continent has yet witnessed. Mothers, with tears of joyous pride half blinding them, helped to buckle on the accoutrements of their sons, and kissed them as they went forth to battle. Sisters and sweethearts, fathers and wives, friends and relatives,—all were represented, and had their individual characteristics in the immense concourse of life which yesterday held possession of Broadway.

Perhaps if there could have risen from the dead one of the old Girondists, after being bloodily put away to repose during the great French Revolution, and if he had been dropped down here in New-York,—by allowing a little for advance in costumes and architecture, he might have seen many curious points of resemblance between the scenes and the impressions here yesterday and those of seventy years ago in Paris. Then the inspiration of Liberty ran through the people, and the most powerful aristocracy of Europe was destroyed. The result of the struggle which cropped out here yesterday and bore its first fatal fruits in the streets of Baltimore, some future historian must record. It was with pride that our City saw her first quota of soldiers departing en route for Washington, to take their share with the troops of other loyal States in the contest now inaugurated. It was such a spectacle as only New York could exhibit—the spectacle instead of being a great pageant merely, having all the grandeur and solemnity of a step in one of those crises of events which involve individual and national life, engraving new names and new dynasties upon the tablets of History. The children of New York, the Seventh Regiment, our pets and pride, eight hundred of our chosen young men with threads woven to hold them, wherever they go, to the million hearts they leave behind, moved down Broadway and started for the Capital. Eight hundred young citizens, each with musket and knapsack, borne along calmly and impassively on a tide of vocal patriotism, making the air resonant with shouts and warm with the breath of prayer. But the dreadful and significant meaning which lay within it all, who can give?

As the call for troops was promptly responded to, even more men being offered than were accepted, there seemed to be no immediate necessity for me to volunteer at that time, though I was satisfied that the time would come and was preparing my mind for it. Your Mother fully comprehended the situation, but our feelings ran too deep to allow much conversation upon the subject.

Both sides were mustering their strength for the final struggle which was now fairly commenced. Regiment after Regiment left the loyal States for Washington and the Capital was secure from danger. In our own County the 14th Regiment was organized with Capt. Skillen for Lieut Colonel. Several young men from Rome joined this Regiment.[17]

On the 18th day of May, 1861, your little Sister was born. We named her Mary Edith. The Mary for both of our Mothers and the Edith for Miss Edith Ball, a young lady friend of ours at Niagara, Canada West, whom we both esteemed very highly. Thus it seemed that while the prospect of breaking up our pleasant little home seemed to become daily more certain, the ties were strengthening to prevent it.

Chicago May 24th 1861

To Wesley & Maria

Dear Children

Our hearts were made glad by the intelligence in Your letter to Lyman which came to hand Monday of this week informing us of the birth of a Daughter and the favorable conditions of both Mother & Child And Your letter under date of Tuesday has this Friday morning come to hand giving us Still further assurance of their favorable Condition. Still we feel great anxiety on their account well Knowing of the many Sufferings the young mother is forced to endure which will find their greatest Alleviate and Support in the Kind warm hearted Sympathy of a Kind and loving husband in this we feel perfect Assurance and gratification.

The new responsibility now committed to your care will for a Short time require much patience and untiring watchfulness—at the Same time will touch many a hidden chord deep in your heart of hearts which will vibrate and tremble with hopes & fears for all time to come—Allow us then to Congratulate you on the Auspicious occasion and trustingly & Confidingly repose in a kind providence which has thus far Succored us and Scattered blessings all along on pathways. Your Mother has not yet Seen Your last letter I will take it home when I go to dinner. She will also Soon write you. I have a little business this morning an thought I would take the liberty of writing you at once and assure that we rejoice with you on this happy occasion—

Lyman came home Safe in due time we were all glad to See him and he was glad to get home once more he had a pleasant trip—business of all kinds is at a Stand Still—our banks and the western Currency is flat how Soon it will revive none can tell. All Eyes and thoughts are turned towards the South and *war* with all its Sad & direful Evils is contemplated with a Stern resolve. A contest Such as the world has never Seen is near at hand and must Soon be decided and he who reposes his trust in an all wise—Just—and beneficent ruler of the Universe will have no fears but that he will Sustain with his own right arm Our Glorious republic—the land of the Free and the home of the brave.

As Wesley Stands in a position liable to be called in to Active Service for the maintenance of the Supremacy of the Stars And Stripes of our Country I will take this Occasion to Say that Should Such an Event transpire I have no fears but that You would meet it with a firm heart and Strong resolve worthy the Son of Our Patriot Siriz. You can most confidently rely on me for the care of Your wife and little one Should our country need you to fight her battles—at the Same time I will Say that I fully appreciate the Circumstances in which You are placed in business and family ties which Surround you. I most ardently hope and present Appearances fully indicate that an Abundance of Volunteers will be offered to meet All the requirements of the government—but I must Close hope you will write often as we Shall feel anxious Your mother will write Soon

And now let us give three
Cheers for the Stars & Stripes &
Nine for the *Mother & Baby*

Truly and Affectionately Yours
E A Gage[18]

A few days after the birth of little Mary we were shocked by the news of the death of *Col. Ellsworth*. He was killed at Alexandria, Va., while in the act of pulling down a Rebel flag from a hotel there. Col. Ellsworth was known to both your Mother and myself. She had seen him in Chicago and I had been introduced to him the year before when he made the tour of the country with his wonderful Company of Zouaves. We felt that we had lost a personal friend. Notwithstanding his youth, he had made a National reputation and the nation mourned his loss.[19] When I went home that night with the sad news, your Mother was Sick in bed. It was a Sad night for us, I assure you.

June passed and July came, and then that dreadful defeat and stampede at Bull Run, July 21st, 22d & 23d, 1861. I shall not attempt a description of the effect this disastrous battle had upon the country, you will read all about it in the books. Congress voted 500 millions of dollars and the President called for 500 thousand men.[20] I now felt that my time was come. At any sacrifice I could not remain longer at home. There was no necessity for me to tell your Mother of my resolution. She saw it in my countenance and said but little, though she trembled at the thought. I talked with my Father [and] his advise was entireley non-committal, he would go or not go if in my place, either or both, but my resolution was taken. The next thing was to find a Regiment that would suit me.

One day my Father brought down a Rochester paper containing the announcement that Col. Chas. B. Stuart of Geneva was raising a Regiment of *Engineer Troops* for service. This branch of service I knew would be suitable to my taste and I opened a correspondence with Col. Stuart upon the subject. The result was his promise of a Captains commission in his Regiment if I would raise a certain number of men.

Charles B. Stuart was an old acquaintance of my Fathers. He had held several positions of trust & responsibility, was an engineer by profession, and had been Chief Engineer of the State [of New York] and Engineer-in-Chief of the U.S. Navy under President Fillmore. He was a man of varied ability but lacked ballance. I shall have occasion to tell you more about him as we proceed.[21]

Now it so happened that many of the members of my company [Gansevoort Light Guard] had often promised that if I would raise a company for the war they would enlist. Whether they intended this for an excuse for not enlisting with the 14th or not I cannot say, but certain it is that when I hung out my advertisement for recruits and

A RARE CHANCE
FOR
NEW RECRUITS!

FOR COL. STUART'S
ENGINEER
REGIMENT!

Wanted, a few able bodied first-class men, to be composed of Mechanics, Carpenters, Farmers, and ordinary Laborers, to fill up the ranks of the Company organized in this village for said Regiment by Capt. WESLEY BRAINERD. This Engineer Regiment is to be one of the best in the service, and as the principal duties in this Corps will be of a professional character a position in it is more desirable than one of equal rank in ordinary Infantry Regiments.

A portion of the Company are now at Elmira, and write home to their friends that every thing is all right. The outfits and uniforms are ready, and are of a superior texture and quality. The time to fill the Company is limited, and those who wish to serve their Country are invited to call at the Armory of the Gansevoort Light Guard, 39 Dominick Street, at the earliest moment and enroll their names.

They will be enlisted for Three Years or the War, and will receive at the close of service a bounty of **$100**, and probably also a Land Warrant.

Men attached to this Regiment, will receive extra pay when engaged in professional duties as an Engineer Corps.

Any further information will be cheerfully given by applying at the Armory as above, by

Capt. WESLEY BRAINERD,

Rome, August 27th, 1861. ***RECRUITING OFFICER.***

A. SANDFORD, PRINTER, CITIZEN OFFICE, ROME.

The original recruitment poster that Wesley Brainerd used to raise a company of men for Stuart's Engineers. Courtesy of the William Hencken Family.

opened Head Quarters at the Armory to receive them they were not to be found. [See Appendix 2, Item 2.]

One man, Mr. George Alexander, had but a few weeks before said that he would forfeit his gold watch if he did not enlist as soon as I would start a company. When the time came, I could find nothing of him or his watch. I must, however, do him the justice to say that a year or two afterward he enlisted and became a Captain, was wounded at Antietam and, so far as I can learn, did good service.

One of my company and the last that I expected, was the first to enlist. He was a fine young gentleman and had inherited considerable property. He subsequently succeeded me as Captain, George W. Folley.

I hung out the American flag at the Armory and commenced to receive recruits. At first they came in slowly enough, but by dint of hard talking [and] with much music by the band, I had at the end of the week secured some 17 or 18 men, mostley hard working honest fellows. Just the material for good Soldiers.

But how was I to break up my little home and what was to become of my Wife and little one? That was the question. The memories of those days are too painful to dwell upon. With a sort of stoicism produced by desperation, your Mother assisted in the packing up of our household goods in boxes and our little home was soon converted into a desolate habitation. The excitement and conciousness of duty stimulated me but, to your poor Mother, all was blank desolation. The agony was too great for tears. It was distressing to look upon her haggard, woe begone countenance as she assisted in packing away the objects of our affection. She was still very weak [and] far from recovery from her sickness but few tears were shed. It was desperation itself.

On Saturday, with my little squad escorted to the depot by weeping friends & relatives, we took the cars for Geneva & thence to Elmira.

Chapter Two

Elmira

Arriving at Elmira with my little squad of men late in the evening, we were met at the depot by Colonel Stuart and a Company of the Regiment who escorted us to the camp, which was situated about a mile outside of the town. Elmira was a rendezvous for volunteers fitting out for the war and the state had constructed wooden barracks on quite an extensive scale for their accomodation. Here we had our first taste of Soldier life.[1]

We were assigned to a barrack and blankets were distributed to us. I was fortunate enough to secure a double one, which I found was none too comfortable as the nights were cool although it was in August. After getting my men packed away in their wooden bunks, I walked out into the moonlight to take a view of the situation. The camp ground bordered upon the Chemung river, a wide but shallow and quite rapid stream of clear water. The large barrack buildings, a dozen or so in number, looked gloomy and cheerless in the still night. All was solemn and I must confess forbidding. The dusky forms of the sentinels as they paced to and fro armed with huge *clubs* in lieu of muskets (for the Regiment was clothed but not armed) worked the boundary line beyond which none might pass or repass except with the commanding officers permission. It was several hours before I felt the necessity of sleep, but finally concluded to bid the world good night and for the first time slept as a Soldier.

It seemed strange enough to be awakened by drum & fife, the merciless "Revielle" unceremoniously dispells the pleasant dreams one is so oft to indulge in when absent from home and loved ones. A wash in the cool river then breakfast [of] cold ham, bread and black coffee, then squad drill until dinner, to be repeated in the afternoon. We did not attempt a dress parade as yet. Every body was in a hurry, coming and going, recruits arriving and being assigned to companies, much writing in the Adjutants office, much grumbling at the rations as well as clothes and shoes that would not fit, friends visiting the camp to bid good bye to their husbands, brothers and sons, here and there a young couple talking confidentially together, and it did not need a second look to be convinced that they were neither brother and sister nor husband and wife.

I was busy getting my men in order, fixing up the innumerable papers required, assisting at the Surgeons examinations &c &c until Monday morning when I left Elmira for Rome, arriving there in the afternoon. I came back to make a final winding up of my affairs, see my Wife and little one off for Chicago and take the ballance of recruits to the rendezvous.

My son, I cannot describe the scene of our meeting when I arrived at our little home. Your Mother was not expecting me on that train and when I entered the house she was giving directions about the things that were being removed from the house. Everything was in confusion, "topsy turvy" would perhaps express it. Seeing me enter the house, it seemed to her (as she afterwards expressed it) as if I had just arrived from the dead. I had been gone but a couple of days and yet it seemed like an age to her. Throwing her arms frantically about my neck we sobbed together and I really thought for a few moments that it would be impossible for me to attempt a final separation. I will not, can not, dwell upon the thought of the scenes that transpired during that Monday and the succeeding Tuesday and Wednesday but to the end of my days I shall never forget them. They were altogether and without parallell the most distressing days of my entire life. Our things were stored at my Fathers house when your Mother and the little one went.

On Wednesday at 11 O'clock a.m., your Mother with little Mary, then three months old, and your Uncle Lyman Gage, who had kindly come in from Chicago for the purpose, left Rome on the cars for your Grand Father Gages in Chicago. I tried to cheer her up and keep my own courage at the proper standard by making myself and them believe that I would run out to Chicago and see them once more before my final departure for the seat of war, that at the most the war would be a short one and I would soon be at home again the better and happier man for having done my duty.

And so they were carried swiftly out of my sight. Alas! I never looked upon that sweet little face again in life and *seventeen* long, long months rolled away before I again met your Mother. I stood alone for a few moments after the cars had passed out of sight, wondering if indeed it was a reality.

The next day Mr. Folley returned from the country with a squad of recruits for my Company and several men from the town joined us. Everything was now in readiness and on Friday morning we took our final departure. A large crowd followed us to the depot and many were the expressions of good will with the wish that we might return safely. It is true that there were but few dry eyes in that crowd and it was with the utmost difficulty that I kept from crying outright. Had *my* wife been there to see me off I know I should have been entireley overcome. My Father was there, it is true, and many kind friends, but I felt that my chief treasures were well cared for and that thought nerved me up. Besides, was I not the *Captain*, and how would it appear for him, the leader, to give way to tender feelings on such an occasion? My pride fortunateley came to the rescue. The band played, the crowd cheered and we moved rapidly away, all of us in good heart but some of us never to return again. [See Appendix 2, Item 3.]

Arriving at Syracuse we were joined by another, larger squad with several Officers destined for the same Regiment whose acquaintance we quickly formed. As at Rome, a large crowd had assembled at the depot to see them off. At Geneva we took [a] steamboat down the lake.[2] It is a beautiful lake and we had a beautiful day. Our anxiety to become proficient in drill prompted us to exercise the difficult manouvers on the promenade deck. We were all enthusiastic and anxious to excell in our new calling so we [took] every opportunity to improve [our] drill. At Jefferson we took cars for Elmira, arriving there in the evening, and as before were met and escorted to camp by a portion of the Regiment.[3]

I now had something like 50 men and others, strangers to me from the surrounding country, some from Pennsylvania, expressed a wish to join. It was soon observed that mine was a popular Company. The prestige of my name as a good Military Officer excited confidence in me. There were so few men in these days who knew anything of Military affairs that my knowledge of drill and discipline proved of immense advantage.

In a few days a young man by the name of Hoyt from Pennsylvania made his appearance with 30 Recruits. He offered to join my company with his men if I would promise him a Lieutenancy, an offer which I did not hesitate to accept. We then held our election (for the Officers were elected in those days) which resulted in placing Geo. W. Folley and Henry O. Hoyt in the position of 1st & 2nd Lieutenants. The Captain could not be commissioned until the company was full. The Sergeants and Corporals were also elected at that time. Our organization was as Infantry, requiring for a full Company: 1 Captain, 1 First & 1 Second Lieutenant, 4 Sergeants, 8 Corporals, 2 Drummers and 1 Teamster, with 84 enlisted men. In two weeks from the time I left Rome my company was full with 6 or 7 to spare. It was the third company completed though several had started weeks in advance of me.

After we had become pretty well organized, my Father, in company with Mr. Sandford, editor of the Citizen, visited us. The result of the visit was published. [See Appendix 2, Item 4.] As I look over these names it makes me sad. Where are they now? Of all these names but six of them ever belonged to the Gansevoort Light Guard & they never promised to go if I would. Those that remain alive are scattered to the four winds. Many left their bones in Virginia. Many died of sickness or wounds in hospital or at home and no considerable number of us will meet again until the last great day.

Our uniforms were of the original Regular Army pattern, grey, substantial, and well made. After we had reached the field, they were changed for the prevailing blue, made of shoddy.[4] In a couple of weeks after our arrival at Elmira we received our tents and went into regular camp. Each company occupied a street with cook tent at the front line and Officers quarters in Rear. At that time we had the regular allowance of tents: one wall tent for the Captain, one for the two Lieutenants and 4 men to each of the "A" tents, so called from their resemblence to the letter A. The Field & Staff had a liberal supply of wall tents and there was one large one for the Hospital. See "Articles of War" for details of camp ground.[5]

Colonel Stuart affected the martinet without the ability to carry it out. His *forte* seemed to be in lumbering up the Adjutants office with a senseless amount of routine, familliarly called *red tape*, to the neglect of more important matters. It required from 4 to 5 clerks continually employed to write passes, collect them, file them away, &c. He attempted to hedge himself about with a great amount of military dignity and he substituted *fuss and feathers* for genuine military tactics. Had we remained at Elmira another month I think the whole machinery of the camp would have been blocked by the additions he was continually making. Until the day he left the service he never learned the Manual of Arms. He could not go though with a dress parade correctly and as for maneuvering the regiment he could not have given a correct order for the simplest movement had his life depended upon it. Yet he aspired to higher military rank and *fully* believed that he was yet to wear the straps of a Major General. Our Army presented many such cases at the commencement of the war, "Political Generals" as they were called, and it was not until we had suffered disastrous defeats that Officers were made soleley upon their merit.

Colonel Stuarts engineering qualifications were much overated. He had held several positions of responsibility and trust, but was famous more for his failures than his successes. He possessed the faculty of appropriating the ideas of others whom he secured as counsellors to his own aggrandisement. By some means or other he succeeded in gaining the title of *General* in civil life, though where it came from has never appeared. Still, he possessed considerable ability as an Executive Officer and had he been more evenly ballanced in mind would have made a passable man but *never* an Officer. In person he was corpulent and heavy [with] grey hair, blue eyes and florid complexion. At this time he was about 48 years of age.

Lieutenant Colonel [William H.] Pettes was a man about 50 years of age. When a young man he graduated at West Point and served as a Lieutenant in the Florida War. He then retired from the service but managed to secure a position under government and superintended the construction of several government buildings such as the custom houses at Chicago & Buffalo. Colonel Stuart found him at Buffalo much reduced in circumstances, but, relying upon his education as an Officer, secured him for Lieut. Col., an error which he ever after regretted. He was of a cold, forbidding nature, ungrateful to his benefactor and patron. He prided himself upon his military knowledge but never attempted to aid the Colonel in his efforts to bring the Regiment up to a high standard of excellence, the very service for which he was engaged. He was a cold blooded, selfish martinet and, although he finally succeeded to the position of Colonel, never did any particular credit to the Regiment or himself. His personal appearence was

Col. Charles B. Stuart. Courtesy of the William Hencken Family.

against him: small in stature, thin, red hair, red face, and red eyes with a voice like a boatswains whistle. He was naturally profane and subsequently acquired the sobriquet of "Mr. God Damn You Sir," a name which I had the honor of originating. [Service records of all soldiers mentioned by WB can be found in Appendix 1.]

Major [Frederick E.] Embick was a young man of about 26 years, about 5 ft. 4 ins., straight as a arrow, black hair and eyes, quick, passionate disposition. He had entered West Point but remained there only one year. His passionate nature could not brook the discipline of that institution and he soon got into trouble which led to his expulsion. He was, however, the best Field Officer we had but the influence of Col. Pettes soon demoralized him and he took but little interest in Regimental affairs. He left us on the Peninsula, was promoted to Lieut. Col. and finally to Colonel in another Regiment but his ungovernable temper brought him into trouble with some General Officer and he finally left the service with a cloud over him.

Quartermaster [Charles B.] Norton was one of those kind of men who went into the Army to make money. He finally became brigade quartermaster, did not account for his receipts satisfactorily and was allowed to leave the service. He then married a rich wife in Washington, went to Europe, quarelled with her and was divorced.[6]

Riches *do not* bring happiness my son, especially when they are not gotten honestly.

Adjutant [Edward C.] James was the son of Judge James of Ogdensburgh, N. York, a young man of brillant talents, well educated but too penurious to be liked by any one. He was a fine writer, fine speaker, and quite a Poet withal. He subsequently became Colonel of the 107 N.Y. Volunteers.[7]

Dr. [Hazard A.] Potter was from Geneva, N. York. He was a Physician of considerable merit and as a Surgeon stood high in the profession. His besitting sin was his love of whiskey. He finally became so inebriated that his discharge was necessary, but after leaving the service he reformed and became an apostle of temperance but much of his usefullness was impaired before his reformation.

Assistant Surgeon [Charles N.] Hewitt was a clever student faithful to duty and took commendable pride in his profession. He succeeded Dr. Potter as Surgeon and remained in that capacity until the close of the war. As a Surgeon he deserved and attained a high reputation.

Rev. [Edward C.] Pritchett was a man of about 45 of Presbyterian proclivities. Like all chaplains in the Army his position was an anomalous one. Not considered as a combattant he had no particular position to fill except to preach on Sundays. He could have done good service in the hospital but he preferred not to, so spent the most of his time in his quarters reading or passing about the camp joking with the Officers. The Colonel made him Postmaster, but on one occasion he refused to deliver our letters to us because it was on Sunday and from that time he became rather unpopular with both Officers and men. He dozed away his time for nearly three years and finally, because none of us who were commanding Battalions would have him with us, left the service in disgust.

Captain [George W.] Ford had formerly been a Sergeant in the famous N. York 7th Regiment. He was well acquainted with military drill and tactics, was systematic, earnest and clean in every thing. His company books and accounts were a model for us all. He was a man of cultivated taste, quite extensive reading, of refined sensibilities, a good companion and in every sense of the word a gentleman of a delicate constitution. We never expected him to survive the hardships of a campaign but, with now and then an exception, he was with the Regiment through nearly all its time of service and finally became Major and Lieutenant Colonel by Brevet.

Captain [Porteus C.] Gilbert was Son in Law to Dr. Potter [and] was a medical student completing his studies in Geneva when the war broke out. He was not cast in a hardy mold & could not stand the fatigue of an active campaign (mentally I mean). He commanded his company passably well in camp where his wife could be continually with him but succumed to nostalgia on our first campaign and threw up his commission.

Captain [Bolton W.] O'Grady was an Irishman by birth, as his name indicates. He raised his company in Syracuse where he was employed in the capacity of a civil engineer. Excessiveley fond of his uniform as well as whiskey, slack in discipline, cowardly by nature, unprincipled by education, his Company soon showed signs of demoralization and was known as "Whiskey D". As might be expected, he deserted his Regiment, his country, and his family upon the first opportunity. I have heard of him since the war and saw him myself when he *denied* that his name was O'Grady, assuming that of *Le Coursey*. A more striking example of total depravity is hard to contemplate.[8]

Captain [John E. R.] Patten was a tall, lean, lank, crooked, good natured specimen of a Dentist. As he was good at heart, of correct intuitions, but of no Military taste or education he was the butt for the Lieut. Colonel, who never missed an opportunity to vent his spleen upon him. He remained with the Regiment about a year.

Captain [Edmund O.] Beers was from Elmira, a straight forward, common sense man and a good Soldier. He had been Captain of a military company at that place so that he soon brought his company up to a high standard of efficiency. A builder by profession, his knowledge of mechanics was often of great service. Capt. Beers finally became Major Beers and was always much respected by all the Officers of the Regiment.

Captain [Ira] Spaulding was a man of about 40 years, of medium hight, blue eyes, bald head, one of the wirey kind that could stand a great amount of fatigue and thrive under it. He excelled in his profession which was that of Civil Engineer. By his own industry he had gained an excellent education and at one time accumulated a large property. He was tenacious of purpose, reticent in the extreme and possessed a happy faculty of obtaining the views and ideas of others without divulging his own. A man of fine tastes and large experience with the advantage of travel and familiarity with men, he was destined, as he eventually became, to be one of the best known of any Officer of the Regiment. Spaulding became my friend and companion. We were more closely allied than perhaps any other two officers of the Regiment and if it had not been for the selfishness which he afterward developed I should pronounce him the best acquaintance I made while in the Army. He knew nothing of military matters when he joined the Regiment and served devoid of any ambition except to perform well the duties of Captain. Indeed, he often told me that that was all he should aspire to. A friendship sprung up between us which was lasting, at least on my part. Whatever knowledge of drill I possessed was freely imparted to him and he relied upon me as his best advisor. He subsequently became Lieutenant Col. of the Regiment and was its virtual commander and after leaving the Army filled many important positions, among others that of Chief Engineer of the Northern Pacific Railway. You will hear much more of him as we proceed.[9]

Captain [John B.] Murray had been a lawyer and a politician. He was a boisterous, rolicking, good natured fellow, generous and brave. He finally left the Regiment but soon came out as Colonel of another one and, as far as I know, did his country good service.

Captain [Walker V.] Personius was originally intended for a farmer, a steady, quiet, rather lazy, unassuming man, always attended to his own business and did not attempt to be sociable with his fellow Officers. Folley dubbed him Capt. *Persimmons*, a name which followed him for years. He was the only Captain of the Regiment who entered as a Captain and came out as a Captain, without disgrace and without a brevet, but during the last year of the war he did some excellent service.

Captain [William O.] Smalley was a butterfly, fond of music and of his ease. His chief merit was in having raised a Company in Geneva, the home of the Colonel. His company would have amounted to nothing had it depended upon him, but fortunately he had several excellent non-commissioned Officers and they kept the company up to a good standard. Having no particular ambition except to draw his pay and have as good a time as possible, he succomed to the pressure when hard times came on, left the Army, speculated in oil, made a fortune, lost it, and died.

Such, my son, were my associates and companions-in-arms. These were the men who commanded when the Regiment was organized. Look at the Register, the Adjutant Generals Report, and mark the changes that occured during the next four years. I have given you a sketch of my peers as they appeared and as they developed in order that you may have a better understanding of events that followed. The twenty Lieutenants I have not mentioned. Suffice it to say that they were men of equal intelligence and general qualifications as any of the same number & rank in any Regiment in the Army. I hope it may be interesting to you as it certainly is to me to look at the photographs of these men that I have preserved in my album.[10]

The time that we remained at Elmira was fully occupied in uniforming and equiping the men. They were examined by the Surgeon, registered and assigned to Companies and drilled. The labor attending the organization of a new Regiment can only be appreciated by those who have passed through it. At last we were ready for the field, ready in everything except drill and Arms. I purchased a new uniform sword, canteen & haversack in the Town. Also a wire bedstead which folded up and was contained in a little bag. We were all anxious for active duty. The mysteries of Guard Mounting and Dress Parade were being mastered but we did not attempt any drill beyond that of the Company.

On Sunday afternoon a carriage drove rapidly into camp and a

Lady and Gentleman allighted in front of the Colonels quarters. The Lady was Mrs. Stuart, wife of the Colonel. In a few moments all was bustle and confusion in the camp. We had received our order to move on to Washington via N. York. Everything was hurriedly packed and the next day the Regiment wound by company front through the principle [*sic*] street of Elmira to the Depot. A great crowd attended us. And so, amid the hurrahs of the crowd, on the 18th day of September 1861, we left Elmira for the seat of War.

Chapter Three

Washington & Halls Hill
1861

Our route from Elmira to the city of New York was an ovation—this is a hackneyed expression but in our case it was literally true. We were favored in having for the most part regular passenger cars, a luxury seldom enjoyed by us in after years. When it was convenient to use a Rail Road for our transportation we were generally packed away in cattle, freight, or on platform cars, but in this instance we were fortunate. At every Villiage & City through which we passed the people turned out to see and cheer us. Flags were flying from every conceiviable position and at night fires in great numbers illuminated the country. The "War Spirit" was at its hight.

Arriving at New York we expected to meet the same demonstrations of welcome but we were disappointed. We met with no reception at all. Things seemed to be moving along in the busy City as though nothing unusual had happened. We were allowed to march down broadway to the Battery without any particular manifestation of joy on the part of the inhabitants. I must confess that our pride was somewhat wounded but we did not consider that the people of that great City had already been surfieted with that kind of display and we arrived at the Battery about 8 O'clock in the morning in no very good humor. Once on the battery we went into a sort of rough bivouac and posted our Sentinels. The Colonel went to report the arrival of his Regiment to the proper authorities [and] we made ourselves as comfortable as possible on that barren spot with the hot sun pouring down upon us and awaited orders.

Now it so happenned that the Regiment of Fire Zouaves, which had been raised in N. York and went out to the war with a great flourish of trumpets promising to do such great things and having accomplished so little except to disgrace themselves, had returned to the City. As they were three months men, their time of enlistment expired soon after the battle of Bull Run. After the death of their Col. (Ellsworth), which event occurred previous to that battle, they had become completeley demoralized and they were hanging around the City awaiting their final pay.[1]

The Regiment was comprised chiefly of the worst characters the city offered, cowards in battle and bullies at home. The arrival of our Regiment of countrymen was a fine opportunity for them to have some sport. They accordingly surrounded our camp and insulted our Sentinels in every possible manner and threatened to drive our Regiment into the sea. Our men took their insults quietley, not wishing to disgrace themselves by coming into collision with them. As the day advanced and the whiskey which they imbibed began to work its legitimate result, their demonstrations became still more offensive. They seized several of our men who had obtained permission to leave camp and amused themselves by tossing them in blankets. Citizens passing that way were taken and served in the same manner. Occasionally they would make a rush and attempt to break through our guard lines.

As our Regiment had not yet received arms, these villains assumed a boldness which they would not have dared had we had our muskets. We armed our men with all the axe handles and clubs we could find and repulsed several of their charges. How we did pray for our muskets. In this manner we passed a most miserable day. As night came on the ruffians retired, promising to return again in the darkness and run us into the bay. About 10 O'clock they commenced to gather again in considerable numbers and were getting very demonstrative when 4 or 5 wagons came into our camp with the muskets. We were happy men then. A few minutes served to distribute the arms and to load them. When we earnestly invited the rascals to "come on" they declined the invitation and, after hovering around the camp a few hours, silentley stole away to their haunts. Thus ended the Battle of the Zouaves. The casualties of the engagement summed up about as follows: killed, none; bloody noses, 19; Black eyes, 11; Badly frightened, 850.[2]

The next morning we were glad enough to leave that detestable place. At about 10 a.m. the Regiment was marched to Pier No. 1 and soon after commenced to file on board the Steamer that was to transport us to Amboy, thence by rail to Philadelphia. While standing in line at ordered arms, one of our men carelessly placed his foot upon the hammer of his musket, his foot slipped, the cap exploded, the ball was driven up through his face, he fell and we thought him dead. He was not very badly injured however, though he was left in hospital in N. York. This was a good lesson to our men, for no accident of that kind ever occurred again.[3]

We had a pleasant trip on the boat & R. R. to Philadelphia, where we arrived in the evening and partook of the hospitality of that city at the famous Cooper Shop, where we were provided with a good plain meal of vituals after which we distributed ourselves around in as comfortable positions as we could find until morning. At 7 in the morning we left Philadelphia with the kindest of feelings for its generous inhabitants.[4]

Baltimore was the chief center of attraction for all the Soldiers passing to the Capitol since the massacre of our troops in April. Not one of the hundreds of Regiments that subsequentley passed through that City but longed for an opportunity to revenge that outrage. Our Regiment partook of this feeling and, had an insulting demonstration been made, the number of Rebel sympathisers in that city would have been considerably diminished. Our muskets were already loaded, the men were cautioned to keep quiet and the ranks were closed up. We were surprised to see considerable friendly feeling manifested as we approached the outskirts of the Town. The waving of hats and handkerchiefs from the doors and windows of the workshops and factories showed what class of people sympathised with the Union.

As we approached the center of the Town all this was changed. The wealthier portion of the population were strongly *Secesh*. As we marched through the principal streets on our way from one Depot to the other these people passed into their houses and stores and when they deigned to look upon us it was not difficult to discover the contemptuous expression of their countenances. But not a word was spoken. The solemn steady tramp of our feet upon the stone pavements was all the sound to be heard on this Sabbath afternoon. But one circumstance transpired to relieve this distressingly sober march. A little boy and girl, perhaps 12 or 14 years of age, stepped forward from the crowd and without a word handed me a little American flag and suddenley disappeared. I handed the colors to Corporal Freeman on the right of the company (we were marching by the flank) who placed the staff in the band of his musket and that little flag was Company property. For two years I preserved it, intending to bring it home should I survive the war, but, upon being promoted, I transfered it to my successor with the charge to him to transmit it to his.

We took cars at the South end of the Town and that evening, just as the sun was going down, we came within sight of the unfinished dome of the National Capitol. What a crowd of associations is gathered about Washington, the Capital of our country. Every man of us, as we stepped out upon the platform at the Depot, thought that it was his special mission to sacrifice his lifes blood, if necessary, in defense of that spot. The last rays of the setting sun were cast upon that magnificent building which but few of us had ever seen and for a time we stood in silent admiration.

We disembarked within the enclosure of the Soldiers Rest and we marched into the barracks where we partook of some small pieces of pork rind, black bread & blacker coffee.[5]

At about 9 O'clock p.m. we fell in line and marched up Pennsylvania Avenue to a large building which had been assigned for our temporary occupation. As we passed by the principal hotels a few hands were heard to strike together which was all the welcome we received. Our Regiment filled the building from garret to cellars. We made ourselves as comfortable as possible until the morning when we were in line bright and early. We marched out 7th street about four miles to Meridian Hill, went into camp, laid out company streets and erected our kitchens, which consisted of tree croches driven into the ground with a cross pole upon which to hang the kettles.

The hills around were covered with troops who, like ourselves, were taking their first lessons in the art of war or, more properley, the business of Soldiering. Looking to the Southeast we could see plainly the majestic capitol building and to the South and West, the hills on the Virginia side of the Potomac, the land we so much longed to occupy. Like all young Soldiers our men were full of ardor and could hardly restrain their impatience when we saw one after another of the Regiments strike their tents and take up their line of march for the "Sacred Soil." [See Appendix 3, Item 1.]

The weather at this season (September) was delightful. The crisp morning air tempted us to exercise and the drills were pleasant beyond anything you can imagine. Already the beneficial effect of camp life was plainly visible and our men began to look hardy and robust. I obtained permission to visit the City and ascended the partly finished dome of the Capitol there. Without the aid of a glass I could distinctly see Alexandria and the Rebel flag flying from Munsons Hill.[6]

We remained at Meridian Hill about one week, which seemed at least a month to us. At last, one beautiful morning, we noticed Regiment after Regiment strike its tents and, to the music of their bands, march off in the direction of Virginia. How the blood *tingled* in our veins when our orders came to march. We were quickley in line and marched down 7th Street, through Washington and Georgetown to the Acqueduct Bridge. Here the road was so blocked up with Artillery, baggage wagons, Ambulances, and Troops of all kinds that we were delayed for several hours. It was quite dark when we finally reached the other side. We passed up through the line of defences beyond Fort Corcoran and halted in an open plain where we commenced to cook our supper.[7]

In a short time we were visited by a number of Officers of the 14th Regiment, which was then encamped near the River. Need I assure you we were glad to see their familiar faces? Some of us obtained permission to visit their camp where we were generously entertained with eatables and drinkables. During our short visit Skillen, Stryker and others regaled us with accounts of the events following Bull Run. Returning to our camp we found our men disposing of themselves for the night and were preparing to do the same when we were ordered to move forward. Proceeding about a mile, we again halted and we all soon rolled up in our blankets on the ground and the nights rest fairly commenced when I was aroused by a summons from Col. Stuart to report to him immediateley. We went to the top of a small hill near by [where] the Colonel called my attention to a long line of fires burning at a distance of about two or three miles and informed me in a very solemn manner that we were now in the enemys country.

"*There*," said the Colonel, "*are the camp fires of the enemy*! I have selected you and your company to do Picket duty tonight as I know that I can trust you!" The Col. said this in a voice trembling with emotion. "Bring up your Company to this spot and I will proceed with you to the line where you are to be posted."

I returned, aroused my Company from their deep slumber and soon reported to the Col. We advanced about half a mile and then deployed to the right and left across the road, [each man] about 50 ft apart and as far as the command would reach. This made my line nearly a mile in length as I had about 100 men. My orders were to allow none to pass or repass without the countersign. If this was not given, I was to *fire* upon them. After giving me these instructions he departed and I paced up and down that line until daylight. The men, fatigued by the march of the day previous, paced backward and forward on their posts without relief until morning. Occasionaly one or another would fall over from downright fatigue. Two or three times during the night heavy firing was heard at the front. Of course, we assumed these to be attacks by the enemy. The great red fires which illuminated the somber heavens with a sort of lurid glare added all sorts of horrors to our imagination.

Towards daylight we heard the clattering of horses feet coming down the road. Hastening to the center of the line I drew up a squad of men across the road and prepared to give the intruders a warm reception. I heard the clashing of sabers and ordered my men to a charge bayonet, expecting the next moment to be dashed upon by at least a squadron of Rebel Cavalry. As the party appeared in sight I challenged them in regular style.

"Who goes there?"

"Friends," was the reply.

"Halt, friends. Advance one with the countersign."

The party halted and an Officer advanced, where something like the following transpired:

"What is this posted here, Sir."

"The outer picket line, Sir."

"The *outer* picket line is it? What fool posted you here and what the devil was he thinking about, with forty thousand of your own men outside of you?"

The party was allowed to pass without further ceremony. I said nothing to the men, but it did occur to me that *Col. Somebody* was wasting his talents in commanding a Regiment simply, when he *ought* to have at least an Army Corps. And it seemed he was of that opinion also. As he was the Senior Colonel of the troops who entered Virginia that day and was placed in temporary command he realy anticipated a Brigadiers commission. Had the Fool Killer passed that way, however, we were not the only ones to merit his attentions, for the troops in front of us had been firing into each other all night and several were killed belonging to the "De Kalb" Regiment.[8] My Company never forgot the Defence of Fort Corcoran, as that nights operations were called, and for years it was one of the standing jokes of the Regiment.

Company C., 50th New York Volunteer Engineers. Courtesy of the William Hencken Family.

We remained in front of Fort Corcoran two days and were then marched out about 4 miles further in advance to a place called Halls Hill. [The army's] outposts [were] established at Munsons Hill, about three miles in advance of us, the Rebels having fallen back from that position on the advance of our forces.[9] Thus our line extended in the form of a semi-circle from the Chain Bridge at our right around via Munsons Hill to Alexandria on the left. As Troops arrived from the North they were assigned positions along this line, organized into Brigades and Divisions and subjected to the strict military drill and discipline necessary to prepare them for the work which was to come.

Our Regiment was brigaded with the 17th New York on the right, the 96th Pennsylvania and the 2nd Michigan on the left. We went into regular camp and commenced a series of Squad, Battallion, Brigade & Division drills which was kept up almost incessantley until about the middle of December.[10]

Our course of drilling was severe, but necessary. General Butterfield, who commanded the Brigade, was a severe disciplinarian and exerted himself to make his Brigade the best drilled in the Army. Before we left them we were really quite proficient in all the military manoeuvers. No one who has not had a similar experience can realize [how] the glorious elasticity produced by pleanty of exercise and plain food soon made new men of us all. Every moment of our time was occupied, there was not an idle minute to dispose of [and] drill, drill, drill, all day long. In the evening we assembled around the camp fire at General Butterfields headquarters and recited our lessons from the tactics, the General himself being the Questioner.[11]

While we were so busy in preparing ourselves for the duties before us the people of the country wondered why the Army of the Potomac did not move and the cry of "On To Richmond" was raised. The humblest soldier in the entire Army comprehended the reasons better than the best informed at home.[12]

The time passed rapidly with us and to me it was pleasant because it suited my tastes. While we were here, the disaster at Balls Bluff occured. The whole country was shocked at the news of that defeat but with us it passed as a trifling incident.[13]

The result of the Defense of Fort Corcoran had begun to develop and the first death in the Regiment occured in my Company. Byron R. Seamans, with several others, was prostrated with Typhoid Fever. He was sick but a few days before he died. He had been the very life of the company, always cheerful, full of fun and jokes. He kept up the spirits of the men by a continual stream of witticisms which kept them in good humor under any and all circumstances. I was with him during his last moments and he requested me to write to a young lady living at Harvard, Mass., by the name of Jewett and inform her of the circumstances of his death.[14]

He was buried with military honors. And here let me say that of all solemn things, a military funeral impresses me most deeply. The ceremony observed on such occasions seem to me very impressive and appropriate. The line of march to the grave is formed as follows: 1st after the Band, playing a dirge, [comes] the Firing Party with arms reversed; then the body; following which the Chaplain; then the comrades without arms; and finally the Officers, lowest in Rank first, highest in rank last. After the body is lowered into the grave, three rounds of blank cartridges are fired over it. Then the procession returns in the reverse order, the band playing a liveley Air. The spot where we burried him soon became a small sized cemetery and I think I could go to the place even after all these years, if only the trees still stand which marked the locality. After the funeral I wrote to the young lady as he had requested [with] Corporal Cook, who was his tent mate, also adding a few lines.

Her reply to our letters is here presented.

Harvard Oct. 20th 1861.
Capt. Brainard & Mr. J.A. Cook
Halls Hill. Va.

Respected Friends

For as such I consider you. I feel that I must respond *immediately* to your *kind* and *sympathizing* letters, which caused my heart to over flow with *sorrow* and yet with *joy*.

To you Mr. Cook my *respected* friend, I would first render my *heart-felt* thanks for your *unspeakable* kindness in favoring me in so *kind* and *sympathizing* a manner of my great affliction, and by complying with your *dear, departed* friends wishes so promptly.

And you Capt. Brainard *honored* & *respected* friend, please accept the *most sincere* thanks of your *humble* servant, and *ever grateful* friend.

Words are *inadequate* to express my thanks to you, *both, strangers* that you are to me, yet expressing your sympathy's for me in such *sweet* words of consolation; Oh! kind sirs I cannot thank you sufficiently, but you will *ever* be remembered by my friends in the *deepest sense* of *gratitude* and *respect*; and *my wish* will be that the *choicest blessings* from "*Above*" may rest on you *now* and *evermore*.

The thought that I shall *never* again see my *dear friend* Byron brings with it feelings which I *cannot describe*. Thoughts which *cannot* be uttered. Oh! would to God I might have *shared* his last resting place, but alas! my time has not yet come. I must *struggle* on through this *cold*, and *dreary* world *alone*. Yes *alone*. For my *nearest* and *dearest* friend is taken from me. *No more to return*. But oh! the sweetest consolation it affords me to know that he was *willing* & *prepared* to die, is worth more to me, than you can describe; and he has died in a *great*, a *glorious* cause, one in which I trust he will be long *honored* and *remembered*.

And now Dear Sirs I have *one* simple request to ask of you, and it is this, that you will *ever* remember me in your *prayers*; that I may have *strength* given me from "*Above*" to endure this *great* & *lasting* affliction with *Christian gratitude* and that I may so live that I may one day meet *him*, my *darling* friend, where *sorrow* & *parting* are unknown.

Again thanking you from my *inmost soul*, for your *ever remembered* kindness & sympathy, and hoping that Heaven will bless you. I subscribe myself your devoted & *ever grateful* friend.

Clara L. Jewett.
Harvard. Mass.

P.S.

Should you ever come to Mass. my Parents & myself would *gladly* welcome you to our home and *rejoice* to meet with those who so *kindly* cared for *my dearest friend*. I shall write to Mr. Isaac Seamans and if he is *not able* to answer, will you extend your kindness a little further by informing me of his health & welfare.[15]

No one can read that letter and not be convinced that she was true to him as a woman should be. But alas, the fraility of human nature. Byron, true to his name, as subsequent events proved, was engaged to be married to *two others* at the same time. Taken altogether, I am not sure but he did the best thing under the circumstances to die. For had he lived, who knows what pain and misery his foolish actions might have caused. Whereas, each of the three, unknown to the others, may now consider his sweet memory as all her own. "Where ignorance is bliss tis folly to be wise."[16]

During this month an incident occured which I will place on record. There had been quite a revival of religion among the men of the 17th N. York. and one Sunday the Chaplain of the Regiment baptised 12 of them.[17] The next day the Chaplain was relating the circumstances to Colonel Mclean of the 96th Pennsylvania. The Col. was a bluff and rather profane man. His chief ambition was not to have his Regiment beaten in anything. So, calling his Adjutant to him, he bawled out, "Adjutant, detail 15 men for baptism immediateley. That d——d 17th Regiment is not going to beat *us*." This was literally true and that was how it occured but I saw the story repeated many times afterward in the papers always attributing the incident to other parties than the true ones.[18]

The ambition of the Colonel of *our* Regiment was to have it said that no profane swearing was ever heard in the Regiment. The Articles of War provide a fine of 17 cts. per oath for every private Soldier and one dollar for each and every offence when committed by a commishoned Officer. Up to this time our Regiment was free from this foolish sin and an oath was rareley heard, but on one evil day all our morality in this respect vanished. The Lieut. Col. (Pettes) was drilling the Regiment. He was a fiery, passionate little man. Something went wrong and out popped a hideous oath [and] the ice once broken, oath followed oath in quick succession. Returning to camp, the spirit was caught by the men and that night oaths rolled from right to left with wonderful volubility. The charm was broken. No more punishment for swearing, for with what propriety could the men be punished for an offence when even the Lieut. Colonel indulged in it unrebuked. Rebuked he was in mild terms by our poor milk & water Chaplain. But the Col., whose duty it was to take the matter in hand and who would have punished severely the private Soldier for like offence, felt powerless to grapple with his Chief Officer knowing full well his own utter incompetency to command the Regiment in the simplest military movements.

No one, however humble his position in life, can ignore the fact that he exerts influence either for good or evil. Here was a striking illustration of the principle, for from that time forward no effort was made except by individual influence to put a stop to profane swearing and I really believe the men were worse in this respect than they would have been had no effort for its suppression been made in the first place. All through the influence of this one man the command fell from its high Standard of Morality.

One day we were ordered out on Picket duty. Our line was established beyond Munsons Hill. General Butterfield himself posted our reserves and gave us our instructions. The Rebels occupied a line of hills about a mile from our line. A deep and wooded valey lay between us [and] we could see them distinctley at their separate Picket stations. What strange emotions possessed us as we saw ourselves face to face with the enemy. The stillness of death pervaded the country between the lines [and] those who lived there, if they remained at home, dared not show themselves for fear of being fired upon by one or the other party.

Our station was at a log house some distance in rear of the front lines. Our men were dispersed in squads of 3 men each [and] secreted in some secure place about 150 yards apart. How sharply we looked out for the least appearance of an attack but none appeared during the day. Occasionally we could see a horseman, evidentley an Officer, visiting the picket posts of the enemy, and an occasional shot from there served to keep our men more closeley secreted from observation. And so passed the day and night came on. It was a long day to us, so full of new experiences.

Major Embick was posted about 1 1/2 miles on my right. About 10 O'clock at night he came in to my reserve where I was stationed, said he had observed a number of men go into a barn in front of his line just before dark and he took them to be rebels. He requested me to take a part of my reserve and with his men go and capture them. Making as little noise as possible, I aroused my reserve from their sleep. With about 30 men we set out following the Major in single file, our arms at a trail lest the glimmer of the moonlight upon their bright surface should discover us to the enemy.

Arriving at the designated spot, we secreted ourselves behind the stumps and trees. There we held a council of war. Was it prudent to attack the barn without higher orders? We were undecided. Suddenly, the sound of musketry was heard upon our right. It came nearer and nearer until it reached us, then our men took it up and blazed away at the stumps, every one of which looked like a rebel, and so it passed on away down to our left. Then followed the general alarm, drum corps after drum corps sounded the "long roll." We could hear them away off in the night for miles to the rear until it seemed that the whole Army was aroused, then all died away again. The whole thing [was] to be repeated at intervals of about two hours until morning.

Once I ventured to reconnoiter the barn. Creeping on my hands and knees to within a few feet of the building I could distinctley hear the voices of men. Returning cautiously, I reported the result of my observations to the Major. He, however, providentley concluded not to order an attack. When, at last, day light appeared, some 15 or 20 of our *own men* scampered out and away to their posts again, having abandoned them probably for a comfortable nights lodging on the hay which the barn contained. The next day we returned to camp the wiser for our Picket experiences.

The army was full of hope and confidence, for General McClellan had said, standing up in his carriage and in response to the multitude who crowded around to greet him,

"Soldiers!! You have made your last retreat. You have met your last defeat"!!![19]

Chapter Four

Halls Hill and Camp Lesley 1861

The monotony of drilling at Halls Hill was occasionally relieved by a review by General McClellan. These reviews by Brigade and Division were both frequent and instruction[al], as thereby the Officers became accustomed to the movements of men in large masses. The review ground was about two miles from our camp. We were practiced in firing by Company, by Battallion and by Brigades [and] to advance and retreat, lay down, &c. all at the sound of the bugle. These reviews were generally witnessed by large numbers of distinguished persons, Military and civilian, foreign ministers, Members of Congress, &c. Many ladies also appeared and the crowds of gay equipages which drove out from Washington on these occasions was a sight to behold. After passing through one of these review days, the exercises of which generally lasted from 4 to 6 hours, you can imagine that we returned to our camp with good appetites and that our sleep was sweet.

Officers who had raised Companies or Regiments had necessarily incurred considerable extra expence for mens board, car fares &c. The Government proposed to refund this to the Officers upon presentation of the proper vouchers. For the purposes of obtaining the money thus due me, I, with several other Officers, obtained permission to visit Washington. I shall never forget how the man in the gold room of the Treasury Department stuck his shovel into a heap of gold coins lying upon the floor of one of the large vaults and asked me if I would prefer gold or greenbacks in payment of my claim. Of course I told him to give me greenbacks. This was the first issue of greenbacks and nearly every body chose them in preferance to gold, as they were of equal value and much more easily carried. Little did we then think that before many months had passed it would require 290 dollars of this paper money to equal in value 100 dollars in gold.[1]

Our regimental Sergeant Major was thrown from a wagon on his way to the city and received such injures that he died in a few weeks. This cast quite a gloom over the Regiment as he was a very popular young man.[2]

On Sundays we were marched up to attend divine service conducted by the Chaplain, who preached to us with his little straight sword by his side and seemed to feel very awkward on consequence. We were always ordered to appear at divine service with knapsacks and side arms. Immediately after the service the Colonel inspected the command in person and woe to the man who appeared with a speck upon his uniform, a button off or unbuttoned or whose shoes did not show the proper amount of blacking. When such a one was found, he was ordered to appear at the Colonels quarters after inspection where a severe reprimand was administered and the delinquent placed on fatigue duty.

Several times during the month of October rumors came of an advance of the enemy. Quickly our tents were struck, baggage packed and loaded and the men drawn up on the color line in marching order, ready to move at a moments notice. After remaining several hours on the line we would be ordered back and arrange our camp again. We subsequentley learned that this was done in order to accustom us to get ready to move at short notice and very useful exercise it was too for new troops.

We began to question among ourselves how long this state of things was to continue, for had we not been raised as an Engineer Regiment? And here we were doing duty as Infantry. At last the order came for us to report to Lieut. Col. Alexander of the Regular Engineers to be assigned to duty as Engineer Troops [and] the order was received with much pleasure. The night before we broke camp the event was celebrated by burning the bough houses with which our tents were surrounded. Strict orders were issued against the proceeding but, in spite of the intense vigilance of the Officer of the Guard, first one and then another of the bough houses were found to be on fire. Perhaps your Father and his two Lieutenants, Folley & Hoyt, may have had something to do with this mysterious burning. I will not say they did not. At any rate, the night was one of much enjoyment to the Officers mentioned.[3]

We bade General Butterfield farewell (he complimented the Regiment upon its skill and proficiency in the military tactics as [well] as its good behavior) and departed for Washington on the 28th day of

October, 1861. The famous Ellsworth Avengers, the 44th N. York, assumed our place in Butterfields Brigade of Porters Corps.[4]

We crossed the Long Bridge for the first time and marched to our ground near the Navy Yard on the East Branch of the Potomac, sometimes called the Anacostia River.[5] Here we commenced the establishment of our permanent camp. Shops were built for the manufacture of such bridge material as our men could make and this became the depot for Engineering material and was nominally the Head Quarters of the Regiment during the entire war. Our camp ground presented a fine view to the East into Maryland and South [into] Virginia. We commenced work at once on our camp ground. Thousands of wagon loads of earth were transferred from the high grounds to the valleys, roads built to the river [and] a permanent Guard House constructed of logs was established. A small house at the South end of the ground was occupied by the Quartermaster and subsequently used for a Hospital. Another little house at the North end of the ground was taken by the Colonel in anticipation of the arrival of his Wife and daughter, his own quarters in large tents near the house. The Field and Staff occupied a semi-circular piece of ground to the Northeast. The Companies vied with each other in beautifying their streets and tents. Everything soon assumed the shape of a permanent garrison and scrupulous cleanliness was the order of the day.

A few days after our arrival and before we had had time to sufficientley anchor our tents, a furious storm burst upon us. As the night came on the storm increased in furey. Folley & Hoyt & myself were rolled up in our blankets on the ground in our tents which adjoined each other. The ground was soft, the rain and wind increased and about 1 O'clock in the night the whole ground upon which our tents stood suddenly started for the Anacostia River. We brought up about half way down the hill and the water rolled over us a foot deep. Fancy our flight, in the darkness and mud and water, our tents blown down over our heads. We crawled out as best we could, singing "A Soldiers life is always gay—Always gay." Almost every tent in camp was blown down that night and it took us several days to right things again.[6]

Here, for the first time, we came in contact with the 15th Regiment N.Y.V. Engrs., Colonel McLeod Murphy [commanding] and as we were brigaded with this Regt. during the war you will often read of it in the course of the narrative. This Regiment was recruited chiefly

"[This] represents the [wooden ponton] bridge in course of construction. The mound of earth on the left shows a little fort we made for practice. You may see me by looking closeley *on the road* about the centre of the picture—looks like a black post." Note the white canvas coveralls the men are wearing to protect their uniforms. Courtesy of the William Hencken Family.

in N. York City and was comprised of many good men but of more bad ones. The Officers as a class were not desirable associates though there were several fine gentlemen among them. Thus the 50th & 15th N.Y. comprised the Engineer Brigade of the Army of the Potomac.[7]

Our camp was named by our Colonel "Camp Lesley," in honor of the Assistant Secretary of the War.[8] Mr. Lesley had rendered material assistance in getting authority from the government to raise Volunteer Engineer Troops, there being at the time much opposition to the measure arrising chiefly from Officers of the Regular Engineer Corps. It was not until mid-winter and much lobbying by our friends and ourselves that Congress finally passed the Act recognizing us as Engineer Troops [and] confirming upon our Officers and men the same rank and pay as those of the Regular Battalion. For nearly a year after the act was passed, those high in Authority would not consent to allow us to wear the Engineer button though the priviledge of wearing the *castle* on our hats could not be denied us. We *won* the right, however, and from that time henceforth kindness of feeling existed between the two corps.[9]

The winter at Camp Lesley passed pleasantly, swiftly and profitably. You may perhaps wonder what we did with our time. I assure you we had none to idle away. We worked nearly two months in getting our camp in order. Then there was the daily Squad and Company drills, the Skirmish drill, in which I was the instructor for the Regiment [and] the baynot exercise for the men and sword exercise for the Officers, the days work always ending with Dress Parade. Guard mounting was the feature of the morning [and] during the evening the Officers *recited their lessons* in Mahans Field Fortifications at the Colonels quarters [with] Field & Staff and Captains reciting to the Colonel, Lieutenants generally to the Major or a Senior Captain.[10]

We constructed a regular fort. The work was done by Companies, relieving each other by day and night as in case of actual hostilities with an enemy in front. Certain portions of each week was devoted to the Ponton drill. In this practice we became very proficient and were frequentley visited by members of Congress and distinguished strangers, The President (Lincoln), General McClellan & Staff &c.[11] Several French Officers also visited us on these occasions and they pronounced our practice superior to the French Pontuniers themselves, which was a very high compliment. [See Appendix 2, Item 5 for a complete description of ponton bridge building.]

We occasionally took trips across the river for practice in the art of making Facines [and] Gabions.[12] The process of making these siege material[s] as well as the Ponton drill I will not stop here to describe. You will find all information in regard to these subjects in Mahans Fortifications and Duanes Manuel for Engineer Troops.[13] These books, with a copy of the Bible and Homers Iliad with perhaps Plutarchs Lives, is all the library a soldier needs. On one of these Facine & Gabion excursions I made the acquaintance of Colonel Naylors Family. I first met the Col. on his grounds when my men were at work. I asked him where I could find a good spring of water.

"Up at my house," he replied. "Please go up there and tell them I sent you and they will give you some good wine also."

This happened to be just the invitation I wanted. I was escorted into the parlor where Mrs. Naylor and her *four* daughters soon appeared. I was not long in making their acquaintance. The ladies were

"[This] represents the [wooden ponton] bridge just after it was swung into position with the different squads in their places. The house under 1 in top of picture is the Widow Naylors. That under 2 on the left is Colonel Naylors house. You can see it by looking closeley." Although the numbers appear in the margin, close inspection does not discern the houses Brainerd is referring to. In the foreground can be seen some wooden pontons rafted together. This is the way they were positioned when a ponton bridge was moved by water, with the different elements of the bridge material stacked on top of rafted pontons. Just behind the rafted pontons is a section of a canvas ponton bridge with a Birago (Austrian) trestle connecting it to the shore. The completed wooden ponton bridge is sixty-five pontons long. Note the chess and balk stacked on the shore at the lower left. Courtesy of the William Hencken Family.

well educated, ladylike and frank [and] we were good friends at sight. The house was situated on an eminence called Mount Henry and commanded a most charming view of the City of Washington and surrounding country.[14] It was one of the most loveley spots I ever visited: to the West, the City of Washington was spread out so that every street and public building was distinctly visible; to the South, the beautiful blue Potomac, Arlington House and the hights of Virginia beyond.[15] After an exceedingly pleasant call I took my departure, receiving an invitation to come as often as my duties would allow me and to bring such of the Officers of my Regiment as I might select. To return the compliment I selected about 20 of our Officers and one beautiful moonlight night, having secured the service of the Marine

Band, we resolved upon a serenade. It was a charming evening and as we floated across the moonlit waters of the Anacostia to the delicious strains of music such as no other band in the country can hope to equal, it seemed like some fairy scene such as I had read about. Arriving at the house, the band played several of their exquisite airs and we were invited in by the Col. to find a house full of Ladies who welcomed us as we entered. So our surprise was not a surprise after all. We enjoyed a social chat, then danced, then marched out to supper, then danced and drank wine and enjoyed ourselves until the "wae sma hours," when we returned to our camp, all testifying to the rare hospitality of the Naylor Family. Long as I may live I shall never forget the kindness I have received from them during that and subsequent years.[16]

Nor was this the only family whose acquaintance we made during that long winter. The Col. had a sister and she was a widow. She had two daughters and they lived about half a mile from the Col. nearer the river. Some of our Officers visited the Widows frequentley and finally, in after years, one of them, Lieutenant Andrews, married one of the daughters.

In Washington City we became acquainted with Mr. & Mrs. Purdy and their two daughters, Mrs. Wallace and Mrs. Thompson, who resided at their Fathers house. Their son John was an Engineer in the Navy. The house was on Pennsylvania Avenue almost within a stones throw of the Capitol grounds. Many pleasant evenings did we pass there. Mrs. Purdy was always anxious to get us up something good to eat and none could appreciate her choice things better than we.[17]

Mrs. Wallace was quite an amateur *Spiritualist* and we often amused ourselves around the table asking questions of the Spirits: one rap, no; two raps doubtful; three raps, yes. On one occasion the Spirits rapped out answers to questions asked giving statements somewhat as follows: that I would be wounded on the banks of a river, which proved true; that Lieut. Hoyt (who was present) would be dismissed from the Service, which proved true; that Col. Stuart and Lieut. Folley (both present) would both leave the service before the war ended, this also proved true; and that I would finally be a full Colonel in the Army, which also came true in due time. I mention these things not to endorse Spiritualism but to give you an idea of our life in Washington, and will simply say that there are more things in heaven and earth than is dreamed of in your philosophy.[18]

It was customary in these days for every Soldier to get his picture

Capt. Wesley Brainerd, late 1861 or 1862, wearing the leggings that precipitated the altercation with Lt. Col. William Pettes. Courtesy of the William Hencken Family.

taken in full uniform to send home to his friends. Here you will find me in gorgeous array ready for the march and fierce as Mars. Mark well those leggins my boy, for they nearly cost your Father a duel one day.

It happened in this wise. I had visited the House of Representatives in company with the Colonel and Mrs. Stuart, returning to camp just in time to bring my Company on the ground at dress parade [and] I did not have time to remove the leggins. At the close of the parade the Officers approach the center, march forward to within a few feet of the commanding Officer, salute and the parade is finished. All went well until the Officers marched forward. Lieut. Col. [Pettes] was in command. After saluting him he signified that he had something to say and we all stood in respectful silence.

"Gentlemen," said he, in his crisp, vinegary voice, "we are not cavalrymen," (one of the Officers had on high top boots) "neither are we 'zoo zoos'," saying this in his most aggravating tone of voice and looking sneeringly at me. "You will never again appear on parade in this dress unless you wish to be placed in arrest."[19]

His manner and tone of voice stung me to the quick. There was no love between us and never had been. He had never supported his Colonel as a Lieut. Col. should and he knew that I knew it. Besides, he knew that I was a friend and confidant of the Colonel and he took this opportunity to insult me. I resolved upon *satisfaction*, returned to my quarters [and] wrote half a dozen letters challenging him to meet me. The language did not suit & I tore them up again until I finally succeeded in getting one to suit me. This I resolved to send him by an orderley when Spaulding, who had seen the mischief in my eye when I left the ground, came in to mediate. He talked with me for an hour or so and then went to see the Lieut. Col. [He] came back and said that the Lieut. Col. would call and see me and offer an apology. Soon after he came in and made the *amends*, such as it was, not by an open, manly apology, but in a sneaking, round about manner. I despised him for his cowardice more than I had felt indignant at his insult and the matter was allowed to drop, but not to be forgotten by either of us as subsequent events will show.

This picture business was carried to a great extent in Washington [and] the men found it to be a good excuse to get permission to visit the city. It was wonderful how they managed to smuggle whiskey into the camp. Every means was adopted to prevent its introduction and still some of the men would get intoxicated every night, become boistrous and have to be locked up in the guard house. It happened one day when I was the "Officer of the Day" that several of the men had obtained leave to go to the city for the purpose above mentioned, taking their muskets with them. During the afternoon the Officer of the Guard noticed that all who returned to camp were very particular to carry their pieces at a Support Arms. An inspection of arms discovered the fact that the barrel of each musket was filled with whiskey, the muzzle & nipple being carefully plugged. Each barrel would hold about a quart. The secret was out and sobrietey reighned in camp for several days.[20]

The arrival of the Colonels Wife & daughter Mary was quite an event for us. She assembled the best of the Officers around her in the little house which she occupied and here many pleasant evenings were passed with music, dancing and social concourse. Our sociables were often graced with the presence of acquaintances from the city, both Gentlemen & Ladies. One evening Mr. Frank Blair visited us. He subsequentley became a Major General of Volunteers and candidate for the Vice Presidency when Horatio Seymour was nominated (and defeated) for President.[21] Here I learned to dance the Lancers, which was then new, and the happy hours passed in that little house will never be forgotten by any of the Participants.[22]

When I think of these days the sad question will arise, where *now* are all those happy and congenial Spirits? It seems but a short time since we were all together there. So vividly are the scenes infused upon my memory and yet many years have passed since then. Some died of disease, some were killed by the bullets of the enemy and of all these bright, happy, hopeful men so full of life and animation but few remain on earth and they are scattered to the four quarters. In every position of life the reflective mind finds many sad recollections, but this is especially true of the Soldier. The friendships which he forms are apt to be devoid of selfishness and consequentley more pure. The changes are also more rapid and sudden than in any other calling, hence to the reflective mind more saddening. But I must not dwell upon this Subject.

The Colonels lack of military knowledge was perfectley wonderful. He did not know the simplest movements and it seemed impossible for him to learn. At his Wifes request I passed many hours with him alone in his tent, long after Taps had sounded and when all in the camp except the guard were asleep, in the vain attempt to teach him the Manual of Arms and some of the simpler movements and orders, that he might at least appear creditably at dress parade. But all to no purpose. As sure as he took command, so sure was he to make some egregious and inexcusable blunder. The Lieut. Col. & Major delighted in his discomfiture for they were ambitious and would not volunteer any assistance and neither would he ask it of them. I felt deeply his disgrace and was willing to do everything in my power to assist him, a kindness which I think he appreciated at the time. No one in the Regiment but his Wife and ourselves had the slightest idea that the Colonel was receiving private midnight instructions from one of his Captains. At one time General McClellan visited us at Ponton drill and remained to see our Dress Parade. When the Adjutant had brought the Regiment to *Present* and had taken his position, the Colonel drew his sword and gave the command to *Order Arms*. Of course the men stood as they were until the Adjutant whispered to him to bring them to a *Shoulder* first. This was glorious for his enemies but dreadfully mystifying to me. His inability to learn anything military resulted finally, as was natural enough, in his complete failure and partial disgrace.

Thus you will see that our Winter in Washington was passed pleasantly and swiftly by. But there was another picture which you have not seen and it was well for me that I did not realize it fully. Your Mother was at home with her little infant. She was weak and suffering from disease of body as well as anxiety of mind. With me all was new and strange and fascinating [but] with her [was] suffering and anxiety of mind which at times amounted to blank despair. The days and weeks and months which passed so rapidly to me, to her dragged their slow lengths along in days of bodily prostration and nights of sleepless anxiety. But no word of complaint came to me from her for she felt that I needed all the courage I possessed to meet the trials which were in store for me. We wrote each other often but it was not until long afterwards that I became aware of the extent of her sufferings that winter. Nor was it altogether loveley with me. I have presented to you the bright side of a Soldiers life. The innumerable privations little and great, the constant apprehension and uncertainty of the future, the anxiety as to the ultimate result of the cause of which we were preparing to fight, the knowledge that henceforth my life was to be taken, as it were, in my hand, with those natural forebodings which *will arise* as to the future of my loved ones should it be my lot to fall, at times pressed themselves upon me and caused an indescribable sadness such as none can realize unless similarly situated.

And so the year 1861 passed away and 1862, with its active events, was ushered in. [See Appendix 4, Item 1.]

Chapter Five

Commencement of the Peninsular Campaign 1862

Thus we passed the time at Washington until the month of March. We received new uniforms of blue in place of our old grey ones and were in every respect fitted out for active service. Among our other duties we were required to learn the use of signals, which had then just been introduced by Major Meyers, to whom we reported twice a week for instructions. His code was adopted by the Army and proved of invaluable service.[1]

During the month of February I was visited by an old acquaintance, Mr. C.C. Coe of Rome, who had adopted balooning as a profession. He presented his plans for the annihilation of the enemy by the use of his baloon, which was to take up large quantities of explosive shells and at the proper moment drop them into the lines of the enemy. A trick operation certainly, especially with a strong wind blowing. No amount of reasoning could dissuade him from his purpose of presenting his plans to the Secretary of war and he finally left in March wounded in spirit because I refused to recommend them.[2]

My Father came to visit me in camp [and] when he saw how my bed was made on the ground with three blankets he concluded that a Soldiers life was better suited to me than himself. We made him comfortable, however, in the little house which was occupied by Mrs. Stuart and her daughter. Every thing was new and strange to him and he enjoyed his visit with us very much.

Grand Reviews were more the order of the day. These movements generally indicated an advance of the Army and, as a consequence, rumors of our sudden departure became frequent. I must not forget to mention what was known as "*The Grand Review,*" at which 75,000 men passed in review before the President, General McClellan and a great crowd of distinguished persons from all parts of the world.[3]

Soon after this and while my Father was still in Washington, I obtained permission to visit our friends of the 14th Rgt. stationed near Munsons Hill. It was a long walk, about 9 miles from our camp, but my Father stood the fatigue like a hero. We crossed the Potomac at the Acqueduct Bridge, passed Fort Corcoran and our old camping ground at Halls Hill. The whole distance was occupied with camps and corralls for animals. Our friends of the 14th entertained us handsomeley and the next morning we started on our return. We had not gone far when we noticed an unusual commotion among the camps and, as we ascended a slight elevation, what should we behold but the whole army in motion. As far as the eye could reach Regiment after Regiment was moving towards the South, knapsacks packed, flags flying [and] bands playing. Artillery, Cavalry, Infantry—they covered the hills like great waves of motion, their bright arms glistening in the sun. It was indeed a glorious spectacle. My Father was enchanted at the sight and I hurried up his walk lest returning I should find my Regiment gone also and without me. But, when we arrived at home, there was the Regiment just as we had left it, not a ripple of excitment visible. At first, I was completeley disgusted for I thought our Regiment had been *forgotten*. This was the move upon Manassas which resulted in the speedy return of the army, for the enemy had evacuated the position leaving nothing but a few "Quaker Guns" to attract our attention.[4] But the days of inaction were over. The advance upon Richmond via the Peninsula had been determined upon and some portions of the Army had commenced the embarkation.[5]

On the 18th day of March we received orders to break camp and report at once to General [Irvin] McDowell near Fairfax Seminary.[6] We received the order with delight and in a few hours were ready to move. All our extra baggage, tents &c. was packed and stored away in our workshops. We called upon and bade good bye to our friends on the river and the next morning we were on the march. It is not to be expected that we could leave our old camp ground where we had passed so many pleasant days without some regret. And so, with mingled emotions of joy and sadness, we marched past the Capitol, across Long Bridge and were once more on the *Sacred Soil* of Virginia. We halted at the other side of the bridge and Mrs. Stuart took her departure. When she left, it seemed that we had indeed entered a strange land and that all our friends were far behind us.

About 2 P.M. we reached the position assigned us and went into camp—but *such* a camp—the ground was covered with stumps and burnt timbers and rather marshy at that. We made no pretentions to a regular camp but *squatted* as best we could. Our wagons did not get up and we went supperless to bed, but this was the only time, for after that our Officers knew enough to have their haversacks full whenever and wherever we moved. The men of arms faired better as they had their rations with them.

The night was chilly and a drizzling rain was falling. About 11 O'clock that night Captain Spaulding came in with a large piece of meat and invited me to share it with him. We cooked it upon the coals of our camp fire and never before or since have I enjoyed such a meal as that. I shall not soon forget the picture we presented sitting on the stumps that night with each a piece of bark for a plate and, with our jacknives & fingers, eating that immense piece of meat.

Of course, we commenced *drilling* at once. We were reviewed separateley by General McClellan and General McDowell. The only remark the latter made was to tell our Colonel to see that every man was provided with an extra pair of shoes. This certainly looked like business. Then the whole of McDowells Corps was reviewed [and] we were allowed to be spectators that day. I was standing with a group of Officers as the Commanding General and Staff rode by. Among them I noticed a large red faced man with a black hat and citizens dress. Instinctiveley, and loud enough to be heard by the whole retinue, I called out, "*put out* that man with the stove pipe hat." The whole party looked around, all but *him*. I did not know at the time that he was the famous D. Russell, correspondent of the London Times [and] better known among us as "Bull Run Russell."[7]

We were drilling one day, as usual, under command of the Lieut. Col. Capt. Ford & myself commanded the two Companies on the left [and] Capt Murray the other. The Col. lost his temper, as usual, [and] he ordered us not to give a command unless first given by himself. The movement from line into column was ordered. We obeyed his instructions literally, and so instead of halting the left of our companies at the proper moment, we let them wheel, and around they went like three tops independently. The Colonel rode down to the left livid with rage but said nothing and they might have kept on wheeling until this day had he not given the order to halt. This circumstance did not tend to increase the Lieut. Col.'s friendship for me and he sullenly treasured up his plans for revenge.

Fort Ward was always remembered by our men as the place where the Chaplain, who also acted as Postmaster, refused to deliver our letters because it was Sunday.[8] This was a stretch of authority which he was obliged to relinquish by order of the Colonel and it lost him the friendship of the entire Regiment.

On the 24th of March, McDowells Corps advanced southward while the ballance of the Army was being transported by water to the Peninsula. On that day we marched as far as Bristoe Station, following the line of R. Road.[9] We were not at all pleased with the idea of being detatched from under the immediate command of General McClellan. On this march I saw for the first time a battery of Gatling Guns. These guns were known in after years during the war of 1870 between France & Prussia as the "Mitrailleuse."[10]

The next day we passed a few miles beyond Bristoe and while on the move were ordered to countermarch and report to General McClellan at or near Fortress Monroe. At this [we] were perfectley delighted. We returned to Bristoe & encamped for the night in a cedar grove. During the night some of my men thought they would try their hand at foraging, a new business to them. How they did it I am not supposed to know [but] in the morning I found some nice fresh pork for my breakfast and asked no questions. Folley told me that a fine horse was waiting for me just inside the guard line and that my men were anxious to see me mounted upon him. It was not my good fortune ever to see the animal, for about 9 O'clock the Major of a Cavalry Regiment which was near by came in and claimed the horse. No one objected to the proceeding, as no one knew how he came there.

A train of cars was brought up to take our two Regiments back to Alexandria. The train was made up of box and platform cars [and] the 15th Regt. was shrewd enough to secure the most of the box cars, which left nothing but the open platforms for us. As we were getting our men on board it commenced to snow, the wind blowing fierceley from the north. The storm increased in violence and coldness as the night advanced. We left Bristoe Station about 8 P.M. and did not arrive at Alexandria, a distance of about 26 miles, until the next morning at 7 A.M. We could have marched the distance in much less time. All that night we were exposed to the pittiless storm, cramped and frozen on these open platform cars, the wind & snow driving with great fury around us and blinding us. Once we stopped, slid down a steep embankment about 60 feet and building a fire with the fence

Brig. Gen. Daniel P. Woodbury. Courtesy of the Library of Congress.

rails managed to thaw out. The sufferings of that night were not surpassed by anything in the whole course of my military experience. Arriving at Alexandria almost dead with hunger & cold, we occupied the houses in the vicinity of the depot and managed to get comfortable again. We were glad enough to get on board the Steamer Louisiana and that night about 12 O'clock [were] happier still when we steamed down the Potomac into Chesapeake Bay.

A pleasant voyage of about 30 hours brought us in sight of Cheesemans Landing. The first time I saw General Woodbury was as we approached the landing. He was standing on the promenade deck in company with Lieut. Col. Alexander selecting through his glass a position on which to run the boat ashore. General Woodbury proved to be a mild, unassuming Christian gentleman, a fine Engineer [and] a talented officer but too retiring in his disposition to attain an exalted position.[11]

Cheesemans Landing is a few miles north of Fortress Monroe.

Col. Barton Stone Alexander. Courtesy of the Library of Congress.

When we landed, the weather had changed and the bright warm sunshine was welcomed by us all. The day was occupied in getting our things ashore as best we could as there was then no wharf there, it having been destroyed by the Rebels. The next day we took up our line of march through the flat, woody, marshy country for Yorktown. We passed near Bethel, where a battle had been fought, and saw many lines of earthworks which had been thrown up and abandoned by the enemy, objects of much interest to us as they were the first we had seen.[12]

We arrived at our destination near Wormleys Creek not far from General McClellans Head Quarters that day about noon and went into camp. Captain O'Grady laid out our camp ground marking the positions of each company in line upon a chip with such euphonious names as "Whiskey D" [and] "Thieving C" in honor of the principal characteristics of each Company. A few days found us in order and ready for the work of the siege which was then progressing. There was enough for us to do [as] we were to make Fascines and Gabions for all the batteries and assist generally in their construction. History has shown that the Army of the Potomac could have assaulted and carried the entire position had we done so immediateley upon our arrival, but I must not enter upon a discussion of what might have been done.

The commanding general had named our position Camp Winfield Scott and his General Order for the conduct of the siege I have thought of sufficient interest to preserve. [See Appendix 4, Item 3.] This order exhibits the predominant traits of General McClellans character—*Caution*. We outnumbered the enemy three to one and yet he makes all his preparations for defense as though they were our equals. But there was a feeling of confidence among the men and a degree of enthusiastic admiration and love for him which no other General Officer in our Army ever attained, for they felt that their General would not sacrifice a life unnecessarily. There was a charm, some magnetic influence about his person, that endeared his troops to him personally, and all who came in contact with him or ever saw him partook of this feeling to a greater or lesser degree. You will see more of the development of this wonderful *idolatry* as we proceed. [See Appendix 4, Item 2.]

You will find in the library a report of the Engineer and Artillery operations of the Peninsula Campaign as written by Genrls Barnard & Berry.[13] This book will give you a good general idea of our duties and operations so I will confine myself to some personal reminicences as they occur to me. One incident will illustrate our Colonels knowledge of military engineering. He had been ordered to detail a party of men to construct a Mortar Battery (this battery was afterwards known as Battery No. 4). Sending for me, he informed me of the order and that he had selected me to have charge of the work.

"And now," said he, "have the Adjutant detail all the *masons*, *bricklayers* and *plasterers* from the regiment and make up the complement (150 men) of laborers." He had the idea that a "Mortar" Battery (so called from the fact that *mortars* for throwing shell are mounted within or behind an earth work) was to be constructed of brick and stone and *mortar* and necessarily required men accustomed to working with *mortar* to construct it. I was so overwhelmed with mortification and disgust at this display of ignorance and felt so humiliated withal, as I had always upheld his character to my associates, that I never mentioned the incident until after he had left the Regiment.

I took command of the party and commenced the battery, which was simply to level down a hill side, throwing the superflous earth into the water (Wormleys Creek) until a sufficient space was made upon which to place our 13 in. mortars, 10 of which were placed in this battery.[14]

One night I was ordered to relieve the Officer in charge of this work at midnight. The night was dark and it was raining. The 74th N. York Regiment had been ordered to report to me for duty [and] the men of this Regiment positiveley refused to work. I informed the Officers that if they had no better discipline than that, it would be my duty to report them at Head Quarters. The Officers used any means in their power to get the men to work but to no purpose. My report of the affair you will find on page 163 of the book above alluded to. No public notice was taken of this act of insubordination but in a few days the Regiment was ordered away. I saw them again in 1864, a mere skeleton of a Regiment. They had been transfered from one department to another in the South and West with no one to take an interest in them until they were reduced to a deplorable condition and this was their punishment for that nights disobedience of orders.[15]

One day it happened that quite a number of cattle were discovered grazing between the opposing picket lines. The men from both sides called them all sorts of pet names to enduce them to come over to them, but as soon as either party attempted to go out for them the other side would fire. At last, Lieut. Hoyt of my company crawled off to the right of the cattle unobserved by the enemy and, by dextrous manoevers, shielding himself behind the cattle, first on one side then on the other, succeeded in bringing them all (17 in number) into our lines unharmed. In a short time the animals were snugly corralled within our Regimental lines and the Hospital was supplied with milk and our company with delicious fresh meat for the ballance of the seige.

Among the cattle were two young steers, which we utilized by foraging a wagon & yok from a neighboring farm house. This little team was used by the company in carrying the Knapsacks and extra baggage (which was not allowed) for several weeks on the peninsula. We also found a store of corn which we had ground at an old mill a few miles to the rear and the way we luxuriated in corn cakes, pudding & milk &c. was the envy of all the companies round about.

Battery No. 4 was soon completed [with] the heavy 13 in. Mortars in position and everything in readiness awaiting the shell. Three canal boats loaded with these had been towed up the York River and left at the mouth of Wormleys Creek during the night. I was ordered with my company to go and bring the boats in. Arriving at the spot at daylight, I found Major Webb of Genrl. McClellans Staff already there with a Squad of men who had succeeded in getting two of the boats in the Creek and out of range of the enemys guns, who were firing vigorously upon the party.

The arrival of my party served to increase their venom and they sent their shells around us in a liveley manner. One of the boats was still out in the river and between us and the battery which was trying to destroy it. Major Webb exhibited considerable excitement as it was unknown whether the canal boat contained any men or not. He had about given her up as lost as it seemed impossible to reach the boat under such a fire and tow her back and into the creek against the stiff wind then blowing.

After consulting with the Major for a few moments I suggested that it would be a good plan to go out to the boat, cut the anchor cable and let her drift ashore, where we might remove her cargo during the night unobserved and unexposed to the fire of the enemy. The Major was delighted with the plan, but asked "who will go out and take that risk?" My reply was that I would and I immediately called for 4 good men. Sergeants [Wallace B.] *Simpson* & [George W.] *Cowan*, & *Corporals Freeman* [Warren] & [John P.] *Strong* stepped forward. I took the steering oar and in a minute we were on our way.

The boat was about a quarter of a mile from us and about the same distance from shore. We rowed vigorously for the boat, the battery in the mean time playing upon us. Soon we reached the craft [and] Corporal Strong ran forward and cut the cable. The boat soon swung around and headed for the shore. I ran down into the cabin but no one was on board of her. The wind being favorable, the craft soon struck bottom about a hundred yards from shore but it seemed a long time to us. Once a huge shell came rushing through the air so near to my head that the compression of the air fairly flattened me upon the deck [and] my impression was that my head was off. The shell struck the water a little way beyond and blew a column of water high into the air. Just as the boat struck bottom a shell struck her in the stern, exploded and carried away the whole after part of the cabin but the cargo was safe as the boat could sink no further. A line was on deck which we ran to shore and fastened to a large cedar tree. Then we rowed back to our position and soon rounded the point out of harms way.

The cargo was unloaded during the night. As soon as I returned, Major Webb ran to me and, taking both my hands in his, congratulated me upon my safe return. Said he, "I shall report your brave action at Head Quarters and you will hear of it again." I did not see Major Webb after that until 1864. Then he was a Major General of Volunteers and I was a Colonel. How he mentioned my name at Head Quarters you will see by referring again to the same book, in which it seems he took all the credit and my name was not mentioned at all.[16]

The seige progressed slowly. The first and second parallell[s] were completed and the order to open fire from our batteries was expected daily but it never came. We were kept busily at work, making new batteries and strengthening the old ones. One day Lieut. Col. Skillen of the 14th Regt. called to see me. Generally cheerful and liveley in manner, I noticed that on that day he was unusually sad. The 44th N. York, who took our place in Butterfields Brigade after we left Halls Hill, had been peculiarly unfortunate, loosing very many of their men by disease and the skirmishes which were continually going on. They had established their burying ground near by our camp [and] Skillen accompanied me to see their ground. Reading the names on the long lines of cracker boxes which served as tombstones, Skillen seemed deepley affected and just as he mounted to bid me good bye remarked that he had now *stopped swearing* and that in future he intended to live a better life. I never saw him alive again.[17]

Chapter Six

Yorktown, White House and Fair Oaks or Seven Pines 1862

About the first of May appearances which a Soldier never misunderstands indicated that matters were rapidly approaching a crisis. On Saturday, the third day of the month, I was ordered towards the center of our line to complete a small battery which was understood to be the last that was to be placed before opening fire upon the town.

We made our way to the position with considerable difficulty as the only approach to the spot was commanded by their batteries and they kept up a continuous and very annoying fire. Reaching the ground, however, we worked like braves all that day and far into the night before our guns were fairly into position.[1]

Once a solid round shot came through the embankment in front of us, passed over my shoulders (I had just stooped for a shovel), rolled forward about 20 ft. and broke the leg of one of the men who stood behind me. Had I been standing upright as I was a moment before, you would not be reading this story from my hands. This was "a close call."

It was the intention of our General to open fire with all the batteries on the following day. The enemy must have surmised his intention, for never before had they kept up such a vigorous and continuous fire upon us. As the darkness increased we noticed [a] very unusual commotion in [the] rear of their lines. Occasionally a bright fire would illuminate the whole heavens and we could see that they were burning buildings. Then the quick report of their guns would cause us to dodge down behind our works again. Then a series of explosions occured one after another [and] masses of burning material would be blown into the air.

Suddenly the explosions became silent and the fires ceased to burn.

Completely exhausted by the days labor and the anxiety of two successive nights in the trenches, I was glad enough to be relieved at 5 O'clock and wended my way camp ward.

Approaching the river I observed our whole fleet of transports, which had for some weeks been laying in the stream, moving upward. "Yorktown is evacuated," passed from mouth to mouth and our joy was mixed with sorrow that we did not have an opportunity to see the effects of our batteries which had cost us so much time and labor to construct. It was a beautiful Sabbath morning, warm and bright. The whole army was soon in motion and Yorktown was soon occupied in force, the Infantry pushing through first, then the Cavalry and Artillery.

The enemy had planted Torpedoes leading to all the approaches and a number of our men were killed by these concealed impliments. Two of our companies were detailed to find and dig up these missels.[2]

Our camp was allowed to remain that day and at 9 O'clock [a.m.] I entered the town with my Lieutenants. How eagerly we examined the works now abandoned but now and then the explosion of a concealed Torpedo cautioned us to be on the watch. Passing through one of the streets I saw a man rise suddenly in the air then drop, a mutilated mass of humanity. The indignation of our men at these works of rebel barbarity was indescribable and had any prisoners fallen into our hands it would have been difficult to restrain them.

I visited General [John Bankhead] Magruders Head Quarters, the fortifications and all other points of interest.[3] A tent was left standing near one of the batteries, entering which we found the table already sett, which showed that the enemy left hurriedly. Here I captured half a dozen napkin rings with the Officers names worked upon them. You will find them among my relics. Entering a hospital tent, I found a book giving the names of the men who were killed & wounded by the explosions of one of their guns a few days previous and which I witnessed from our battery No 1. The first thing that attracted my attention was some writing on a sand bag at one of the entrances. I cut it out with my knife and preserved it as a memento of the occasion.[4]

General Butterfield was at once appointed Provost Marshall and was arresting every one who attempted to carry away anything, so we had to conceal our plunder as best we could under our coats or in our pockets.[5]

The forces on board the transports under command of General

[William Buel] Franklin moved up the river to White House Landing, a detachment of the 15th Regt. accompanying them for engineer purposes.[6] A large force also moved in the direction of Williamsburg where a battle was fought, an account of which you will read in the histories. We could distinctly hear the sound of the cannonade but had no idea that our forces were so near meeting defeat until long afterwards.[7]

Leaving our camp at Yorktown we took up our line of march following the right or South bank of [the] York river, highly elated with the prospect of soon quelling the rebellion. It rained almost every day and every night, rendering the roads almost impassable for ordinary travel, but when our heavy Artillery and baggage wagons came upon them they were soon converted into quagmires.

The Regiment was divided into Sections and each Section allotted a certain number of wagons. The animals soon began to give out and die and hundreds of wagons were tipped over and abandoned. Here was where our little ox team came into play. The little fellows would drag their wagons loaded with Soldiers knapsacks and muskets through the mud as though they were used to it. This relieved the men of so much weight that the little oxen received the blessings of every man in the company and I believe our men would have gone without meat for a month before they would have allowed the death of the little fellows.

This march up the peninsula has passed into history as one of the most tedious and exhausting of the war. Officers and men alike were covered from head to feet with mud, presenting anything but a military appearance. At night we were glad enough to rest upon the ground making no attempt to raise our tents.

The second night of our march we bivouaced in a field near the road. Folley, Hoyt & myself had taken one tent from the wagon (Company Officers were then allowed *one* tent only) and spread it upon the ground. We were soon asleep as I supposed. Several hours later I awoke and hearing Folley & Hoyt whispering in earnest, of course listened to get the subject of their conversation, as they evidentley did not wish me to hear.

Folley it seems, soon after we had lain down [and] being satisfied that I was asleep, arose quietly, went down to the Quartermasters corrall, mounted a horse and rode out into the night in search of plunder. He was determined upon capturing a horse for his own use. As soon as the moon arose he could discern objects at a considerable distance on the road. Observing away in advance of him a horse standing by the roadside, he approached cautiously, sure of his prey. But as Folley put it "the goll damned old horse I was riding wheezed so that you could hear him a quarter of a mile," and just as he was nearing his coveted prize out came a negro from a hut nearby who mounted the animal and rode as rapidly as he could urge him away. Folley put spurs to his horse in hot pursuit but after a chase of several miles the poor old quartermasters animal gave out and Folley was obliged to give up the chase. He returned to camp very much disappointed.

Hoyt in the mean time, with the assistance of Corporal Freeman [Warren], had succeeded in capturing *two* fine horses, one of which I overheard him say was intended "for the Captain." They were brought into camp and fastened nearby. This was the substance of their conversation, every word of which I heard as they supposed me fast asleep.

Reveille aroused us at 5 O'clock and we were soon ready again for the march, but the two splendid bays were "non est." Some adroit thieves from another regiment had stolen them from the stealers. Then the secret was let out. There was no resource left but to continue the march on foot. Engineer Officers were allowed horses if they desired to purchase them, hence the desire of my Lieutenants to mount themselves as cheaply as possible. After this escapade of Folleys, I informed him that he need not take any more chances of that kind on my account and of course, as I was in duty bound to do, had to administer a reprimand for leaving camp without permission. Besides, I had an ambition to lead my Company through one campaign on foot and it was my boast for years afterwards that I *marched* through the entire Peninsular Campaign. I have always felt that my experience during that campaign was of immense service to me ever after, for an Officer should know by experience the actual fatigue of the Soldiers to properley lead a Regiment.

Speaking of horse back riding reminds me that I knew nothing of this art when I entered the Army and subsequentley felt rather awkward when the Colonels Wife invited me to ride with her one day at our camp in Washington. I postponed the ride upon one pretext and another until, after several weeks of practice in the early morning hours unobserved by anyone, I became quite proficient and finally accepted her invitation. This is one of the manly accomplishments which I hope time and opportunity may be offered you to acquire.

We dragged our weary way up the peninsula through the interminable mud until we arrived at White House. The enemy in the mean time moved rapidly up the James and entered Richmond, which they had surrounded with impregnable fortifications. One night we entered the great wheat field consisting of about 11000 acres. The property of General Lee, the wheat was breast high and the whole Army was encamped in it leaving room for another of equal size to encamp beside it.[8]

We remained at White House Landing for several days. The weather was getting very warm [and] the continuous rains and hot sun had raised a miasma which was telling with fearful effect upon our men. Several of the Officers also came down with chronic diarrhea and fever and two of them, Lieuts. Yates and Perkins, were sent home completeley enervated. The former, Lieut. Yates, died before his arrival at N.York. The steamer which carried them, the Daniel Webster, was loaded down with sick and well do I remember their woe begone countenances the morning I assisted in carrying our Officers on board.[9]

From the White House (so called from the House which stands there and in which General Washington was married) we marched over to Bottoms Bridge on the Chickahominy.[10] Here we laid an Austrian Tressle Bridge one rainy afternoon, which [was] left in charge of a detatchment under Lieutenant Hine and that night we marched all night reaching Gaines Mills at daylight. Here we went into camp and commenced our operations near Mechanicsville and but 7 miles from Richmond itself.

Our camp was situated in a grove about 1/4 of a mile [north] from the river on a rising ground which overlooked the flats extending on each side of the Stream [with] the rebel batteries in plain sight on the opposite side. The Chickahominy itself was a small and sluggish stream except in time of freshets, when, as it has its rise in the mountainous regions to the West, it was subject to rapidly overflowing the banks on each side.

A portion of our Regiment had been fitted up with a Ponton Train. This detatchment consisted of Companies F & C, the former under command of Capt. Spaulding, the latter my company. Capt. Spaulding, being senior in commission, had command of the detatchment. The regulars under Major Duane also had one bridge train in their charge. We, Spaulding & myself with our companies, commenced the construction of a crib bridge on the extreme right. The work was very laborious as we had to cut the logs for the bridge at a considerable distance from it and haul them to the place with our teams. The amount of work we did on the Chickahominy was simply enormous. We completed this bridge and the approach on the north side, but when we commenced the South approach, the enemy opened upon us with such a vigorous fire that we were obliged to abandon the position. This bridge was never of any practical use to the army.

Appearances indicated a movement in some direction and that before long. [See Appendix 4, Item 4.] One night during a fierce thunder & rain storm the whole of Porters Corps passed by our camp. It was so dark that one could not see their hand before their face. The flashes of lightning revealed the troops in solid column moving rapidly by. The next day the battle at Hanover Station was fought resulting in a victory to our troops, though not on a large scale. They (Porters Corps) returned to their old position the following day and I was never able to learn that the victory resulted in any substantial advantage. It was understood by us that McDowells Corps should have cooperated in the movement but they did not.[11]

On Friday, May 30th, long lines of troops filed down towards the river and crossed on Sumners Bridge, sometimes called the "grape vine bridge", so called from its resemblence to a grape vine in its crooked windings. This bridge, situated about 1 1/2 miles down the stream from our camp, was constructed in a rude and unsubstantial manner by General Sumners Corps. It was frail and rickety but saved the Army. Caseys Division was already across and had taken up its position nearly opposite to us.[12]

That afternoon General Woodbury called me to accompany him on a reconnaissance. His Aide, Lieut. Cassin, was then with him.[13] He dismounted and we three crept cautiously towards the stream and then along its bank sometimes up to our knees in water, jumping from log to log, always in a stooping or creeping posture in order to avoid the attention of the enemys Sharp Shooters, who were posted on the opposite side. After two or three hours of this kind of work, General Woodbury indicated the position which he had selected for me to construct a bridge that night. I waded the Stream, which was then not much wider than a common street, and found the water about up to my middle. Observing well the position, we returned to camp thoroughly Saturated with mud and water. This was what General Woodbury termed a "still hunt."

Soon after we returned, the sharp rattle of musketry on the opposite side informed us that the skirmishers were already at work. My orders were to get my section of a bridge in readiness, move down in rear of Gaines House, where I would be concealed from the view of the enemy, and there to await for orders [and] remain ready to move at a moments notice. I moved my command down to the position indicated and there remained. Night came, dark and cloudy, still no orders. I laid down on the ground and fell asleep. About 11 O'clock an Aide from General Woodbury came to me and imparted the order to move my train down to the river and commence laying my bridge. Taking a lantern in my hand I went ahead of my train in order to guide them over the narrow and ill defined path which no one knew but myself. Having passed over about half the distance, an Officer came up to me and said that Col. Pettes wished me to report to him immediately. This was the first intimation that I had had of his presence. Halting the train, I passed to the rear and found him in a towering passion because I had not reported to him. His insulting manner added to the responsibility which I already felt was more than I could stand and I informed him that the command of the expedition was now in his hands and that the responsibility no longer rested with me.

"Then, Sir," he roared out with an oath, "I order you, Sir, to go ahead and lay the bridge."

"That, Sir, is just what I should have done if you had let me alone," I replied and again took my position.

The field which I had reconnoitered but a few hours before was now covered with water and all my landmarks were obliterated. I could mark the position I had selected only by a clump of bushes. The water was rising rapidly and running like a mill race. Before I could get the first tressle in position it had raised 18 inches. The fates or something else had ordered this sudden rise in the Stream evidentley to thwart our movements.

My men worked with almost superhuman energy. The roaring of the water drowned the voice of command and all depended upon individual effort. Oftentimes a tressle would be completely upset and the men who were upon it swept rapidly down the Stream. It was simply impossible to construct that kind of a bridge under the circumstances.

General McClellan, supposing that the bridge would be completed by 1 or 2 O'clock [a.m.], had ordered Porters Corps to cross before daylight. The troops moved down into the plain to find it covered with water from 6 to 12 inches deep for half a mile from its banks. General Martindale and several other General Officers came up to where I was at work with all my might and inquired impatiently when the bridge would be completed. I could only answer as soon as possible. Then Col. Pettes would come forward with another volume of oaths but not a suggestion could he make whereby the work could be expedited. And so everything went in the most heart rending confusion, for every one felt the importance of getting troops across to save those from utter annihilation who were already there and completely cut off from the main body of the army.

I soon discovered the cause of the singular conduct of Lt. Col. Pettes when I saw him go so often to Capt. O'Grady, who always handed him his canteen which contained something beside water. Taken altogether it was a fearful night. I can never tell half the torment I suffered. I verily believe that that night added two years to my age. Glad enough was I to be relieved at daylight.[14]

The next and succeeding days was fought the Battle of Fair Oaks, in which owing to the causes I have mentioned, less than one half of our Army met the shock of the whole Rebel host and the only wonder was that they were not entireley annihilated.

We stood in our positions all next day listening to the battle which raged so near us but powerless to aid. We could hear the cries of the men as they charged and recharged, [see] the smoke of battle rising in a dense column above the trees, but not a man could we lend them. The old grape vine bridge below still stood, but owing to its feeble condition but few men could cross it.

Our troops under Casey were at first routed but rallied again and brave old Sumner saved the day, for, as you will see by the published accounts, our men held their own notwithstanding the odds against them. All accounts agree that had the ballance of the Army been able to cross, Richmond could have been taken without a shadow of a doubt, for after their repulse the enemy fled back and through the city panic stricken.[15]

Was the sudden rising of the Stream a direct interposition of the *Almighty*? I think it was. *He* did not design to have the war end until his own good time.

Chapter Seven

The Seven Days Battles and End of the Peninsular Campaign 1862

Thus ended the Battle of Fair Oaks or Seven Pines, as it was sometimes called. Our troops on the South side of the Chickahominy fortified their positions and both sides seemed to await for the other to take the iniative for the next movement. In a day or two the stream which had given us so much trouble at a critical moment subsided into its original channel and we commenced to build more bridges to connect the two wings of the Army.

I was ordered to take a detatchment of men down towards the Grapevine Bridge for fatigue duty. On the way we met a party of Officers uniformed and equipped as I had never before seen. *Gorgeous* is the only word that will express it. Silver and gold covered the laphels and skirts of their coats, caps and sleeves. My party was armed with picks and shovels, the muskets being slung upon the back. As we approached them they drew off to the side of the road. One of them who seemed to be the chief and whose uniform fairly sparkeled in the sunlight, his breast all covered with hanging medals, lifted his hat, revealing a pleasing countenance, burning with intelligence. In that position he remained until we had passed. This was General Prim, the Spanish Soldier of renown who was on his way back from Mexico to Spain and who had stopped over to visit the Army of the Potomac. His reputation as a soldier and a scholar was already fully established but subsequent events in Spain placed his name among the first of Statesmen. The prominent part which he played in connexion with Spanish affairs and his tragic death are matters of history.[1]

Now I must tell you another *horse story.* On returning to camp from fatigue duty one evening, a party of my men discovered an old grey horse lying beside the road where he had evidently been left to die. The men gathered around him and concluded if possible to save the old fellow. By propping him up from both sides they finally succeeded in getting him into camp where, by careful nursing, in a few days the old fellow was able to stand upon his feet. Lieut. Folley took him in charge and a most valuable animal he proved to be. "Old Frank" had a friend in every man in the Company. Gentle, intelligent, an easy rider, affectionate and fast when it was necessary to be, it is no wonder that he became a great favorite. Lieut. Folley rode him until we arrived at Washington where I bought him for $40 and kept him during the ballance of the war. When on long marches I always rode "Old Frank." He was as easy as a cradle and always knew when I was asleep and conducted himself accordingly. When we bivouacked, "Old Frank" would be turned loose to graze at pleasure but he would never go beyond the length of a lariat rope and after eating the grass within the circle would come and lay down beside me. On *drill* or *parade* I rode another and more *showy* animal [and] on these occasions my Orderley rode "Old Frank."

A few days after the Battle of Fair Oaks I obtained permission to visit the field. It was the first I had seen and the sights of that day shocked and sickened me. The dead had been only partially buried. Bodies of men and animals were still exposed to the scorching rays of the sun and the stench arising from these partially decomposed and hideas masses was perfectley horrible. While I was roaming over the field moralizing upon the dreadful horrors of the war who should I discover but my old acquaintance, Elder Vogel of Rome. He was sitting on the steps of an ambulance [with] his sleeves rolled up resting from his labor, for he was now an Assistant Surgeon. From my earliest boyhood I had known him as a Baptist Minister. He was a man of considerable notoriety in our part of the Country. Had I seen my own Father in that position my surprise could not have been greater. Having visited the principal parts of the field, I returned to camp that evening weary and heartsick.[2]

Soon after this I was ordered to join Capt. Spaulding and with our 2 Companies move to the South side of the Stream where we commenced the construction of the largest bridge built across that river which was known as the Woodbury & Alexander Bridge.[3] The construction of this bridge was no small undertaking. First, the Stream had to be cleared of logs and underbrush to enable us to float our material to the spot. My company was detailed for this work [and] my men soon became regular "Water Dogs". The dense woods bordering this sluggish stream was filled with black snakes, lizzards, wood ticks and other unpleasant vermin. Working in the water up to our breasts for two or three weeks prostrated many of the men with what was known as the Chickahominy fever, for which there seemed to be no antidote but *quinine* in immense quantities. The Army lost far more men by this disease than by the bullets of the enemy.[4]

After the completion of this bridge I was ordered to build another just above. Company K assisted us in the construction of this bridge under my charge. It was 1100 ft. long exclusive of the river, which was spanned by 4 Ponton boats built in regular form. Large trees from 1 to 3 feet in diameter were felled, cut to the proper length, split in two and laid as corduroy [and] this, when covered with brush and earth, completed the approaches. This was known as the Cavalry Bridge and was second only to the Woodbury & Alexander Bridge in size and amount of labor expended in its construction. It is hardly mentioned in any of the reports as very soon after its completion commenced the 7 Days Battle and the enemy very soon got possesion of it. It was of no practical benefit to our Army.[5]

One morning I noticed a great commotion among the men in one of the Company streets. Going down to the spot, they showed me a huge black snake which they had just dispatched, having found it under the blankets on which they slept, where the reptile had crawled during the night probabley in search of warmth. I often amused myself by firing my revolver into *clumps* of them as they were gathered in the trees above us. A shot into the center of the group would cause them to dart in every direction like the lightning from Jupiters hand.

Twice while on duty clearing out the stream my little wooden pipe was knocked from my mouth into the water and twice did I pay 1 dollar in gold to the man who dove for it and returned this precious relic which orginally cost me 37 cents. As it was smoked by me during the entire campaign, I have thought it worth saving [and] you will probably find it among my things.

A whole month passed away in this manner, at the end of which our communication between the two wings of the Army were ample for any emergency. Gradually the main body of the Army was transferred to the South side of the Stream until nothing but Porters Corps was left on the North side. Thus we had about 20,000 on the North and about 75,000 men on the South side of the Stream. Every day we expected orders to advance. Once I crawled out to our outer picket line and enjoyed an unobstructed view of the spires of Richmond. If we were undecided what course to pursue, the enemy was not.[6]

[General T. J. "Stonewall"] Jackson, by a series of rapid movements, had defeated our detachment in the Shenandoah Valley and was already in rapid movement for Richmond when McDowell Corps was dispatched [but was too] far away to intercept his movements. The authorities at Washington were alone responsible for this action, which was ordered contrary to the advice and protest of both McClellan and McDowell.[7] Suddenley, on the 26th of June, Jacksons forces formed a junction with the ballance of the enemys Army and they pounced down upon our 25,000 with their 75,000 men. We had 75,000 men on the South side of the Stream against their 25,000, but they were so adroitly handled by McGruder as to completely deceive our Generals as to their numbers.

During the night Troops were hurried over to the North side to reinforce Porter and a continuous stream of wounded were returning. The troops immediately in front of our camp had been removed to the South side so that there was no force in our front nearer than Mechanicsville. We heard the firing off in that direction but, as it ceased soon after, we concluded a slight attack had been made and repulsed. We had already transferred the rest of our camp equipages to the other side of the Stream and were drawn up in line preparatory to moving leisureley away when suddenly the *Rebel Yell* was heard and the woods near us was filled with Rebels making rapidly for us. We were but two companies while they numbered several thousand. On they came pell mell, firing and screaming like so many devils. This was Jacksons Corps, which McDowell was trying to find a hundred miles away. If we ever took a double quick we took it then. The bullets flew about us but did no material damage.[8]

Martindales Brigade was drawn up in line about a quarter of a mile to the rear of us. We soon placed them between us and the Rebels and made our way across the river. We lost a couple of teams and wagons but more than anything else we deplored the loss of our little Ox team, which had been stolen the night before by a Vermont Regiment. Martindales Brigade held the Rebels in check and then commenced the Battle of Gaines Mill. Our troops, outnumbered 3 to 1, manfully held their own until darkness closed in upon the combattants, when a change of position was affected.

Early daylight found us on our way to White Oak Swamp where we were to prepare a passage for the army over a little stream which flowed through its center. Orders were given to give us the right of way and everything drew off to the sides of the road allowing us to pass rapidly through.[9]

At Savages Station we saw our wounded to the number of 2500 men, [many] suffering the torments of amputation while a large party was burying the dead. These wounded were finally abandoned to the enemy owing to a lack of transportation to get them away.[10]

Sumner was getting his troops into position to check the enemy as they approached. Jackson was delayed nearly all day by building the bridges across the Chickahominy but Magruder attacked Sumner with great impetuosity. He held his position with a stubbornness that marked that old brave, so that Magruder, assailing his position in successive charges till dark, met only bloody repulses. Thus, stout Sumner stood at bay, while, thanks to the barrier, the mighty caravan of artillery and wagons and ambulances moved swiftly & silentley through the meloncholy woods and wilds all day and all night without challenge or encounter on its winding way to the James.

During the night the rear guard also withdrew across White Oak Swamp. In our hurry to construct a bridge for the passage of the Army we had left our camp equipages on the North side of the swamp [and] all day and night we stood at our posts guarding the structure against accident while the surging masses passed over the bridge or forded the Stream. The 14th [NY] Regt passed us and an ambulance containing the dead body of my old friend Skillen was hurried rapidly over. They burried him a short distance beyond in a silent loveley place.[11]

Suddenly we heard great cheering along the lines which gradually approached us and then the General appeared, followed by his numerous and brilliant Staff. As he crossed the bridge, the pleasant smile upon his countenance betokened anything but a sense of defeat. Of course we cheered him lustily as he passed and he disappeared along the lines.

At early daylight our rear guard came rushing across the bridge [and it was] then we remembered that our camp equipages were still on the other side a half mile away from us. We hurried to the place but the enemys pickets were already in view. Then there was hurrying to and fro. I secured my papers and blankets and abandoned everything else. The rest did the same and we beat a hasty retreat for the bridge which was already on fire. Three of my men who lagged behind in the endeavor to save more of their effects were gobbled up by the enemy. We managed to get across the bridge which in a few minutes after became impassable.[12]

Our train was soon upon the march southward but I lingered behind to see our troops forming in a long, dark line to resist the passage of the stream. The line was soon formed with several batteries placed so as to command the road leading to the bridge. Then for the space of half an hour all was as silent as a Sabbath morning. A column of smoke ascended from the bridge which we could distinctly see in the ravine which lay before us. Then a party of horsemen appeared in the clearing beyond [and] we could see that the woods were rapidly filling up with dark masses of men on each side of the road. Suddenley a battery was run out into the clearing and in an instant the shot began to fall about us. One of our batteries was almost annihilated at the first fire. Then with their characteristic *yell* the masses of men rushed from each side of the woods into the road and moved rapidly in the direction of the burning bridge. Now was the time for our batteries to do their work. Simultaneously they opened fire directly at the masses of moving men. I could see what appeared to be portions of human bodies flying in the air. A few minutes of this dreadful fire answered the purpose. The road was again clear, with the exception of the prostrate bodies which were left lying there and which seemed to me almost to cover the entire road. Remembering that I had no business there, I left the spot and overtook my command, which was then two miles away.[13]

The passage of White Oak Swamp and the Battle of Glendale insured the integrity of the Army, imperrilled until that hour. Henceforth, all felt confidence in the ultimate success of our movements whatever they might be.

A march of a few miles brought the pursuers again in contact with the Army, which was found occupying a commanding ridge extending obliqueley across the line of march in advance of Malvern Hill. On this admirable position General McClellan had concentrated his Army [and] prepared to receive final battle.

We pushed on rapidly towards the James, our ears filled with the sound of the cannonade at Malvern Hill on our flank and rear. The day was bright and the stiff breeze from the west cooled our fevered temples, making the march an exhilerating pastime. Our command halted not until we reached the very bank of the James into which our men and animals plunged and rolled to their hearts content. We struck the river at Haxhalls house, some distance below Turkey Bend where the fight had already commenced.[14]

We were allowed to finish our bath in the river and then drew off a few hundred feet from the river and bivouacked. General McClellan occupied the house near by. That night we were permitted to sleep [a] sweet sleep, the first we had enjoyed for seven consecutive days and nights. At daylight next morning we were up and doing.

A portion of our command was retained in camp to reorganize our schattered Ponton Trains while the ballance was detailed to go into the different roads leading to our position and cut down trees, falling them across the roads in such manner as to render them impassable for a force attempting to flank our main position. About 8 O'clock in the morning and while the fight was progressing, I saw General McClellan with General Porter and their respective Staffs ride by us towards the front. General McClellan was afterwards accused of being on board one of our gun boats all that day, but to my certain knowledge he was not, although he did go on board one of them for a short time to consult with the commander.[15]

All day long we worked like beavers in the woods, at times exposed to the fire of the enemy but miraculously we escaped unhurt. I caught several glimpses of the battle going on around us but was too much engaged in my own work to pay much attention to it. The day resulted in a great victory for us and the complete demoralization of the enemy. At evening I returned with my command to our position of the morning to find our ground covered with the wounded and dying brought in from the field. Recognising among them many of my acquaintances, I remained among them doing what I could to alleviate their sufferings until 10 O'clock that night when I sought my blankets. We all knew that we had gained a great victory and expected on the morrow to [be] ordered to advance towards Richmond.[16]

I had hardly layed myself down when an orderley summoned me to the quarters of the Colonel. He was sick and despairing [and] had a presentiment that he was going to be killed and gave me his orders to attend to his matters in case of such an event. For the first and only time to my knowledge the Col. pulled out his flask and requested me to drink with him. Then saying a few words of a cheering nature I retired, but not to sleep, for again, as soon as I had touched my blankets, we were suddenly aroused with orders to *move immediateley*.

We were soon on the road in the thick darkness, but what was our surprise when we found that we were not *advancing* but still *retreating*, whither we knew not. We were fleeing with no man to pursue for the Rebels were going the other way as fast as possible. The night was pitchey dark and the roads were blocked with Infantry, Cavalry, Artillery & baggage wagons in one dense mass and then it commenced to rain. At daylight we had not moved more than two miles and all that dismal day we alternated between marching & halting, the drizzling rain making the roads almost impassible. About 4 O'clock we wallowed through a small stream, up through a ravine and once more stood upon high ground. Here I saw General McClellan alone on his horse covered with mud from head to foot. He was directing the position of the lines to cover our position at Harrisons Landing where we now were. As the troops crossed they were at once assigned their positions, some filing to the right, others to the left. We moved on about half a mile and stopped to cook our supper, the rain still falling.

In a short time the General came by still alone. As he passed the troops the scene was most impressive. Squads would assemble around him, throw their caps in the air and cheer themselves hoarse, while with tears in their eyes and streaming down their rough and grimy faces they showered upon him all the endearing names they could think of. They seized his feet, his legs, and would not let him pass until he had said a few cheering words. Then he would pass on a few pases and the same scene would be repeated again and again. I heard one of his little speeches which ran as follows: "Soldiers: you have done nobly. I shall now give you some rest. We have plenty of hard tack and vegatables at the river. I am glad to see you all. God bless you!"

That night we moved down to the Harrisons House, which had been abandoned by the family, leaving every thing in its place. Had they remained at home all would have been safe but abandoning it as they did the Soldiers rushed in and, before a guard could be established, the fine carpets, beautiful pictures and substantial furnishings were soon distributed among the camps. This was the home of the Harrison Family. Rebels though they were it was a sad sight to witness such wanton distruction but they had "sown the wind" and were now "reaping the whirlwind."[17]

That night we encamped in a mud puddle and it was the most uncomfortable night I had yet experienced. The next day we moved our camp to higher ground. My tent was within a stones throw of that of General McClellan. It was now the 4th day of July and as the band at Head Quarters struck up the "Star Spangled Banner" the tears would come to my eyes in spite of all efforts to prevent them. Thus ended the Seven Days Battles and [the] *change of base* so famous in history.[18]

Our Army was soon established in an intrenched camp at Harrisons Landing and here we remained, swetting in the hot sun in the hottest place in Virginia (as the Richmond Enquirer said), fighting flies, digging trenches, cutting down trees, fortifying our position, giving quinine to the sick and burying the dead until the 16th of August, 1862.

The Army was crowded into a small space [and] consequentley the Regiments & Brigades were encamped in dense masses. The weather was intollerably hot with but little rain. In consequence, much sickness prevailed. We had not had any vegatables for a long time and the scurvy broke out. Had it not been for the efforts of the Sanitary Commission I know not what would have become of our Army.[19]

Our Suttler succeeded in getting a barrel of Onions, which vegatable is one of the best anti-scorbutics in existance. I remember paying one dollar for two of them. Speaking of the Suttler reminds me that ours was a stick. He cared nothing for the health of the regiment, he only lined his pockets well and so instead of getting useful & healthful things he expended all his means for cookies and other worthless jimcracks which he sold to the Soldiers at enormous profit. It was therefore with no compunctions of conscience that I witnessed some of my men slyly roll away some half a dozen barrells of his cookies down to the company street where they quickly disappeared while a crowd of noisy Soldiers occupied his attention in another direction.[20]

Body lice made their appearance in countless swarmes. We all had our hair shaved close to the head. Flies darkened the air with their numbers [and] there was no protection from them. Our poor horses stood and kicked themselves to death in the vain endeavor to keep them off. Mosquitoes tormented us during the night rendering sleep almost impossible and taken altogether, our position was far from being comfortable. Under the circumstances it was natural to get homesick. Some few leaves of absence were granted but not many. I made the attempt to get one but failed.

Colonel Stuart was taken quite sick and grandly *lost his voice*, a peculiar disease from which he did not recover for many months. He obtained leave of absence with powers to recruit men for the Regiment. I well remember the morning I took him in a ponton boat and rowed him down to the wharf to take the steamer for home. It was a sad time for me, but a happy one for him. Once he *forgot* that he had lost his voice and spoke up quite naturally but the next moment subsided again into a whisper. However, he did good service in recruiting. [See Appendix 3, Item 2.]

I returned to camp in a sad mood as my most inveterate enemy, the Lieutenant Colonel, was now in command of the Regiment and I was at his mercy. We had an occasional review of our decimated battallions but the contrast presented by their thinned ranks as compared with former reviews saddened rather than inspirited the men.

While our Army had ceased active operations, the enemy were more active than ever, had strongly fortified Richmond and, leaving a sufficient force to protect the place, took the iniative by dispatching a large Army for the invasion of the Northern States. Of course the authorities at Washington were terrified [and] peremtory orders were given for the immediate return of the Army of the Potomac.

On the night of August 15th we received our orders to move *at once*. We had 3 hours to cook rations for the men, but the poor Officers were provided with nothing. I started on that march with nothing to eat but my haversack full of hard tack, more than half of which was wormy. We picked out the worms with our fingers and ate the ballance. The diarrhea, which until this time had affected about every one but me, had at last settled itself upon me and my strength was rapidly giving out.

We started out about midnight and on all that march I had nothing to eat except my wormy hard tack, green peaches and green corn with no salt. Every time we halted to rest we would jump over into the corn fields which bordered the road & pluck our arms full of the corn, start a fire and roast what we could before the order came to "Forward" and nibble our corn while marching. We came across numerous peach orchards also but the fruit was only half grown and had not yet begun to turn its color—this was our desert. A fine diet one would think for chronic diarrhea but it was all we had and it must be eaten or nothing.

Our Regiment was divided in two: one portion under Capt. Spaulding followed the road near the river and built the long bridge, 2100 ft in length, across the mouth of the Chickahominy; the other portion, including my command, taking a more circuitous route into the interior.[21]

The sun blazed an unmerciful heat down upon us. Water was very scarce and we suffered terribly from thirst. When a well or stream was reached, the men fought over its possession like so many fiends. Kearneys Brigade followed us and their men straggled ahead to secure the wells for the use of their comrades when they should come up. As our men arrived and endeavered to get water it required all the tact and skill of the Officers to prevent bloodshed. I saw Kearney on this march for the last time. His impetuous nature could not brook defeat and he did not hesitate to condemn our General and his policy.[22]

The roads were filled with dust from 4 to 6 inches in depth [and] the marching of the men through it raised it in clouds which filled our eyes, nose and ears and almost suffocated us. It was so hot that it burned the feet and both of mine were blistered completeley.

At last we arrived at Fortress Monroe, our bodies so completely covered with dust and sweat that one could not recognize the color of our faces or uniforms. Grimy, dirty, hungry [and] exhausted we debouched upon the plain back of the fortress and cried aloud with joy as the cool breezes from the Ocean struck us. Folley, Hoyt & myself purchased a boat load of *watermelons* from a Negro and did not stop eating until they were all gone. But my diarrhea was gone, cured by green corn, green peaches & watermelons. After remaining here a couple of days getting washed and rested, we again shipped on board of a Steamer [and] left that delightful spot with many regrets. We were soon at Alexandria with no incident worthy of mention except that 3 of the men fell overboard and were rescued with much difficulty.

Chapter Eight

Second Bull Run, Antietam, Harper's Ferry & Berlin 1862

The Peninsula campaign had ended. The Army of the Potomac, or what remained of it, had again returned to near its original position but now it was transferred to another commander. General Pope was now to control its destines. He came to us unknown except as we had read his flaming bulletins reflecting severley upon our beloved commander [and] ascerting that *his* Head Quarters would henceforth be "in the saddle." General Popes command of the Army was brief and disastrous, resulting in the disgraceful defeat known as Second Bull Run.[1]

McDowells Corps was again united with the Army of the Potomac. The troops around Washington, who up to this time had not seen active service, were added to the command. The troops were hurried forward from Alexandria as fast as they arrived in detatchments of Divisions and Brigades to meet the enemy, flushed with victory, who were now menacing Washington. They met on the old ground on the plains of Manassas.

General McClellan, divested of his command, remained at Fairfax Seminary during these disastrous days. Our two Regiments of Engineers, the 50th & 15th, were the only troops that were not ordered away from him. We were marched into the line of fortifications and there remained until the battle was over.

For two days we saw the long lines of baggage and ammunition wagons moving to the front. We heard the noise of the battle and stood anxiously awaiting tidings but none came. The third day disclosed our defeat [and] the long lines of wagons and thousands of stragglers returned to the line of defences bringing tidings of our misfortunes. Our Army was whipped in detail [as] there was no head to guide affairs. All was a mass of chaos, confusion, misunderstanding, defeat.[2]

On the night of the 31st of August, I was on duty in command of an advanced line of Pickets. The night was chilly and our camp fires burned with unusual brightness. I was out about a mile from our front with orders to halt all parties of stragglers approaching and hold them in custody until daylight when they might be returned to their Regiments. It was also feared that the enemy would follow up his advantage to make a descent upon the Capitol.

At about midnight a cavalcade approached escorting a single ambulance. Within the ambulance [was] the dead body of the gallant Phil Kearney. I had seen him but a few days before on our march down the Peninsula, so full of life and animation. Could it be that he was dead? Yes, he had gone to his last account, but his memory will ever be held dear by those who knew him. He was indeed a "preux [valiant] chevalier" in whom there were mixed the qualities of chivalry and gallantry as strong as ever beat beneath the mailed coat of an olden Knight. Like Desaix, whom Napoleon characterized as the man most worthey to be his lieutenant, Kearney died offering a heroic breast to disaster.[3]

On the 2nd of September the Army was withdrawn back within the lines of Washington. Lee, abandoning direct pursuit, began his march for the north of the Potomac. It was now evident that Pope was not the man to command an Army [and] the country as well as the Army turned its eyes to the neglected McClellan as the only man who could save them in this emergency. Pope was therefore removed and McClellan reinstated in command. The Army received the news with evident joy, confidence was again restored and order soon took the place of chaos.[4]

With the re-organization, the Army commenced the Maryland Campaign. Our Regiment was ordered to Washington. We marched the 7 miles through the hot dust with light and cheerful hearts. As we entered Washington we presented a most comical appearance. Covered with dust so as to be scarceley distinguishable, our uniforms old and ragged, our horses poor, scrawney and scraggy, we presented a striking constrast to our fine appearance when we [had] marched through the same streets 5 months before. We marched up to our old camp ground and commenced fitting up a Ponton Train for field service.

The next day Folley & myself obtained permission to visit the city. We went straight to Willards Hotel determined upon enjoying *one good, square meal.* We sat down to the table with appetites whetted by months of hard and scanty fare. Commencing at the head of the bill of fare with the intention of eating it straight through, we presented a unique appearance at that table filled with the elite of the country, Ladies & Gentlemen dressed in the hight of fashion. Our dusty and worn uniforms, our bronzed faces and short cropped hair attracted the attention of those present, who gazed at us with astonishment not unmixed with admiration & sympathy. What a dinner we did eat. You can never know until you have lived for months as we had done on pork & hard tack and scanty rations at that, to be suddenley transferred to the table of a first class hotel.[5]

In a few days we were on the way to Harpers Ferry. The Army, meanwhile, [was] in advance of us on the route to Antietam. Our train stretched out upon the road nearly two miles in length. Our command consisted of 4 Companies divided into Sections, each Section having a portion of the train under its special charge. The roads for the most part were good but an occasional bad place would detain the whole train and it was not until after midnight of each day that our whole train would be in and bivouacked together.

The country through which we passed presented a striking contrast to the desolated Virginia we had so recentley left. Rich fields of grain waved luxuriantly in the bright sun, houses and fences [were] intact, evidences of thrift and comfort on every side, cattle grazing indolentley in the fields unconscious of the fact that an Army was passing by. The roads for miles were lined on each Side with fruit trees laden with ripe fruit, with pleanty of running brooks of clear cool water, too. The air [was] fresh and pure and balmy. What wonder that our hearts were filled with delight and that we sang as we marched along "Sweet Maryland, My Maryland." From every house the national flag was flung to the breeze and the inhabitants came to their doors or to the roadside to greet us with pleasant words and assurances of good will.

It was a beautiful day when we passed through Fredrick City. General McClellan and the great bulk of the Army had passed through but a few days before, receiving on their way a perfect ovation from the inhabitants of the town. Just beyond the city we approached a beautiful mansion situated at the summit of a high hill overlooking an extensive and beautiful country. At the gate of its entrance leading to the mansion stood a beautiful girl of some 16 or 18 summers. Her flaxen hair hung in ringlets over her shoulders and her sweet blue eyes reminded me so forcibly of a certain person whom I had known in my younger days that I could not resist the impulse to stop and speak with her. Two sable attendants, or more explicitly two "darky women", stood beside her, each with a pail of milk from which the fair maiden aforesaid would ever and anon dip the lacteal fluid with a tiny cup, which she held in her ivory hand, and with the smile of a sweet angel, offer to the weary Soldiers passing by. I think that cup of milk the sweetest and best I ever tasted. I could stop but a moment, long enough however to learn from her that she was the daughter of Colonel, afterwards General, Kenley who commanded a Brigade of Marylanders.[6]

All this time the battle of Antietam was being fought, that terrific battle in which victory and defeat was so evenly ballanced that at last our victorious commander did not dare pursue the checked but not defeated enemy. He has been blamed for this but people forget what was at stake in case of defeat and of which he alone felt the responsibility.[7]

On the 19th of September we arrived at Harpers Ferry. The place had recently been captured by the Rebels with 12000 of our men as prisoners. General White had foolishly turned over the command to Colonel Miles who in turn turned it over to the enemy when by withdrawing to Maryland hights opposite he might have held it until doomsday against the whole Confederacy. After their defeat at Antietam the Rebels had evacuated the place and were fairly out of sight when we arrived.

The picturesque little Villiage of Harpers Ferry lies nestling in the basin formed by Maryland, Bolivar and Loudoin Hights, which tower into almost alpine sublimity. A line drawn from any one mountain top to the others must be two miles in stretch yet rifled cannon crowning these hights can easily throw their projectiles from each to [the] other, a sort of titanic game of bowls which Mars and cloud-compilling Jove might carry on in sportive mood. But Maryland Hights is the Saul of the triad of giant mountains and far o'ertops its fellows. Of course, it completeley commands Harpers Ferry, into which a plunging fire even of musketry can be had from it. While therefore Harpers Ferry is the worst military trap, lying as it does at the bottom of this rockey funnel, Maryland Hights is a strong position, and if its rearward slope were held by a determined, even though small force, it would be very hard and hazardous to assail. However, all these advantages were thrown away and the surrender of Harpers Ferry has passed into history as one of the most disgraceful events of the war.[8]

Our approach to the Town was along the north bank of the river, a narrow and torturous cut into the mountainside just wide enough for a team to pass but too narrow for two to meet and pass each other. Before us the road was blocked with Artillery by which we were detained until nightfall. As the darkness fell we commenced to ford the rapid stream to an island near its middle where we parked our train. In order to reach the spot, we were obliged to station our men at equal distances along the bank and in the Stream with the water up to their middle, each bearing in his hand a torch of pine knots. In this way we succeeded, after several mishaps, in getting our men and bridge material concentrated upon the island.

Large fires were lighted, around which our chilled and tired men stood or lay until the morning. It was a weird scene: our little island all ablase, casting the flickering lights across the waters which roared around us [and] were reflected back from the lofty, somber, solemn mountains on each side of the stream. The wind whistled down through the gorge with a sad and mournful sound [and] it seemed as if the ghost of Old John Brown still haunted the place of his historical exploit. The very engine house where he made his last and heroic defense [was] but a few hundred feet from us on the Virginia side of the Stream.

The next day we built a bridge across the river, [with] a part resting upon the island and many of the abutments spanning the rocks which showed their ugly protruberances above the surface. Sumners Corps soon passed over and took up its position on Bolivar Hights to the rear of the Town. Then we crossed our own teams and wagons and established our camp on the hill in rear of the town, or rather to the west of it & between it and Bolivar Hights. The line of Officer tents being near the bank of the Potomac [and] this bank being about one hundred feet high and nearly perpendicular, we had a splendid view from our camp of the river and mountain scenery.

During the day we were visited by the brave old Sumner at our bridge. He spoke cordially to Spaulding & myself and seemed so much pleased with our bridge that he gave an order to his commissary to find us a barrel of whiskey for the use of our men who had been so long in the water.[9]

Our bridge completed and camp established, we had camparativeley little to do and so enjoyed the season of rest offered us, as indeed did the whole Army, which was now encamped around that country from Sharpsburg on the West to near Berlin to the East and

Harpers Ferry to the South. Both Armies in fact felt the need of repose and, glad to be freed from each others presence, rested on their Arms, the Confederates in the Shenandoah Valley in the vacinity of Winchester and the Army of the Potomac near the scene of its late exploits amid the picturesque hills and vales of Southwestern Maryland. General McClellans Head Quarters were situated in the beautiful valley east of Maryland Hights.[10]

As for the engineers, we enjoyed our rest most thoroughly. Our camp was situated on a high bluff exposed to the winds as they came swooping down through the gulley made by the Potomac and we had no wall tents with us. At least I had none, my only covering being the *fly* belonging to my tent, which we left in Washington. This did very well during the day but at night it was very severe. However, we managed to enjoy ourselves.

I shall never forget how, on these clear, cold October nights, our little party of Officers used to gather around the camp fires, bundle ourselves well up in our blankets & overcoats and *sing* by the hour. Captain Folwell, who subsequently became the President of the University of Minnesota, introduced what was then a new song to us, the "Tombigbee River." We would all join in the chorus,

"Then row away row. O'er the waters so blue
Like a feather we'll float in our gum tree canoe"

All of which was very beautiful to us, if to no one else.[11]

One afternoon as I was laying on the ground under my fly reading Homers Illiad, which I always carried with me, who should make his appearance but Philo Judson, your mothers uncle. He it was who married us and I had not seen him since that time. He had but recentley been appointed Chaplain of the 8th Ill. Cavalry and was on the way to join his Regiment.[12] Learning of our whereabouts he had sought me out and glad was I to see him. This being his first experience of camp life he found our bed (the ground) rather hard that night, but Folley and I placed him between us and being well covered with blankets he rested some but did not sleep much that night. The next day I furnished him with a horse and we rode the 10 miles to Sharpsburgh, where he joined his regiment and I returned to the Ferry. I did not see him again until he came to see me after the Battle of Fredricksburgh. [See Appendix 3, Item 3.]

We had our usual drilling to perform and an occasional review. Once President Lincoln, in company with General McClellan, reviewed the Troops on our side of the river. I well remember how sad he looked as he rode by within a few yards of where I stood. For some reason or another they had given him a very small horse and his leggs almost touched the ground. This was the last time that President Lincoln ever reviewed the Troops with McClellan in command.[13]

Our forces commenced crossing the Potomac soon after, through the town and across the Schenandoah, filing up the mountain sides [with] a long line of glittering steel and blue uniforms.

On the 27th of October we broke camp leaving Capt. Personius with Co G. in charge of the bridge and took up our line of march for Berlin, 7 miles below Harpers Ferry. We were here reinforced by another company or two from our Regiment at Washington, who brought additional bridge material for the construction of a bridge at that point. We also received a large number of recruits, the result of Colonel Stuarts efforts while on leave of absence.

We had soon completed a bridge across the Potomac at Berlin, the stream at this point being about 1100 ft in width. Soon after, the main portion of the Army commenced crossing and for two days and nights we were kept on constant duty guarding against accidents while all this immense body of men, animals, wagons, [and] cattle crossed in one continual stream. We saw the last wagon, the last straggler safeley across and away and then awaited patiently for orders which we daily expected.[14]

The campaign had commenced with a vigor which promised success. The weather was charming, neither cold or warm, clear, gentle and bracing. The roads were in excellent condition [and] everything seemed to favor a speedy and favorable termination of the War. Gradually the rear of the Army became so far separated from us that we could get no news from Head Quarters and, as the Country between us was infested by bush whackers, we were entireley cut off from news.[15]

Our time was chiefly occupied in drilling our new recruits, the number of which having increased our rank and file to the proper dimensions, entitled the Regiment to an additional Major. The question was, who should it be? Unquestionably, the choice lay between Spaulding & Myself. Spaulding was my senior in rank in date of commission but he had not raised a man for the Regiment while I had recruited over one hundred. Spaulding knew little of *military* matters strictly speaking, while that was my *forte*. On the contrary, he was an *Engineer* of skill and experience and this was my weakness. He was

Ponton bridge across the Potomac at Berlin, Maryland, November 1862. Courtesy of the Library of Congress.

also learning rapidly and his executive ability was by far superior to my own. We were friends and not rivals: both wanted the position and both felt that the other ought for some reasons to have it. The position had been *promised* me months before but then how did I Know but it might not have been as secretley promised to Spaulding? At any rate, I was not much surprised, though considerably disappointed, when intelligence was received of Spauldings appointment. He came to my tent with the letter himself and inasmuch as it was him and not someone else, my ambitious hopes were soon satisfied, as there were two more vacancies yet to be filled when the number of men in the Regiment should justify it and though I knew I could not be the senior Major, yet to be 2nd in favor was my reasonable expectation. Spaulding was therefore promoted in place of Major Embick, who had been promoted to command another Regiment. [See Appendix 3, Item 4.]

Before going any further with my story I must tell you about *Johnny*, My *Johnny*: John E. Jones. Johnny was so intimateley connected with me in the capacity of man servant that it may almost be said that whatever applies to me applies to him also, for wherever I was to be found except on the field of battle, there was Johnny also.

Johnny was a Welshman. In Rome, where he lived for many years, he was generally known as "honest John." He had been in the employ of Mr. Elmer for seven years and upon his failure in business he (Mr. Elmer) recommended John to me and I had him in my employ for about two years before the breaking out of the war.

During this time I had become very much attatched to him on account of his faithfulness and attention to my interests. When he found that I was going into the Army, his ambition was at once aroused to accompany me. Having no family or near relations, his whole affections seemed to center upon me and he was ready cheerfully to go where I went or do whatever I might bid.

At first I intended to make a Soldier of him but he could no more learn to keep step to music or with others in line of march than could a wild Kangaroo. When the mustering Officer, Capt. Tidball, came to muster in my company he would not accept Johnny on the ground that as several of his teeth were gone he could not bite a cartridge. Now Johnny could bite cartridges enough to keep a whole platoon firing but the mustering officer thought he knew best and so Johnny could not be mustered as a Soldier. But he could be my servant and as such he was installed and so remained for four long years.

Johnny was known all through the Army among my acquaintances as the *model* Servant. I could fill a small volume with the sayings and doings of Johnny. Far and near he was known and quoted by the Officers. He seemed to live and move and have his being in me, faithful as a Spaniel. Johnny was never known to get between his blankets no matter how late the hour or what the circumstances until after he had first seen me snugly stowed away where he would tuck me in, place a cup of water at my head and roll himself up to sleep, ready to rise good naturedly at any call.

He never lost his temper and seemed always happy when he thought me comfortable. My very wish his pleasure, he seemed to have been given me by a kind providence expressly to administer to my comfort. Is it any wonder that a strong attatchment sprang up between us which has always existed & which will exist as long as we live? I cannot do justice to the subject of this sketch but as you will often notice that I speak of "Johnny," you will in future better know to whom I then refer.[16]

Day after day we waited for orders and wondered [why] they did not come. Gradually we accumulated our material and put things in readiness for instant departure, feeling that our orders *must* come be-

Northland College
Dexter Library
Ashland, WI 54806

John E. Jones, Wesley Brainerd's servant during his time in the army. Courtesy of the William Hencken Family.

fore long. Thus the beautiful Autumn days wore along, but no news from the Army. We were completely isolated, perhaps forgotten. This wonderful quiet proved to us to be the calm before the storm. Suddenley we received vague rumors of the removal of General McClellan from command of the Army. Then the news was confirmed and we were informed that General Burnside was then in command.[17]

Time passed on and we were full of anxiety when one day late in the afternoon came the order for us to move *immediateley* to Washington with one entire bridge train. The order had been written *seven days* before reaching us and therein consists the secret which seems never before to have been revealed of all our subsequent woes, misfortunes, and defeats. Seven days the order written and we [were] still at Berlin, 48 miles from Washington. Five days before we received the order to leave Berlin we *should have been in Washington*. Like men made of steel we sprang to our work and before the sun cast its shadows across the Potomac our entire command was well under way and rapidly moving into the darkness in the direction of Washington.

On the very day of McClellans removal, the order above mentioned was issued but during the confusion incidental to his sudden departure (as by his peremptory orders he was forced to do) the message was laid aside or held in abeyance until the designs of the new commander should take shape. Hence this *fatal* delay which cost us the loss of a terrible battle and plunged the whole country into depths of dispondency and gloom from which it did not recover until the final termination of the war. "Verily how great a smoke, a little fire kindleth."[18]

The disastrous consequences of this change of command in the midst of an active campaign cannot be measured. Of all the accounts published of the causes of our defeats at Fredricksburgh none have yet given the *true* reason as I have herein stated it.

We marched rapidly until late that night and were on our way at an unusually early hour the next morning. Arriving at the nearest railway station, Spaulding went on in advance of us to prepare for our arrival. And so on the 16th day of November, 1862, we were again near our old camp ground by the Navy Yard, fitting up as rapidly as possible our transportation for the lower Potomac.

Chapter Nine

Preliminary Operations Before Fredricksburgh 1862

On our Arrival at Washington we learned that the advance of the Army was already at Fredricksburgh and that the whole main body was rapidly advancing thitherward. Our orders were to fit up two Ponton Trains as rapidly as possible, the one to go by water, the other by land, Acquia Creek being their destination. Night and day we labored to get our trains in marching order. Five hundred greene, wild mules were to be subdued and harness fitted to them. Our men knew little of the ways of the mule but the Negro Teamsters seemed made especially for that purpose.

In about 3 days one of the trains was dispatched [in rafts] down the Potomac in charge of a company or two from the 15th Regt. without animal transportation. On the 19th day of November, 1862, we left Washington, crossed the Long Bridge and were once more in Virginia. It rained a dull, heavy, sleeting rain, just enough to make us all feel lonely, cheerless [and] desolate. We marched off down to the right of Alexandria, across a little stream known as Hunters Creek, into the flats beyond and, as night approached, bivouacked in a field wet but fresh and still greene, notwithstanding the lateness of the season.

I now had another horse, a bay of splendid bearing & grace but a terrible hard rider. "Old Bill" was his name. I was often glad enough to exchange the *style* of the bay for the *ease* of the gray. These two horses were my companions for the ballance of my service in the Army.

The next morning, Nov 20th, we were on our way bright and early, or rather early but not bright, for soon the sky became overcast and the rain commenced falling in earnest. The road followed along in the valley beside the Stream mentioned. This Stream soon overflowed its banks and covered the road so that we could not tell whether

Engineer ponton train on the move. *Frank Leslie's Illustrated*, January 3, 1863. Courtesy of the Library of Congress.

we were in the road or in the Stream until we occasionally upset wagons, Pontons, men and all into the channel. Sometimes a dozen teams and the men accompanying them would be floundering over their heads in the water at one time. In this way we *navigated* all that cold, raw, November day and we did not go into bivouac until nearly 11 O'clock that night. Then we stood or lay around our fires all night drying ourselves, the rain, however, still pouring. We made 8 miles that day.

The morning of Friday, Nov 21st, was like its predecessor, dark and rainy. I am particular in regard to dates as so much importance was attatched to our movements that memorable week and the succeeding month. As soon then as possible we commenced pulling our train through the mud, having left the valley and stream the night before.

About noon we passed Pohick Church, an ancient brick structure in rear of the Mount Vernon estate. Here Washington was accustomed to attend divine service when at home or on his Plantation. Riding up to the building, I dismounted and entered. There was nothing of the church within the bare walls except the platform of the pulpit and the sounding board above it. Ascending its pulpit I cast my eyes over and around this sacred place. There were no windows remaining, no floors even, for they had been torn up and burned by the soldiers of both Armies. In the center of the room was a party of three of our cavalry-men silentley cooking their coffee over a fire built upon the ground, the smoke of which raised up and passed out of the holes in the walls where stained windows once were. I fancied the spirit of Washington visiting this spot where he had once worshipped so peacefully, now so dessolate. But the house of Washington was safe, both parties having too great reverence for his name to commit any depredations there.[1]

We trudged on all day and far into the darkness until 10 O'clock at night when we reached the Town and river of Occoquan. The Town was deserted. The Stream is of considerable magnitude and empties into the Potomac a few miles from the Town. The approach to the Stream was by a narrow, crooked and rocky road which was very difficult for us to pass on account of the lengths of our Ponton boats & wagons. We lay down in and beside the road and awaited for daylight.

Saturday, Nov 22, the rain ceased falling and a freezing temperature took its place. The Stream was rapid and much swollen. The Army was waiting for us and obstacles seemed to accumulate as they had never done before. The situation was *extremeley* unpleasant.

Spaulding & myself held a consultation. We had strained every nerve to hasten through our ponderous train and still we were 7 days behind. Messengers continued to arrive with orders for us to "hurry up" and inquiring anxiously the whereabouts of Sickles Brigade, which seemed to have been lost.[2]

We decided to build a bridge across the Stream, cross it ourselves and divide the train, a part going by water (all that would float) and the ballance by land. Capt. McDonald rode back to Alexandria for a Tug to tow the bridge material from the mouth of the Occoquan down the Potomac and at 10 O'clock on Saturday night we had built our bridge, crossed our trains [and] made the Pontons into rafts upon which was placed the other material. Spaulding, with one half the command, started off with the Pontons and bridge material down the Stream, leaving me with the other half of the command, all the animals and with nothing but the empty wagons to make my way over land through the mud as best I could to Belle Plaine. This arrangement proved to be our salvation for had we started into the country beyond the Occoquan with our train as orginally made up we would not have worked through in a fortnight.

The next day, Sunday, Nov. 23rd, I started with my command out into the unknown country with no guide but my pocket map. [We] pulled through the mud and at dark had made about 12 miles and went into bivouac 2 miles from Dumfries, a notorious Secession hole and rendezvous for Bushwackers. A portion of the 1st Maine Cavalry was posted here.[3]

[On] Monday, [Nov.] 24th, [we] started bright and early with one company of the Cavalry for our escort. Several times during the day we expected attacks from the roving bands of the enemy who were hovering around us and made ready to give them a warm reception but they kept at a safe distance, fearing that we were too many for them. Couriers still continued to arrive with orders for us to press forward with all possible speed and still inquiring for Sickles Brigade. We tugged and pulled all day, not halting to feed the men or animals, and when the darkness became so thick that I could no longer see a face ahead, I gave the order to halt and we dropped down *in our tracks*, having overcome about 16 miles during the day.

[On] Tuesday, the 25th, as the sun came up, I found that we had stopped within 3 or 4 miles of Belle Plaine, which I could distinctly see. We were not long in joining our companions. At 2 O'clock that day our whole Train was loaded and on its way to Falmouth, where we arrived that afternoon. Had we not divided as we did at Occoquan, we would not have reached our destination in at least a weeks time. It rained copiously the night of our arrival.[4]

The enemy were strongly posted in the Town of Fredricksburgh and our Armies were facing each other on each side of the Rappahannock, 400 ft wide. The day after our arrival we saw the NY Papers for the first time since leaving Washington [and] the first item that attracted our attention in the paper was that Major Spaulding and all the Officers with him had been placed in arrest for *delay* in bringing forward the Pontons. As this was the first intimation we had had of being placed in arrest, we went up to Head Quarters to ascertain its truthfullness. The result was that we were informed that in order to learn the news one must go away from home and we indulged in a good laugh, the first we had enjoyed for many a day.[5]

Of our march from Washington to Fredricksburgh much might be written. Seldom has a body of Soldiers ever struggled harder or more faithfully than did we to get to our destination in order to save a movement of the Army from disastrous failure and for which we received no credit whatever, not even an honorable mention. [This] owing to the stupor produced upon the public mind by the overwhelming disappointment which soon followed. On the other hand, the whole country, ignorant of the facts, were blaming us for not being at Falmouth before we had received our orders to leave Berlin. The facts which I have narrated were only partially disclosed at an official investigation which followed the disaster at Fredricksburgh.[6]

We were now at our destination and the whole face of things was entireley changed. Instead of finding no obstacle but the river to prevent our onward march, the Enemy was strongly posted on the opposite side and adding to the strength of their position hourly. He had checked us, that was evidint. What was our commander to do, that was the question. The Army was full of hope and confidence.[7]

The next day, Nov. 26th, General Woodbury sent for Spaulding & myself to accompany him on a reconnaissance down the Rappahannock River. The party consisted of the General, two Officers of his Staff with orderlies, a detatchment of 10 mounted men from the 4th Regular Cavalry [and] Spaulding & myself. We took a back road from the river to prevent being seen by the enemyes Pickets and were soon outside of our lines.

You will remember that I once said General Woodbury was famous as a "still hunter." He showed his capacity again that day. Every now and then we would approach the river as near as we dared with our horses, then leaving them in care of the orderlies we would cautiously approach the river and look over the ground on the other side. It was evident that General Burnside was at a loss *where* to cross and it was now our business to find a good position. It was very singular to us that day that when a good position [with] hard ground [and] easy approaches was found on our side of the Stream, that the very reverse would appear on the other.

About 4 in the afternoon we reached a house on the bank of the Stream where lived a Mr. Walker [who] was at home with his family. The General asked him if he enjoyed duck shooting, to which he replied that he did. We were now about 10 or 11 miles below Falmouth and from this point began to retrace our steps.[8]

After we had returned a couple of miles the General asked me to go with him alone and leaving the ballance of the party back some distance we approached the river. The General then pointed out to me an old sunken canal boat on the opposite side and calling my attention to the ground beyond said that it was, in his opinion, a good place to cross, but, said he, "In order to be sure of the nature of the ground beyond, I would like to have you return to Walkers house, demand his boat, for I know he has one (I now understand why he asked Walker if he was fond of duck shooting), come up here with the tide, reconnoiter the ground thoroughly and report to me in the morning. I will not *order* you on this duty," said the General, "unless you feel perfectly willing to undertake it."

I replied that I was only too happy to be chosen for the duty and that I would do my best. "Look out for a picket post over in that corner of the woods," said he.

We then returned to the party [and] the General gave me a corporal and five men of the Regulars. The Officers all bid me good bye and we separated, they for the camp and I to return on my errand.

When I reached Mr. Walkers house it was about dark. The man and his wife seemed very much alarmed at our approach. I told them to keep quiet, treat us fairly and not attempt to escape and they need have no fears. I then had the Corporal post a sentinel at the house and one at the barn, where our horses were already feeding. The Lady commenced at once to cook a supper for us.

While this was preparing, I took Mr. Walker aside and asked him where he kept his boat (I knew if I asked him if he had one he would probably deny it and so my policy was to assume that I knew he had one). After some little hesitation, he admitted that he had one and took me to the spot where he kept it, but he had no oars. [It] seemed the boat would not possibly hold more than I, [as] it was nothing but a little boat used for duck shooting. However, we made some paddles and after the supper that the good woman gave us, or rather the good supper that the woman gave us, I quieteley waited for flood tide and at 11 O'clock was ready to move.

Leaving the guard at the house I took two men with me in the little boat and with our paddles headed up the stream. We all three sat in the bottom of the boat which was now brought down nearly to the level of the gunwales. Silentley we paddled up with the tide, keeping on the shady side of the stream, for the moon was up and frequently came out from the clouds with wonderful brilliancy.

Reaching the point opposite the picket post, where I could plainly see the men standing around a fire, we waited for several moments for the moon to get behind a cloud and when at last it hid its face, sending a dark shadow across the stream, we paddled our canoe across as if our lives depended upon it, as indeed they did. Reaching the opposite shore, we floated along close under its shadow and passed the Picket not 20 feet away. There must have been about half a dozen of them and we could hear every word they said as we floated quietly by almost under them.

By 12 O'clock we had reached the canal boat [and] here I landed, taking one man with me [and] leaving the other in charge of the boat. We walked out cautiously and returned, [as] I was satisfied that this was no place to cross as the water, now at high tide, had overflowed the whole bottom and I waded up to my knees before I returned. Near the bank I discovered a road and thought to follow it at least a ways.

Remembering what the General had said about the picket, I stooped low and took slow steps, taking good care to keep close to the bushes, revolver in hand, my man about 10 ft. behind me. In this way we progressed until suddenley some one called out in the darkness, "Who goes there?" and I heard the click of his musket as he made ready to fire at the first object that should meet his vision. Did my hair stand on end? I cannot positiveley affirm but there certainley was a very peculiar feeling about the scalp.

We stood perfectly motionless and almost breathless for what seemed a good half hour (perhaps it was but a few minuites) and then, one step at a time, slowley retreated backward until we finally reached the boat.

The same caution in eluding the pickets had to be observed on our return and we finally reached the house in safety at about 4 O'clock in the morning. In order that Mr. Walker should have no opportunity to report my proceedings to the enemy, I took an axe and demolished the dear little boat which had borne me so safeley, knocking it into a hundred pieces. Then, after cautioning the family against making known to any one what they had seen that night, I *raised the seige* and dashed away for camp, reaching there about 6 O'clock in the morning.

I at once reported to General Woodbury the result of my expedition [and] he then gave me a note addressed to General Burnside and requested me to report to him immediateley. The note said simply,

> "Dear General, This is the Engineer Officer (Capt Brainerd) who visited the position within the Enemys lines last night, I send him to you as requested 'at once.'"
> (Signed) Woodbury

When I arrived at Head Quarters the General was already awake and attending to business. The Sentinel allowed me to pass to his tent unchallenged. Armed with my letter, I made straight for his tent and *scratched* and was at once bidden to enter. There were two General Officers within who at once retired as the General approached me and so I was left alone with the commander of the Army.

Requesting me to be seated, he drew up his camp stool close beside me, read my note and requested me to tell him all I knew. I shall never forget that interview with General Burnside. His manner assured me and captivated me. I thought and felt that I was having a private interview with a personal friend. He was so frank, so generous, so unreserved and unassuming. One felt at ease in his presence at once and I sat and chatted with him for a good half hour, told him all I knew about the ground, at which he seemed to take a deep interest. It was evident that he wanted to cross the river below. Thanking me for my kindness (when I felt that I ought to thank him instead), the interview was closed. I then returned to my quarters and slept all the ballance of the day.

The next day it was decided to make still another reconnaissance on a larger and more extended scale and still further down the river. I was selected by the General to command the expedition. This time a regular life boat, which had been brought down for the purpose, was provided. My escort consisted of a Sergeant, a Corporal and 8 men from the 4th Regular Cavalry. And a splendid sett of fellows they were too, [as] having been *picked* from the Regiment for the purpose, they felt proud of the distinction and I felt every confidence in them.

Our boat was placed upon a wagon drawn by 4 mules and then covered with brush so as to entireley hide it from sight. Should the enemy chance to see the party with the most powerful glass, it would have been impossible to mistrust that a boat was under those bushes. We got off in good season and early in the afternoon without any extra effort arrived opposite Skinkers or Buckners Neck, 12 miles in a direct line but more than 20 by the route we came from the City of Fredricksburgh.[9]

Besides General Woodbury & Staff, General Hunt, Chief of Artillery, accompanied us. Leaving my men with the boat well concealed, we, Generals Hunt & Woodbury & myself proceeded to reconnoiter the position. A casual glance over was enough to convince us that this was a most magnificent position for a crossing. Could we once obtain a foothold on that high ground opposite, the whole rebel Army

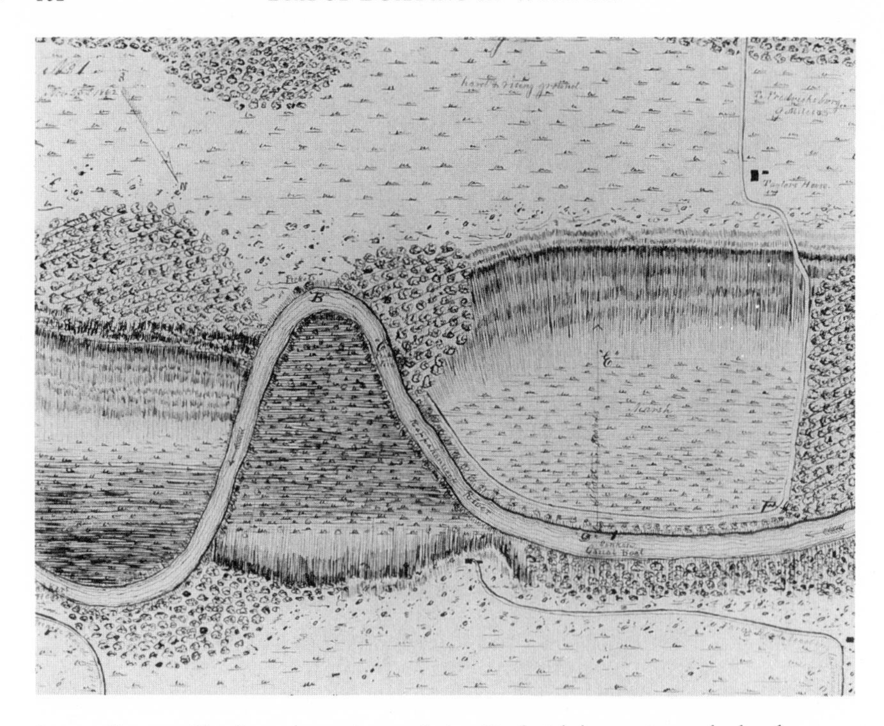

Brainerd's map of his first reconnaissance below Fredericksburg prior to the battle, November 26, 1862. B & C, pickets avoided by Brainerd; D, sunken canal boat where Brainerd landed; E, area looked over by Brainerd, marshy at high tide; F, picket post pointed out by General Woodbury and where Brainerd was challenged. Courtesy of the William Hencken Family.

would not be able to dislodge a single division. Besides which, our Gunboats, which were but a few miles below at Port Royal, could be brought up to Berry Plains and with our Artillery posted in favorable positions on this side, [the Army's] crossing could be completely covered should the enemy discover the attempt in time to gather a force for resistence.[10]

The only questions now to be decided were, whether the enemy were in any considerable numbers beyond and if the ground was sufficientley hard to admit of the passage of Artillery. This was my mission.

About 5 O'clock I was left as before, alone with my men, with a good bye and earnest wishes for my safe return. This time however, I had an Ambulance.

We saw some very fine sheep grazing near by. A remark from me

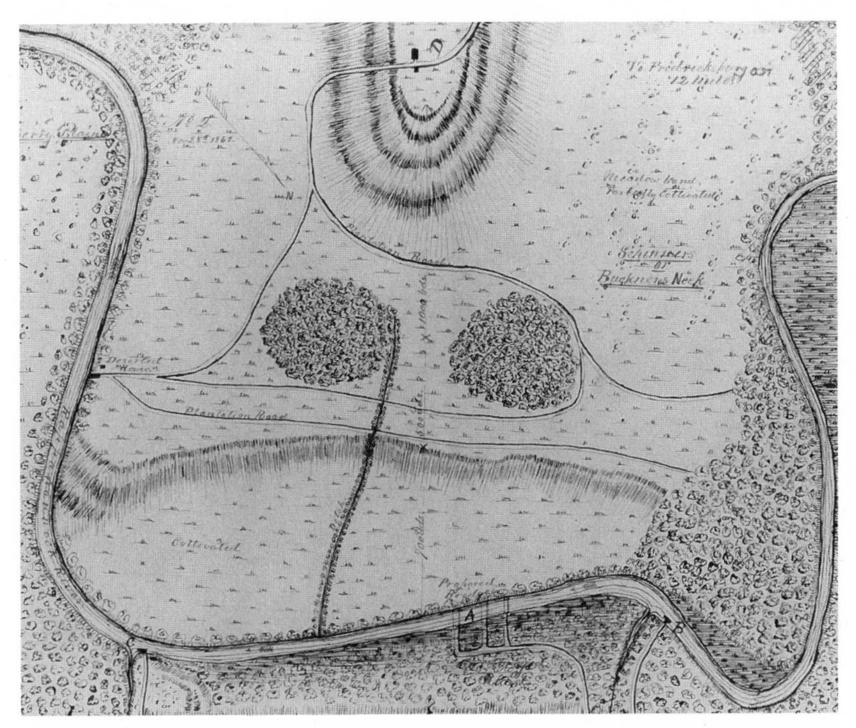

Brainerd's map of his second reconnaissance at Skinker's and Buckner's Neck, November 28, 1862. A, point at which Brainerd landed with his reconnaissance force and the point selected as a possible site for crossing bridges; C, where Brainerd left his force for the initial reconnaissance of the neck with Generals Woodbury and Hunt; D, the high ground and the objective of the mission. Brainerd makes no mention of B in the text although there is a B clearly marked on the map. Courtesy of the William Hencken Family.

that some of that sheep skin would be fine to muffle our oars with was enough to bring in three or four very fine fellows in about as many minuites. Mutton chops tasted good that evening in the cool, crisp air, for it was now the 28th of November, and the thick wool skins of these bloodied sheep were soon wrapped around the oars.

Then we waited patientley for darkness and the tide. At preciseley 10 O'clock the boat was shoved from the old wharf into the water. Eight men besides myself were soon seated and I took the stearing oar whaleboat fashion. Two men were left with the Ambulance. The men who pulled the oars were excellent rowers. So softly and gently did we glide through the water with our muffled oars so niceley feathered, that I doubt if you would have heard us had you stood upon the bank ten feet away.

Arriving on the South shore I left two men in the boat and taking the other six with me started across the field. The order of march was as follows: first myself, then the Sergeant, then the Corporal and then the other men, all in single file and ten feet apart. Each one carried his revolver in his hand ready for instant use and all were to watch my motions. When I laid down, all laid down, when I raised up they all did the same, and in this cautious manner we proceeded across the fields, stepping high to avoid the ruslling of the grass and laying low whenever the moon came out from among the clouds. We all knew that it would be no funny matter for us to be captured within the enemys lines while acting in the capacity of spies, nor did we intend to be taken alive, for we knew too well what mercy we might expect.

Off to the left was a deserted house, which we reconnoitered very carefully, fully expecting to find a picket post there, but not a solitary living thing could be found. There it stood, dismal, alone, deserted. Then we made for the eminence, our grand objective point. There was a house upon its summit and one or two small shanties at some distance from the house. When we had approached to within several hundred yards of the house, a dog began to moan and howl at a most dismal rate. So acute is the sense of hearing with these animals that he had heard our footsteps in spite of all our precautions. I came up with my men and surrounded the little house which proved to be occupied by a couple of negroes, as we could distinctly hear them talking within. Then we lay around that house for nearly an hour, hoping that one of them might come out and give us a chance to gobble him and make him show us the roads. But after a while, fearful of making an alarm lest there might be pickets in the neighborhood and the whole object of the expedition frustrated, I drew my forces away by signs and jestures and resumed the journey inland.

After travelling perhaps a half or 3/4 of a mile we came to the main road [and], hearing the sound of horses feet in the distance, we all lay down close to the fence by the side of the road, where soon a squad of cavalry passed us. This then was the patroll and it was quite evident that no picket line had then been established so far to their right.

From this moment we felt at our ease. I had found all that was required of me. We had also found an excellent position to cross the river and there was no enemy to prevent. Once across and in possession of this position, we would take the enemy by the flank and rear and annihilate his Army. But in order to do this immediate action was necessary. Not a moment was to be lost.

I returned with my party and deliberateley paced off the distances. I found my men at the boat anxiously waiting our return, rowed back to the place where we had launched the boat and found the men with the Ambulance. [We] cooked our coffee, killed a nice young steer and had him dressed and in the Ambulance and were well on our way back to camp before 6 O'clock in the morning.

I reported to General Woodbury and he said that he would see General Burnside. I felt so confident that this was to be our position to cross and so anxious that General Burnside should know of all the particulars, that I could not refrain from expressing my anxiety about it to General Woodbury again, but he seemed very busy working out some mathematical problem on a piece of paper and treated me so indifferently that I did not dare intrude my ideas further.

The next day, however, I was called and again we visited the position. General Hunt had two of his Officers with him. We remained all night a short distance below where I had launched my boat at the house of Mrs. Gaines, who was alone with her house full of children, her husband being in the Rebel Army. This Gaines was a tenant of Dr. Taylors, as venomous a Rebel as ever lived, whose house was about a mile distant.[11]

That day (Dec 1st), after the two Generals had satisfied themselves with observations and had turned homeward, I was riding at some distance in rear of them and, happening to turn around, I discovered a man on a black horse about a half mile in rear of us skirting a belt of woods who by his peculiar actions excited my suspicions. I was soon convinced that the man was watching us and rode back with the intention of capturing him. He did not attempt to escape but faced me boldly and impertinently. On demanding of him his business there, he coolly informed me that he had been over to see his sister near by. I told him that that story would not take and that he must come with me. Upon this he produced his protection papers from General Patrick, Provost Marshall of the Army of the Potomac. Of course I could do no more and suffered him to pass on towards his house after some pretty sharp and not very polite or loveable expressions on either side.

This man was Dr. Taylor, the owner of the estate, and to this man

I attribute the loss of the battle which soon followed. For, at the time of my reconnaisance, there was not a rebel soldier on or near Snickers Neck and within five days thereafter the ground was covered with them. Who carried the information of our operations to the enemy? Who else but he, the only white man who had observed our movements, a thorough Rebel, too old to fight in the ranks but keen enough and strong enough to obtain protection from us, watch our movements and find means of conveying his information across the river.

When all secrets are divulged it will be seen that this man carried information of our doings to the enemy and our disaster followed. Had it not been for his protection papers I should have taken him with me to camp.[12]

CHAPTER TEN

Fredricksburgh 1862

The result of my second expedition across the Rappahannock was communicated to General Burnside by General Woodbury. At least I supposed it to have been, for the next day or the day after a large force was detailed to go down to the place selected and prepare the way for the Army by cutting down trees and making corduroy roads. The men were allowed to make as much noise as they saw proper, which alone was fatal to the enterprise *if* the General ever had any serious intention of crossing there.

Indeed, it has long been a question in my mind whether General Burnside fully realized the value of that position or had been fully informed of its superior advantages, for that by allowing those boistrous demonstrations to be made when the work could just as well have been done quietley, it soon became the general belief that he intended the whole operation as a ruse, intending to draw the attention of the enemy in that direction while he perfected his arrangements for the real attack from our extreme right.[1]

None of *our* Officers could believe that General Burnside would throw away such a splendid position as the one which I described to them. Nor could I believe it, until on the morning of December 9th, Spaulding informed me that we had orders to meet General Woodbury at the Lacy House, in order that we might be shown the position which had been selected for our bridges and familiarize ourselves with the ground. What was our amazement when we were informed that the General had decided to force a crossing directley in front of the Town.

Well, we went down to look at the position: Spaulding, Ford, McDonald & myself, and after we had pondered the situation well we all came to the conclusion that we might now return to our quarters and with great propriety execute our last Wills & Testaments. With the enemy so strongly posted on the opposite side of a river 400 ft. wide and for us to attempt to lay a Ponton Bridge right in their very faces seemed like madness. But, as that was the *order*, there was no fault finding or hesitation.

There was scarceley a word spoken as we returned to our camp. Then that old soldiers song which I used frequentley to repeat came up in my mind and I sang out in the best voice I could command,

> "O why should we be meloncholy boys,
> Whose *business* tis to die."

in which they all joined heartily and our spirits were once more restored.[2]

That morning two more Companies (J & F) of the Regiment joined us, having marched up from Belle Plaine where Lt. Col. Pettes was with the Head Quarters of the Regiment. Our Detatchment consisted of Companies C, H, J, F, A, & K. Company G was still at Harpers Ferry and Companies B, D, & E were at Washington and Belle Plaine. Colonel Stuart was still absent in N. York, regaining his health and recruiting for the Regiment. During the afternoon of the 10th, I rode down to see the position with Capt. Perkins, who on our return seemed particularly infused with gloomy forebodings. Poor fellow, it was his last ride.

We moved our Train down to within a mile of the river just before dark and as we passed through the different Divisions of the Army every man knew that the moment for action had arrived, for the *Pontons* were moving for the river.

Evanston Ill.
Sunday evening
December 10th 1871—

And now my son, having just assisted your Mother in tucking you away in your little bed upstairs with your dear little Sister and while your Mother has gone out to evening church service, I will employ the time in writing, for you, what transpired to me nine years ago this very night and two years lacking 5 days before you were born.

We moved our train down near the river during the day (the 10th), halting back of the stream in a ravine far enough away to conceal us from the observation by the enemy.

After partaking of a hearty cup of coffee at about this time in the evening (the church bells are tolling now), Spaulding and myself with our Surgeon, Dr. Hewitt, mounted our horses and rode toward the river. Nearly one quarter of a mile from the stream we came to a ravine in which the Dr. selected a position for his hospital tent. He remained to superintend its erection while Spaulding and myself rode on. We went down to the Lacy House [and] saw no one that we knew, only a few men in civilian clothing evidentley belonging to the Quartermasters Department and an Officer or two lolling about.

We then rode out to the rivers bank. All was quiet, peaceful, calm. Who would have dreamed that a *terrible tragedy* was soon to be enacted there? The moon shone with a quiet light which was reflected from the sparkling snow which scarceley covered the ground. [Reflections from] little fires here and there on the opposite side of the river glimmered and sparkled from the ice which bordered the stream. On our side of the Stream everything was as quiet as if no enemy or Army had an existance.

Spaulding and I stood and gazed upon the scene for some minuites without exchanging a word. We took a good view of the landmarks so as not to be mistaken when the moon should go down. We rode up to the Phillipps House to learn if General Sumner, in whose Grand Division we were to operate, might have any orders or suggestions for us. Upon the announcement of our names, we were shown by an Orderley into a large room on one side of the hall and very soon thereafter the door opposite was opened and General Sumner approached us with a smile on his dear, frank old face and a hearty hand shake. He replied to our question for orders that he had none to give us and had nothing to add but his earnest wish for our success and his belief that we were about to achieve a great victory. As he retired to the room opposite we could see that it was filled with Officers who had been summoned to receive their final orders from the old chief of the Right Grand Division of the Army. This was the last time my eyes ever beheld glorious old General Sumner.[3]

We then rode back [to] where our train was bivouacked and were soon on our way to the river. We marched as quietley as possible so as not to disturb the enemy, who were expecting us to strike somewhere but could form no idea of the spot. We proceeded cautiously and noiselessly to the rear of the Lacy House and there, in compliance with our instructions, waited for the hour of *One*. The moon was supposed to go down at that hour but my recollection is that it was bright enough to discern objects at some distance all through the night.

It was cold and raw, the night air laden with a sort of haze or fogg [that] penetrated to the bones taking all the *vim* out of the body, especially as we were not exercising. The ordinary Soldier going into battle is stimulated by the exciting nature of his surroundings but with the Engineer Soldier the case is often entireley the opposite and this was one of those occasions.

I had made up my mind that morning when we went down to the river and were shown our positions that escape from there with life was extremeley problematical and now, as the time approached for the opening of the grand tragedy, my mind passed through a variety of emotions which I apprehend must be experienced by those who are conscious of the near approach of their final dissolution. I retired to a room in the Lacy House, where by the light from a fire in a grate I could see to write and taking out my diary wrote these words:

> Dear Father.
>
> To night the grand tragedy comes off. If I am killed and should this meet your eye, please accept it as my "good bys," remember me to all my friends and I beg of you to look after the interests of my Dear Wife and little Daughter. And so, my dear Father, Good bye.
>
> Wesley
>
> Lacy House opposite Fredricksburgh, Dec 10th 1862"

Having done this and offered a short but earnest prayer to the "God of Battles," I returned to my command. If in considering the possibilities of that night I had done so with more or less dread and fear, as I must confess that I did, from that moment and thereafter those feelings departed. My mind was serene for I felt that there was no escape and had made my peace with my Maker in expectation of the worst possible result.

As I stepped out into the clear, crisp air, the bell in the tower of the principal church of the Town tolled out slowly and solemly the hour of *Twelve*. There was another hour to wait and it seemed an age before the bell again struck, *One*, but when it did strike, my men were all in line and at the word of command we marched towards the river.

The Ponton wagons where hauled to the crest of the hill about 200 ft. from the bank of the stream by the teams, which were then unhitched and taken to the rear. As we approached the crest of the hill overlooking the stream, two Regiments of Infantry approached us, one from the right and the other from the left. These were our supports. They moved silentley down to the river bank and stood at ease waiting for us to commence operations. Our Arms had been left in stacks back at the Lacy House.

Soon we had all our boats unloaded from the wagons and were passing them down the hill, 150 ft. across the level ground and into the water at a liveley rate. Then the little fires on the opposite side of the River were one by one extinguished but not a word was spoken or a shout made.

Our Artillery began to arrive rapidly and it was not long after, when casting my eyes up the hill in rear of us, I saw a long line of black cannon unlimbered, shotted and ready, their muzzles pointing over us, waiting to send forth their "winged messengers of death" at the first demonstration of the enemy.

Our men sprang to their work as if their lives depended upon their efforts. The time passed rapidly by and I heard the Town bell strike *Two* and my bridge (for I was in command of the first one, Capts. Ford & Perkins of the 2nd) was fast assuming shape. Our Squads were so thoroughly drilled and so well did each man understand his duty, that it was scarceley necessary to give the words of command.

The bell sounded *Three* and we were near the middle of the river but the ice had prevented a rapid construction. The fogg and mist which covered the River had now become very dense. We had completed the tenth abutment or length, making the bridge 200 ft. long (the Stream at this point was about 400 ft. wide). I was standing at the extreme outer end of the bridge encouraging my men, when, happening to cast my eyes to the shore beyond just as the fogg lifted a little, I saw, what for the moment almost chilled my blood. A long line of arms moving rapidly up and down was all I saw, for a moment later they were again obscured by the fogg. But I knew too well that line of arms was *ramming cartridges* and that the crisis was near.

I said a few encouraging words to my men & walked deliberateley to the shore. Here I met General Woodbury and explained to him what I had seen. He remarked that he too had caught a glimpse of the same but, said he, "Captain, do the best you can and when they open fire upon you, as they probably will in a few moments, retire to the shore in as good order as possible." With these words the General moved off in the direction of our Artillery and I went out again to the end of the bridge.

Soon the bell sounded the hour of *Four* and in an instant we were in the midst of a perfect storm of bulletts. Our Artillery on the crest of the hill in rear of us quickly responded, their shells roaring and screeching over our heads. Our Infantry on our right and left returned fire also.

The battle was now fairly opened. The bulletts of the enemy *rained* upon my bridge. They went whizzing and spitting by and around me, pattering on the bridge, splashing into the water and thugging through the boats. Where were my men? They did not require any command to fall back in good order for had any such order been given not a word of it could have been heard in that pandemonium. Every one started for the shore end of the bridge. Some fell into the boats, dead. Some fell into the Stream and some onto the bridge, dead. Some, wounded, crawled along on their hands and knees and in a few moments all of us were off the bridge, all except the dead. The storm of lead continued.

We had no protection, not the simplest earthwork. The Regiment of Infantry, our supports, were alike unprotected and were rapidly melting away. It was simple murder, that was all. We instinctiveley dropped upon the ground and so most of the bulletts passed over us. Here we lay, our faces burried in the snow and mud while the balls whistled around like bees let loose from a hive.

There was one large tree near the bank behind which many of the men & Officers were standing & lying for protection. When I got ashore the only thing I could find that promised any protection was a bundle of lashing rods about a foot in diameter and 18 or 20 inches long. This I placed on the ground and lay down close behind it. Soon five other men were in line behind me, all depending upon that little bundle of rods to protect us from the bullets which spatted and sputtered into it in the most venomous manner. Here we lay in this uncomfortable manner for fully an hour.

When it began to get light I heard Spaulding calling for me to come behind the tree which was about 20 ft. away. There was a slight lull in the firing of the enemy who had taken refuge in the rifle pitts and houses opposite. There on the ground near us lay Capt. Perkins, the cape of his blue overcoat up over his head. We called but he did not answer. We went to him, pulled down the cape from his face and there he lay, dead. We thought he had been shot in the mouth as the blood was coming from there, but it appeared afterwards that the ball had passed through his neck.

Spaulding, Ford & myself then tried to form our men to make another attempt at building the bridge. While we were forming them, a very difficult task to perform under the circumstances, General Woodbury coolly walked down the hill to see us and returned unharmed amidst a perfect shower of bullets. He told us that General Burnside had said that our bridge *must* be completed at whatever cost and that the whole Army was waiting for us.

The second attempt was soon made and soon over. We rushed on and were speedily driven off again. Our situation was now dreadful in the extreme. [With] no possibility of finishing the bridge under a galling fire that 70 pieces of Artillery could not silence and no possibility of retreat or relief, as there were no Troops in the Army that understood our duties or that could relieve us, there was nothing left for us but to die, or so it seemed to me.

It was now about 7 O'clock in the morning. After great difficulty we succeeded in forming another party for another attempt. I was to go on first with ten men, lashers, and upon my signall Ford was to follow with twenty chess carriers and Spaulding was to follow him with ten balk carriers. At the signal we started on a double quick, ten men besides myself. Our Artillery opened with all their might to keep the Rebels down but they were prepared and resumed the firing with greatly increased ferocity.

When I reached the end of the bridge but five of my men were with me, the other five had either been killed or were wounded and were crawling off. Corporals Kinney & Dunlap were both wounded in the shoulder as they were lashing the balks and fell in the boats. Then Blakeley was killed and Bascomb wounded, leaving Wilcox & myself the only ones unhurt. Wilcox stooped over, lifted the last balk to complete the abutment, shoved it about half way out and fell dead at my very feet. The ball passed through his body. I was obliged to stand straddle of his body while I lifted the balk, shoved it out to the boat in which Dunlap & Kinney were and the two Corporals, in spite of their wounded condition, fastened their end while I fastened mine.

I was now the only man of 11 who remained unharmed and 20 ft. more of bridge was ready for the chess (flooring). I turned around to give the signal to Spaulding to send forward the Chessmen and saw him beckoning me to come back. It was evident that, seeing the fate of my party, they had concluded not to sacrifice 20 more men in the attempt to lay the chess between the boats.

Just then my left arm swung around over my head and down. A

The bombardment of Fredericksburg by the Army of the Potomac. Courtesy of the Library of Congress.

stinging sensation as though I had been hit with a bar of hot iron followed and looking down to my left I saw a stream of blood about the size of a lead pencil flowing to my feet. Solitary and alone, the last of my party, I walked back over the bloody bridge to the shore, the enemy meanwhile used their best endeavors to bring me down but without success.

When at last (for it seemed a long while) I reached my comrades on the shore, I began fully to realize that I was wounded. The blood was leaving me very fast for an arterey had been severed. Spaulding & Ford succeeded in getting off my overcoat but there was no time to be lost in getting off my uniform coat so they took their knives, haggled off the sleeve near the shoulder and bound up the arm as tight as they could with a pocket handkerchief.

Then they insisted that I should be taken to the rear. As my arm was still bleeding profuseley and we found it quite impossible to stop the flow, I at last consented to go. To go up that hill was to take one chance in a thousand for life. To stay there was swiftly to bleed to death in a few miniutes. It was concluded best to take the chance. Sergeants Simpson & Cowan volunteered to go up the hill with me and so, with an arm over each of their shoulders, we three started up the hill.

We made about one quarter of the distance very well, as I had considerable strength left, but for the ballance of the distance, each step seemed to me my last. Loss of blood had so weakened me that I could with the greatest difficulty lift one leg after the other and I lay almost a dead weight upon my brave supporters. Three men going up the hill so diliberateley attracted the attention of the *gallant* Mississippi Riflemen on the opposite side of the river and they resumed their fire upon us with increased malignity. My privates overcoat having been removed they saw that I was an Officer and seemed determined to kill me. The bulletts flew around and about us, struck the ground ahead of us, spat into the ground behind us, whistled beside and between us but not a ball struck us.

It was a long, long journey. When at last we reached the top I was little more than a lifeless lump of lead. Our Artillererymen, seeing us comeing, attempted by rapid firing to keep the rebels down but they seemed well protected in the cellars and behind the heavy stone walls of the houses. As we reached the summit we met our Ambulance men with a stretcher and I was rolled over on it more dead than alive, lifted upon the shoulders of four stout men and carried all bloody (the handkerchief which bound my arm had got loose) and faint, almost insensible, to the Lacy House. [I was] carried into a large room and sat upon a chair. The men then left me to go for others.

The room was filled with the wounded, dead and dying. Some were crying, some groaning and others were too far gone to do either. Some of the crying were wounded Officers, crying for *grief* that their men were being sacrificed, slaughtered for *naught*.

Here in this house which I had left so quiet, so peaceful a few hours before, there now arose the most horrid sounds from the crowd of freshly wounded Soldiers. The windows shook and the solid walls of the old mansion trembled with the reverberations of the cannon which surrounded it and were belching forth upon the town.[4]

The fatigues of the night, the loss of blood, the excitment of my escape up the hill, the horrid sights and sounds that now surrounded me was too much for my poor nature to bear and I fell from my chair to the floor, helpless, exhausted, almost insensible. In this condition I lay until a Surgeon came around and, seeing my condition, bound up my arm again, stopping the flow of blood or I would certainly have bled to death. He gave me some stimulus but my stomach was too weak to bear it and the stimulus was soon rejected.

After a time our Ambulance returned and I was taken away from that horrid place by kind hands to our own hospital. Here at last was peace and safetey. I was placed in a tent with quite a number of our wounded but there were no complaints, no grumbling. All were cheerful and happy notwithstanding their sufferings for we were now under the kind and competent care of our own excellent Surgeon, Dr. Hewitt.

Captain McDonald was brought in wounded about the same time with myself, the ball passing through the left arm near the elbow joint. In his case the bone was seriously damaged, his sufferings were acute and his arm was never much use afterwards. Johnny, my ever faithful Johnny, whose anxiety for my safety during my absence could not have been increased had he been my brother, was now by my side, so happy that he might again minister to my comfort. Our beds were on the ground with a little straw underneath, then a blanket and a blanket over us, but mine could not have been sweeter had it been of rose leaves and soon I fell away into a long, sweet sleep.[5]

Meanwhile, how fared our men at the bridges? For some time after I had been taken away they remained where I left them, protecting themselves as best they could behind the tree and one solitary tool wagon which had been left on the ground. Lieut. Folley succeeded me in command of the company.

Here they were with no chance of service or escape until Spaulding managed to get word to General Woodbury suggesting that some Infantry be sent across in our boats to drive the enemy away. Finally this plan seemed to reach Head Quarters and volunteers were called for. Our men meantime were waiting patientley in the boats ready to pull them across when they should come down. Two different Regiments volunteered, came down as far as the top of the hill overlooking our men and the river and *backed out*.

Then a long pause ensued. More of our men were wounded by the Sharpshooters across the river. At last a N. York and a Michigan Regiment made their appearance. They came down the bank on the *double quick* but before they were loaded into the boats many were wounded and many more ran back up the hill again.[6]

Captain (then Lieut.) Robbins, with his crew of 4 men, shoved off in the first boat. Each Ponton contained about 18 or 20 men. The Infantry all lay flat down in the bottom of the boats. No heads ap-

Amelia Maria and Mary Brainerd. Courtesy of the William Hencken Family.

peared above the gunwales except those of our men at the oars and the Officers in command.

Captain Robbins boat was in the advance. When he had reached about the middle of the Stream the Officer in command of the Infantry raised up his head and *requested* Robbins to *return*, as the balls were comeing through the boat uncomfortably near him. This Robbins refused to do but kept steadily on his course. Then the Officer positively *ordered* him to *turn back*, threatening to use violence if he did not. At this, little Robbins drew his revolver and, bringing it to bear upon the Colonel, told him to lay there quietley or he would use it. In this manner the first boat crossed the river at Fredricksburgh.

As soon as the first boat had touched the shore the firing ceased and the Rebels tried to make their escape but in a few moments a large number of our men had reached the shore and rushing up the hill captured quite a number of them. The remainder made their way through the Town to the main body beyond.[7]

Our own men soon completed the bridge and the one above it and our Army commenced crossing into the Town. Then the Engineers, leaving a small guard to care for the bridges, returned to camp and for the next few days were allowed that quiet and rest to which they were so justly entitled. [See Appendix 2, Items 6 and 7.]

The next day, the 12th of December, I was visited by Philo Judson, also by Majors Beveridge and Ludlam of the 8th Ill. They had heard that I was killed and were glad enough to see me alive.

The day following, Uncle Philo Judson visited me again. I was sitting up then [and] he asked me if I had heard from home lateley. I told him [that] the last time I heard that my little daughter Mary was not very well. He looked at me with a peculiar look. "She is better now," he said, "*far* better."

Little Mary was dead.

[See Appendix 3, Items 5, 6, and 7.]

Chapter Eleven

After Fredricksburgh: Leave of Absence and Return 1862–1863

Our Army had gained a foothold in the Town on the 11th. All day the troops poured across the bridges for there were two other bridges below ours which were completed with comparativeley little opposition early in the morning. One of these was constructed by two Companies of our Regiment, Cos. F & K, [and] it was at this bridge that Captain McDonald was so seriously wounded. The other one was laid by the Regular Engineers some distance below with no opposition whatever. One was also laid by the 15th Regt. without opposition but this I think was after my bridge was completed.

Why did General Burnside not allow troops to cross at these lower bridges? Why, instead of allowing these three bridges to remain unused awaiting for the completion of the 4th & 5th before giving the order for the Troops to cross, gaining many precious hours of time in which many thousands of troops could have been crossed and, comeing up by the flank, drive the enemy away from the position in front of us or capturing them? This question has never been satisfactorily answered.[1]

There was considerable firing during the day but not of a general character. The next day, Friday, the 12th, the body of Captain Perkins was removed to the rear to be delivered to his friends for burial in N.York. The firing was heavy and continuous and troops were crossing on all the bridges all day and all night.

Saturday, the 13th, the battle commenced in earnest, or rather it culminated on that day, for the fighting had been going on since Thursday morning. But on the 13th the noise of musketry was absoluteley frightful, one continuous, ceaseless roar. McDonald and I could stand it no longer [and] we asked the doctor to let us go out on the crest of the hill beyond the hospital and see the battle. To this the Dr.

positiveley refused to give his assent. Soon observing him busy with some freshly wounded from other regiments we stole away from his observation like two school boys and were soon on the crest of the hill. But this was not near enough to suit us, so, weak as we were, we more crawled than walked along until we were up to the bank of the river or rather the elevation back a little distance from the Stream which overlooked the whole scene.

For miles on our right and left we could see the line of battle. I doubt if any field of battle of our whole war presented such an uninterrupted view of so large a part of the army when in action. The city of Fredricksburg was surrounded by a line of hills, or rather one hill in the shape of a semi-circle, reaching for about 8 miles back from the river and around again to the river while the Town lay in the valley or plain. Inside of and far below the circle this line of hills was fully fortified and manned and this was the position our army was mainly endeavoring to carry.

We could see long lines of blue advance while the ground was dotted all over behind them with blue coats as men fell. Then the line would falter, then halt and then fall back, leaving the field all covered with the dead. This we saw repeated time and again, from far to the right to far to the left our men could get so far but no further. It was a glorious and yet a sickening sight.

Occasionally a random shot would reach over near where we stood. Quite a number of the wounded and some General Officers had gathered here when who should come up but our Doctor Hewitt. Having reasoned with us for some minuites without effect he quietley went to Genrl. Woodbury, made known his request and the General good naturedly but firmly *ordered* us back to the hospital.

That afternoon we buried the dead of our regiment in regularly made graves with wooden coffins made of rough boards. A regular coffin and a regular grave was a luxury that few of our Soldiers besides those of our Regiment were honored with. Towards sundown that night the firing was the most terrific that I ever heard before or afterwards in my military life and made such an impression upon my mind that I think I shall never forget it.

On Sunday morning the firing commenced very briskly again but did not hold up as fierceley through the day. Who can tell of our anxiety all during those terrible days? All night long for two or three nights the wounded were being brought to our hospital for first attention and were then sent to the rear for further care. Of all the wounded men that I saw or heard utter a word not one but said he was willing to suffer all he was suffering and *more* if we only gained the victory. I can never get away from my eyes or clear of my nose the sights of that hospital and the smells of supurating wounds.[2]

Our men fought braveley if ever men fought well but all to no purpose. It was not intended by the Almighty that a victory should be gained by us. Sunday there was a lull in operations and on Monday I signed my application for leave of absence for 40 days. It was properley drawn up and signed by the Surgeon. It was approved by Spaulding and I took it myself to General Woodbury. He signed the request but stipulated 20 instead of 40 days. It then went to Head Quarters and was returned *not approved*. This was cheering intelligence for a wounded man who had never missed a days duty for 16 months and, as may be immagined, I felt a little downhearted.

That same day, Monday, there were unmistakable signs of a retreat and as soon as darkness covered the face of the earth our Army commenced recrossing. A dense mass of men, horses, Artillery & the various *impedimenta* of an Army poured over until after daylight when the last of the Army except a few hundred stragglers were across. The enemy were so close upon us that our men did not have time to dismantle the bridges regularly. Casting loose the end towards the foe the bridges were allowed to swing around by the force of the current, one end being held fast until they touched on our side. They were then dismantled and taken away.

The enemys advance guard standing on the other side of the stream cooly observed the operation, not choosing to fire upon our men while so many of our heavily shotted cannon pointed to their faces. Some of the stragglers came to the river after the bridges were cut off and begged pitiously to be taken over. A few pontons were kept running for this purpose as long as possible [but] the ballance of the laggards were left to their fate.

That night we could hear the people in the church opposite us (the same whose bell had before tolled the hours) singing, praying, and praising God for the victory he had given them and invoking his aid for the total annihilation of the hords of northern barbarians who had dared to invade their country. The prayers were *earnest* and *sincere* and what strange emotions filled my breast as I reflected that at that very moment so many thousands of *our* friends were then imploring the same Almighty God to protect *us* and destroy *them*. Alas, poor human nature.

Having failed in the attempt to obtain leave of absence from Army Head Quarters the next step was to get to Washington, as there were so many wounded that proper care could not be given them in the field and the City of Washington was amply supplied with comfortable hospitals. This our Surgeon accomplished in due season and on Wednesday the 17th we were sent in cars to Acquia Creek and from there by Steamboat to Washington. Arriving there at 4 P.M., Capt McDonald, Johnny and myself put up at Murrays Hotel on Pennsylvania Avenue.

Who was to blame for the disastrous failure at Fredricksburg? Investigations were commenced and testimony was taken [but] nothing ever came of it except that General Halleck and the *Pontons* were finally in some indeffinite manner held accountable.

"Our Pontons in Virginia," written by Col. B. S. Alexander to the NY Herald in our defence, was called out by the congressional investigation. Probably not a tenth part of the people who condemned the *Pontons* & Engineers [with] promiscuous abuse from one end of the land to the other ever read this. As we were in the service all we could do was bide our time & suffer in silence, it being a grave offence for an Officer to write anything for publication except what passes through Head Quarters.[3]

I suppose the matter would still be under investigation if General Burnside had not promptly stepped forward and assumed the entire responsibility and blame. But everybody knew that he assumed more than he deserved.[4]

Would you like *my* opinion of the matter? You shall have it. I attribute the failure to several causes:

1st, The removal of General McClellan in the midst of a brilliant movement, the consequent confusion and change of plan by the new commander and the opportunity this confusion gave the enemy to move from the dangerous position in which he was held. It is a question whether Lees Army would not have suffered a terrible defeat within ten days had the order of McClellans removal not been made;

2nd, The mistake that was made in not crossing large bodies of men at Falmouth and United States Fords when our advance arrived at the river and taking possesion of the hights beyond, which could easily have been done, instead of sitting down and waiting for *Pontons*, which were then at Berlin without orders to move;

3rd, The magnificent mistake of the delay after my reconnaissance at Skinkers Neck, when a prompt and active movement would have crossed the Army at that point, gained an impregnable position, taken the Enemy by the flank, forcing him to change his front and all without the loss of a man on our side;

4th, The delay in waiting for the 4th & 5th [bridges] on the morning of the 11th after two or three bridges were completed before any Troops were allowed to cross, thus giving the enemy an opportunity to discover our real intentions and a good half day for the concentration of his forces, which we could have prevented by throwing a large body of men across the bridges as soon as completed (before daylight) and by charging in mass upon his center, severing his line and fighting him at great advantage to ourselves with an absolute certainty of victory.

There may have been other reasons before and during the battle itself. General Franklin may or may not have been to blame. General Burnside says the country will never know how near we came to a great victory. I dont think it ever will. One of the saddest reflections in regard to the whole affair is to think of our poor President Lincoln and how the people poured abuse upon him. Lincoln, Stanton, Halleck: they are all dead now.[5]

Having arrived at Washington, the prospect brightened materially of getting home during the season that I would necessarily be unfit for duty. The wounded came in from the front so rapidly that all accommodations for them, immense as they were in Washington, were soon completeley exhausted. The authorities were glad enough to give furloughs to the men and leaves of absence to those of the Officers not seriously wounded in order to make room in the hospitals for those who were.

It was not until the 24th of December that I succeeded in getting my leave of absence properley signed. In the meantime, Capt. McDonald & myself received calls at our Hotel from many sympathizing friends including Mr. & Mrs. Moore, who were formerley from Rome. Captain McDonalds sufferings from his wound were so intense that mine seemed nothing in comparison. The streets of the City were filled with the wounded or their friends who came from home to administer to their comfort or perhaps to perform a still sadder office. Those five days of waiting seemed an age to me.

At last on the 24th my papers came. I had seen that important personage, the Paymaster, who paid me for two months, leaving two months salary yet due me. That night at 5 P.M. I left Washington hoping to reach my Fathers house in time to spend some of the Christmas day there but we missed a connexion at N.York so instead of reaching Rome at 3 P.M on Christmas day, I did not get there until 10 P.M. that night.

There had been quite a number of *Prominent Citizens* at the cars at 3 O'clock or so my Father informed me. As I did not come on that train they assumed that I would not make my appearance until the next day and so I missed a reception. When I did return there was not a solitary individual acquaintance at the depot to do me honor, but, as I did not know that any demonstration was intended until informed of it the next day, my feelings were not injured.

I walked slowly up through the dark streets of my native town toward my Fathers house, the house of my boyhood and early loves and cherished memories, with peculiar emotions after sixteen long months of exciting Army life. He knew my footfall as soon as it sounded upon the doorstep and in a moment the door was open and Father and Son grasped hands.

The next day I remained at home all day and had quite enough to do to see the callers that came to welcome me back again and make inquiries for their sons, brothers & husbands in the Army. Everyone had a word of kindness for me and I felt almost repaid already for the sacrifices I had made.

On Saturday I went out to make calls upon families who had relations in my company and on Sunday attended church. Colonel Stuart called on Sunday and I was obliged to recite again all the particulars of the battle to him, as I had already done to others. He was recruiting for the Regiment and was very successful in raising men. He left for Albany that evening after insisting upon my stopping at his house in Geneva.

On Monday I left Rome, arriving at Geneva at 4 P.M. Here I was met at the depot by Mrs. Stuart, who gave me a most cordial welcome. That night the Colonel returned from Albany. My visit there was made very pleasant by the entire family. So many things had to be talked over and so much to be said about Captain Perkins, who was Mrs. Stuarts Nephew. She could not reconcile herself to his death.

Just as I was about to step onto the cars at the depot the next morning, the Colonel handed me a letter and my Commission as *Major*.

GENERAL HEAD
QUARTERS, STATE of NEW
YORK.
Adjutant General's Office.
Albany, Dec. 29th 1862

Major Wesley Brainerd
50th Regt NY.V Engineers
Dear Sir

It affords me great pleasure to forward your Commission as Major in the 50th NY Vol Engineers, a promotion so richly earned in the terrible battle of the 11th before Fredericksburgh in which you were so severely wounded in the gallantest & thrice attempted effort to build the Ponton Bridges in front of that City.

May this "New Years Gift" be highly valued by you & yours hereafter as the Just reward your brave & skillful conduct has received. With my best wishes in your Future success in your honorable career & in your health & prosperity.

I am, dear Major,

Your Friend & Colonel
Charles B. Stuart
50th Regt NY VEng

The Commission was not entirely unexpected as I had been notified that my name had been sent forward but I was agreeably surprised to receive it just at that time. Lt. Col. Pettes, with his usual *good feeling* for me, had witheld his consent to my nomination for the office for more than a month before the Battle of Fredricksburg but after that event he dared not refuse it longer.

I arrived at Chicago at 10 1/2 P.M. on the last day of 1862. Your Uncle Lyman met me at the depot. We jumped into a Carriage and arrived at home, 157 West Washington St, an hour before the year departed. Need I say that your Mother was awake and anxiously awaiting my arrival? Sixteen long months separated and now we were actually in each others presence again. It all seemed like a dream, but the joy of our meeting was sadly toned down by our recent affliction.

Two weeks before my arrival home the body of our little Mary had been carried out and deposited in the receiving vault of the Cemetery.[6] Her little Spirit had flown to the arms of her Father above in the land where sorrows are not known. It was not for me to see my little daughters face again in life. Sixteen months before, after I had kissed the little 3 month old infant, I had begun to form plans for the future of our little one. So recent and so sudden had her removal been [that] now I could not free myself from the impression that she had been purposeley taken from the house by one entrance while I was entering by another.

But, if my grief and disappointment was great, what think you must have been the agony of your dear Mother during that terrible week. Just think of it, little Mary was burried on the 7th and news of my wound and of the terrible battle then raging reached her on the 11th and the papers were filled with such vague intelligence that for several days she knew not but my wound was a fatal one. What a terrible week that must have been for her, sick herself, down with fatigue and anxiety, now childless and perhaps a *Widow*. My sorrows and my distress seemed as nothing in comparison.

A few days after my arrival we rode out to the cemetery and there I saw the dear little form, so sweet, so loveley, such a noble looking child, the sight of which in life had been denied me. The flowers around her little body were just beginning to fade and wither. The least perceptible yellow tinge had overcast her countenance [but] not enough to produce any disagreeable sensation. So I fastened the image of my little first born indellibly upon my mind and then the casket was closed forever.

My leave of absence expired on the 9th day of January, 1863, but upon application to Dr. McVickar, who was stationed in Chicago, I obtained an extension for 30 days. The time was profitably employed in getting rested & rejuvinated, visiting quietley at home with your Mother, Grand Father, Grand Mother and Uncles Lyman & Lloyd. The time passed pleasantly and all too rapidly for me, knowing as I did that before me were greater hardships and dangers as any I had yet passed through.

It was decided that your Mother should accompany me when I left for the East and take a season of recreation and rest. Accordingly, our preparations all completed, we left Chicago together via the Michigan Southern RR on the 15th of January at 7 P.M. Your Uncle Lyman went with us to the Depot and we embarked during one of the most terrific snow storms that had been known for years. At Cleveland we were detained from 10 1/2 A.M. until 8 P.M. on account of the storm. I wonder if your Mother will remember the Cuffs she lost at Toledo when we changed cars? They were subsequentley recovered by the liberal use of the telegraph.

We passed through Buffalo and arrived at Byron on the 17th, were met at the Depot by Cousin John and driven to his house four miles away. The girls: Mary, Jennie, & Debra, were all at home and there we had a splendid visit though the storm continued for several days.

We enjoyed our time among our relations at Byron until the 20th of January when we left there for Geneva. Arriving at Geneva in the evening, we were met and entertained by Colonel & Mrs. Stuart and daughters in a very hospitable manner. During our visit there the Colonel and I drove out to Seneca Falls and saw Captain McDonald, who had also obtained leave of absence and was doing as well as could be expected with his wound, though the bone of his elbow had to be re-broken [and] re-sett. He was suffering much pain with his wound while mine gave me comparativeley little trouble.

On the 23rd we left our good friends of Geneva for Rome, stopping one day at Syracuse to visit your Mothers relatives there. We arrived at Rome on Saturday the 24th and were entertained at my Fathers house. Many of our old friends and acquintainces called on us during our stay, which was extended to February 3. The days were all too short for us as the time approached for my second departure. It really required more fortitude to sustain me in this second leaving than it did at the first, for now I knew something of what I was to expect. All was romantic and uncertain before, now it was a dreadful reality.

Arrangements were made for your Mother to travel and visit during the Summer [as] it was necessary for her to have a change of air and to recover her overtaxed energies. We braved each other up as much as possible and finally the day of my departure came. Your Mother, my Father & my Step Mother accompanied me as far as Utica and there I left them.

On looking back upon my past life it occurs to me that the saddest events of the whole period were the frequent leave takings which it seemed to be my destiny to experience. I can never attempt to describe the loneliness [and] the desolation which I have felt upon these occasions. As I left all that was dear to me behind, I felt more than ever before how extremely brittle is this thread of life. The probability that we might never meet again now seemed exaggerated to an absolute certainty.

Some friends had placed in my charge a young lady, Miss Clark, who was returning to her house in N.Y. City. The conversation of an agreeable companion served in a measure to dispel my gloomy forebodings. At Poughkeepsie we stopped for supper and as we stepped again on board the train my charge dropped her pocket book. The train started as I got off to endeaver to recover it from the snow and before I could get on again the train was under such a rapid motion that by the merest accident I succeeded in getting a foothold upon the last platform. My young ladies Father met her at the Depot in N.York and she was as happy on her arrival home as I was sad at leaving mine.

At Philadelphia I stopped over a day and visited some of the land marks that were so dear to my memory. [I visited] my friends at Norris Locomotive Works (where I had learned my profession), my old friends the Wilsons and passed the evening at Rev. Dr. Brainerds, who with his good Wife had always taken the warmest interest in my welfare.[7]

From Philadelphia to Washington was but a short ride and I arrived there on the 5th and called upon my old friends, the Purdys. Had the prophecy of two winters ago been fulfilled? The *Spirits* (as we jestingly called our table rappings) then said that I was to be wounded on the bank of a river. So far the prophecy had been fulfilled but what of the future? We shall see.

On Saturday the 7th I took the Steamer to Acquia Creek and reported to Lt. Col. Pettes one day *before* my leave of absence had expired. Had I been one day *too late* he would have gladly embraced the opportunity to have had me court martialed. The next day, Sunday, February 8th, I joined my old comrades at a new camp, having on my way thither reported in person to General Woodbury at his Head Quarters.

I found the camp situated a few miles to the north and west of Falmouth and not far from a station on the Acquia Creek R.R. called Stonemans Station. The ground selected for our winter quarters had pleanty of wood to last us all winter with proper management, good drainage, an isolated position and pleanty of room with good water. What more does a Soldier need to make him comfortable?

My old companions gave me a hearty welcome and I must confess that as soon as I felt that I was again actually in the field the desponding feelings which during my absence were uppermost now almost entireley vanished. Immediateley on my return, Spaulding (who had his papers ready) left for his vacation of 30 days and I, as Major & Senior Officer, assumed command.

Chapter Twelve

Winter on the Rappahannock 1863

During my absence at home, General Burnside, anxious to retrieve the desaster of Fredricksburgh, determined upon another movement, the particulars of which I will not attempt to recite as I was not an observer. The movement was known as the Mud March and resulted in a rediculous failure. He determined to cross suddenly at Banks Ford a few miles up the river from Falmouth and strike the enemy on his left flank. No sooner had the movement fairly commenced than the rain began to fall and continued to fall in torrents until the roads became absoluteley impassible. Entire Regiments were employed in the vain attempt to drag the batteries of Artillery through the mud.

The enemy discovered the movement and reached the point to which our Army was trying to force its way long before we had arrived in any considerable numbers. They seemed to enjoy the joke hugeley and painted rough signs which they elevated for us to see on which were inscribed such sentences as, "Why dont you come over Yanks" and, "Stuck in the mud." This time the movement or its failure could not be attributed to the Pontons, for our men had succeeded in getting several of our cumbersome Pontons on the spot selected for crossing long before a single piece of Artillery could be got near the position.[1]

General Burnside discovered that any further attempt at crossing was useless and ordered the Army back into camp. He then resigned the command of the Army and was succeeded by General Hooker. The particulars of his resignation and the action of his Division Commanders which hastened it, you can read about in the histories.[2]

The next day after my arrival I rode on to Head Quarters and was mustered out as Captain and mustered in as Major. I now commenced anew in a new field of operations. The first thing in order was to get fairly to house keeping. Thanks to my own faithful Johnny this was a matter of very little trouble to me. My bed was made, stove up, mess chest filled and every thing in smooth running order in a very few hours. For sociabilitys sake our Officers had formed a *mess* and I therefore temporarilly abandoned my domestic arrangements to join them. While we remained in camp near Stonemans Station, as our camp was called, this mess was kept up and I look back upon it as one of the pleasantest periods of all my Army life. [They were] genial, witty, whole souled, generous and brave Gentlemen. The long, dreary, snowy, sleety, sloppy, disagreeable days and cold, windy, dark, dismal nights are almost forgotten and quite shorn of their prominence in my memory of that winter as my heart warms at the recollection of the many happy hours we passed in each others society. Dear old Comrades in Arms! where are you *now*? Like shadows, your familiar faces flit across my memory and it all seems like a dream.

Our chief occupation was to make ourselves as comfortable as possible for the winter and get our Trains in order. Drilling for the time was omitted [but] the necessary Guard Mounting was of course attended to promptly and strictly. By General Woodburys orders, 500 mules had to be received by our command in exchange for as many horses, which were turned over to the 15th. These animals were to be corralled, harness fitted and drilled to their duties. Wood must be brought and our commissary and Quartermasters stores hauled from the Station or from Falmouth. The mud was so deep and the roads so badly cut up that it required two full teams to haul half a load. But by the 22nd of February we were very comfortably fixed.

I shall never forget this 22nd day of February, 1863. It was Sunday and a terrible snow storm was raging, as it had been all the night. The wind whistled and moaned through the trees, carrying the snow in blinding sheets. My tent shook and flattened but could not break loose from its secure fastenings. It was Washingtons birthday and of course a day of rest. I took occasion to write a long, long letter to your Mother and had just finished the 8th page when suddenly the booming of cannon along the line in front brought us to our feet in great haste. I was certain that an attack had been commenced in force and was about to give the necessary orders for my men to fall into line when the idea occured to me that the cannonading was intended as a salute in honor of the day and such it proved to be. A hundred guns were being fired in its honor but [there was] no response from the lines on the other side of the river. The evening was passed in true Soldierly style around the camp fire in company with Ford, Folwell, Hewitt & others.

As soon as the storm had ceased, we commenced our drills again to which I added another of a different nature, viz. that of practicing the men one hour each day in the science of splicing rope, tying knots, &c, pertaining to pontonnering, all under the direction and instruction of Captain Folwell, who had qualified himself especially for that purpose.

We had our non-commissioned Officers drill from 6 to 7 in the morning, then the Officers drill which was under my charge after breakfast, then company drills during the forenoon and Battalion drill and bayonet exercise by Battallion during the afternoon. These were also placed under my command. Then Dress Parade and the days labor was completed. The command attained a very high state of perfection in the various Battallion [and] Company drills and the Bayonet exercise of the command was considered, by the numerous Officers who visited us for the purpose of witnessing it, as unequaled by any other Regiment *except* the 5th N. York in the Army. Our Ponton drills were not neglected, though the motions had to be gone through on dry land and recitations in Field Fortifications were kept up during the evenings by the Officers.

An occasional ride over to visit some acquaintance in another Regiment was all the outside variation we had but the time did not drag on our hands, at least it did not on mine. Two long letters a week were to be written to your Mother, a custom which I followed religiously during all my army life, never passing by the time to write unless it was absolutetely *impossible* to adhere to it.

But it was *glorious* that [the] out of door exercise from morning to night and then the excitement and *uncertainty* all lent a sort of real but indescribable charm to the life. The tendency of all this physical exercise was to develop fully the *manliness* of ones nature. Never since and never again will I feel that bouyancy, that peculiar elasticity, as if every muscle was a *steel spring* ready for instant action. I suppose we can never have these feelings but once in a life and not then unless developed by a life similar to the one we were leading.

Colonel Stuart returned to the Head Quarters of the Regiment at Acquia Creek about the first of March, having been absent since the July previous, a period of about seven months. He had recruited quite a large number of men for the Regiment, including a Band. The Band was sent up to our camp and a few days afterwards the Colonel visited us in person. He did me the honor to occupy my quarters during his stay with us. He occupied my bed, which was made in the usual manner of camp beds, viz. 4 crotchets driven into the ground [with] poles laid across and covered with leaves or pine branches on which was spread the blankets. Johnny and I took up our lodgings on the ground.

I quietley intimated to the Band that perhaps the Colonel would be gratified by a seranade and hinted the same to the Officers. So at about 11 O'clock, much to our *surprise*, we were awakened by music from the Band. The Colonel, of course, felt much complimented but as he was not feeling very well, asked me to respond in his behalf. I did the best I could but having been chiefly instrumental in getting up the seranade and then to do the responding also was almost too much for my risibility. The joke was highly relished by the Officers who were in the secret and we had many a quiet laugh over it afterwards.

The Colonel remained with us but a few days during which I accompanied him as guide to all the positions of interest around about in spite of the snow and sleet which fell continually. He then returned to Acquia Creek. Meanwhile we were busy getting our complement of material and animals for a movement of the Army. We made a requisition upon the Quartermasters Department for our two trains and supply train that required 120 teams of 931 Animals and 236 men for drivers.

Stringent orders were received from Army Head Quarters reducing the allowance of baggage for Officers and everything indicated earnest work ahead. Another indication of an early movement of the Army, and one on which we generally depended, was the fact that we had now finished our camp and there was no more to do except to enjoy it and keep it in order. Then the March winds were beginning to dry up the muddy roads so that it was now possible to get our supplies through without having to double the teams.

On the 15th day of March we witnessed a most wonderful phenomenon. The day was cold and a heavy wind was blowing, the dark clouds flitted past and over us, every now and then discharging flurries of snow which completeley filled the air. At the same time loud reports of thunder were distinctly heard and vivid flashes of lightening darted across the sky in the vastness above. It was a wonderful sight and filled us all with awe and admiration.

During the night the *long roll* was sounded and we were all aroused expecting an attack from our extreme right, to which numbers of our Troops had been passing for several days, showing that our Commander felt a little uneasiness from that quarter. The attack did not come however and all settled down quietley again. People at home had an idea that the Army was doing nothing that Winter because no battles were fought but could they have seen the marchings and countermarchings, the shifting of position night and day of large bodies of Troops, or could they have known the arduous duties performed by these men in guard duty alone with such an extended front to guard, they would probably have thought differently as all civilians did who visited us.

Fortunateley we, as Engineer Troops, were exempt from general Guard or Picket duty, having only to guard our own camps and property. For this exemption I could never be too thankful and so we all felt. If there is any one thing more disagreeable than any other thing in a Soldiers life [it] is the performance of Picket duty. Occasionally we were called upon to perform it but very seldom. I used to pity the poor Infantry as they trudged back and forth by our camp through the slush and mud on their way to relieve the Picket.

As the weather became better, the Officers of different Divisions sought amusement in various ways [and] Steeple Chase became quite a popular sport. At one of these which was largeley attended, two horsemen came into square collision, the sound of the horses heads as they struck each other could have been heard far across the field. Both riders were thrown together with such force as to mortally injure one, who died the following day, and cripple the other for life. This ended *that* kind of amusement. An order from General Hooker positiveley forbade any more Steeple Chasing or Horse Racing in the limits of the Army. Several men of Company H died during the month of March, otherwise the health of the command was good.[3]

One naturally concludes that familiarity with such numbers of death scenes as one witnesses in the Army hardens the heart and dulls the sensibilities. Such indeed may be the effect on many men and to a certain extent I was also influenced by it, more from a sense of duty than any thing else. But I could never witness a military funeral without experiencing emotions of the deepest and most inexpressible nature. I used to think in those days that when I died, could I be buried with military honors my *spirit* would be satisfied.

I shall never forget how on one sunny afternoon returning alone from a visit to a neighbooring Regiment some three miles away, my horse walking slowly along through the muddy road, my ear caught the sound of that well known dirge, The Dead March, plaintively swelling and again dying away in the clear, calm, crisp air. Looking in the direction from whence came the concord of sweet sounds, I saw the procession slowly ascending a small hill to the place of burial. The sun was just going down and cast a peculiar combination of colors upon that very spot.

I stopped my horse and sat quietly in the sadle for many minutes fairly revelling in the beauty of the scene, my eye entranced by the loveliness of the landscape and my ear charmed by the ravishing melody and sweetness of the tune which the band played while the little group of Soldiers were forming a circle around the grave. There was a pause for a few moments then the quick rattle of musketry as the fairwell salute was discharged over the grave. Forming into line and at a quick march the burial party were soon out of sight. Slowly, "old Frank" and I left the spot and for fear of breaking the charm we walked all the way to camp, which was not reached until some time after darkness had fallen. This may seem a trifling circumstance to you but the older you grow the more will you realize that lifes sweetest recollections as well as the most profound and lasting impressions take rise in just such unexpected and seemingly unimportant incidents.

The Engineer Brigade was henceforth to be commanded by General Benham. General Woodbury was relieved of command in the Army of the Potomac and sent to Key West on the coast of Florida. We never heard of him again as an active participant in the war and about a year afterward he died. General Woodbury was an excellent scholar [and] a fine Engineer but not a man to command men in the rough experience of actual war. His nature was modest, retiring, studious & reticent to the last degree. He was not a man to excite enthusiasm among his troops but a braver man never lived and when he died those who knew him said, "The world has lost a Christian Gentleman and the Army a devoted, faithful, exemplary Officer."[4]

General Benham, his succeser, was as nearly an opposite character in many respects as could well be immagined. Gruff, blunt, severe, a Martinette of the old school and a good Engineer Officer if confined entireley to the construction of permanent fortifications for harbor defences. [He was] proud of his West Point training and jealous of every other Officer in the Army, every one of whom both in the Rebel

Brig. Gen. Henry W. Benham.
Courtesy of the Library of Congress.

[Army] as well as our own, he knew from their youth up. Selfish to a superlative degree, sensuous, fond of his "toddy," kind to those whom he wished to patronize but unmerciful to those whom he happened to dislike—such was General Benham. He came to us from South Carolina with a record not entireley clear. But everyone knew of his disposition to push [and] drive and we trembled and sighed as we thought of the gentle Woodbury.[5]

Our detatchment was reviewed by our new General for the first time on the 25th of March. My impression was that the General was just the least bit intoxicated. I judged so from the manner of his riding. O, Lord, thought I, what is to become of us now.

The month of March passed by leaving us still in the mud. Once I rode over to the camp of the 8th Ill. Cavalry 9 miles away and saw my old friends Uncle Philo Judson, Major Ludlam and Major Bevridge. Once we took our band over and serenaded the Officers of the 146th N.Y. Regt. in which were many acquaintances from N. Y. state. Of course we had a pleasant time of it. In this manner we passed the intervals between storms until about the first of April when our new

Headquarters of Brig. Gen. Henry W. Benham, commanding engineer brigade near Falmouth, Virginia. Drawn by Sgt. John S. Vernan. Courtesy of the William Hencken Family.

General desired to have us remove our camp. So we broke up one camp with all its associations and moved to a spot near the river south of the Phillips house.

It was with sad hearts that we left our old camp ground. After all, as in after life we are so apt to forget the disagreeable features of the past and remember only those of a pleasant nature, so did we feel in leaving this old ground where we had made our home during the winter. The associations of the winter, taken all in all, had been pleasant. The future was all uncertainty.

We were soon temporarily fixed in our new camp for we knew full well we were not to remain long there. A few days afterwards there was a review of the cavalry at which President Lincoln attended and soon after that a Grand Review of the Army. The President was also present at this Review, which was remarkable for the absence of *civilians*, who generally managed to get passes to the front to witness the Reviews.[6]

The ground on which the review took place was in easy sight and almost within cannon range of the hights occupied by the enemy beyond the Rappahannock. What must have been their feelings as they witnessed our solid columns wheel and march past the reviewing Officer hour after hour as they did nearly all that day? As the review was drawing to a close they gave us one round of shell from their nearest battery as much as to say, "Come on, we are ready for you."

After the review we all knew that nothing more remained to be done before a move than to get our rations and wait for favorable weather. Meanwhile, General Benham was busy planning and, as he said, re-organizing the Engineer Brigade. His orders came thick and fast, every one complicating still more the machinery of motion.

General Order No. 7 revealed our new commander in his true character and the threat at the end of it told us what we might expect of him. Had we never had any experience in the duties required of us such an order *might* have been appropriate, but to issue an order of that kind to Officers who had had ten times the experience in that line than he had and who knew of its absurdities and impracticable details accompanying the whole thing with a threat, insulted every Officer in the Brigade. It was amusing to hear Lt. Col. Pettes swear between his teeth at being ordered to perform the duties of *Chief Wagon Master*. The Colonels of both Regiments were entireley ignored. [See Appendix 4, Item 5.]

The Paymaster (always welcome) now put in his appearance as he generally did just *before* a battle. My *friend* Pettes, as a last shot of his malignity, had reported me Absent Without Leave because I had not returned at the expiration of my time, though he was aware that it had been extended. He had the satisfaction of giving me considerable trouble before I could get my pay, for it was necessary for me to be acquitted by a Court of Inquiry, my name having gone on the books at Head Quarters as Absent Without Leave. This took a couple of days to get the Court summoned and *five minuites* of explanation on my part produced an immediate acquital of all blame, though I had not been in arrest. Then I drew my pay from the paymaster. Poor Pettes, this was about the silliest of all his little souls performances.

We were all ready to move, where, no one knew. General Hooker kept his own councils. None but Gen. Butterfield, his Chief of Staff, knew of his intentions. But a fearful storm came on which drove us all inside our shelters for nearly a week. When it cleared away there was here and there a Brigade missing then a whole Division gone, then another and another, then a whole Corps was gone. In this way disappeared the whole V Corps and then the XI & XII and the II in like manner. Those who remained knew not where they had gone, [but] we soon learned that they had moved away to the right. It was now evident that the long looked for movement had at last commenced, that the campaign was opened and that our movements were being managed by a skillful if not a masterley mind. General Hooker had brought the Army up to a high state of efficiency & organization and all had the utmost confidence in him. [See Appendix 3, Item 9.]

It was now the 27th of April. The detatchment under Spaulding & myself [was ordered] to move to near Banks Ford where General Hooker was trying to make believe he intended to cross. General Order No. 7 was of no account [as] Cols. Stuart & Pettes were both at Acquia Creek and Spaulding, Ford, Beers, McDonald, McGrath & myself had things our own way!

Early on the 28th of April vast bodies of troops commenced to move by our camp. We had orders to hold ourselves ready to move at a moments notice. At 10 O'clock it commenced to rain. Was it an evil omen? At 1 O'clock we were on the march and reached Banks Ford at three in the afternoon. We commenced at once making demonstrations to attract the attention of the Enemy such as cutting down trees, opening roads to the bank of the River, measuring distances with a tape line, &c, in order that they might see us with their glasses and think the crossing was intended to take place there, while in reality the main body of the Army was moving far to the right and that very night, the V, XI & XII Corps crossed the river away up near Richards Ferry and then coming down crossed the Rapidann and were secureley lodged in the enemys country.[7]

We worked all night at Banks Ford and succeeded in attracting the attention of the Rebels, who concentrated in large force opposite. Somehow or another I had an idea in my head that I was going to be killed that night and, although presentiments were not common with me, I could not get rid of this one. So I made my will again on a piece of paper in lead pencil by the light of a camp fire. The principal *property* to will being my horse and equipment, which I desired to be sold to pay Johnny what might be due to him.

As I said, we worked all night. At 7 a.m. I rode back to report to General Benham for orders and found that another detatchment of our Regiment had constructed a bridge below the city on which the VI Corps of General Sedgewick had crossed and was in position to attack the enemys right.

I returned to our camp at 2 p.m., having ridden 18 miles with orders to move *immediateley* for United States Ford. We did not get our train up within distance of the Ford until after 12 O'clock that night and dropped down in our tracks without a mouthful to eat to get what rest we could before morning. No fires were allowed as here we intended to cross in reality. We found General Couch near by and reported to him. His Corps was waiting for us to build a bridge for them. It rained all night and we were uncomfortable enough.

In the morning the Enemy evacuated their strong position opposite to us when they found both their flanks turned. We slid our pontons nearly a quarter of a mile down a narrow road to the river and in two hours from the time we commenced had a bridge completed. At three in the afternoon the Troops were crossing. So with Sedgewick on our left, Meade, Hancock, Sickles & Howard on the right and Couch in the center, our entire Army was across the Rappahannock without the loss of a single man.

Chapter Thirteen

Chancellorsville May 1863

The crossing of the Rappahannock River by the Army of the Potomac previous to the Battle of Chancellorsville may rightfully be considered as one of the most ably conducted movements of the entire war and I doubt if history can show a parrallell case. Here was a large, well officered Army with a 400 ft. wide Stream in its front with every available ford well guarded and fortified with works of great strength. [An enemy] vigilantly watched every movement of our Army, who for months had confronted them and whose object they knew to be to get across the Stream. And yet, so completeley were they bewildered and deceived by the skillful manoevors of our forces and so entireley ignorant were they of our *real designs*, though for days the army had been in motion in plain sight, with all these natural advantages in their favor, yet, the entire Army of the Potomac had made a successful crossing without loosing a man in the attempt and were now making the attack on their right and left and centre. It seemed to me then and has ever since, a most wonderful achievement.[1]

The day our forces were all across I heard General Patrick, the Provost Marshal, say in the presence of several Officers that our forces outnumbered those of the enemy two to one and in order to prove his assertion he read to us from a paper which held a complete statement of every Corps, Division, and Brigade in the Rebel army, their numerical strength and position. The document was an exact copy from the Adjutant Generals report of a few mornings previous. How it was obtained General Patrick of course knew and, of course, I did not.[2] It did seem that success was now an absolute certainity, so far as human judgement could go. The chances were ten to one in our favor but, if "Man proposes, God disposes," as the result proved.

A good sound sleep on Friday night, May 1st, 1863, and Saturday morning broke bright and beautiful, the weather now very warm. During the morning I mustered three Companies and while Reynolds V Corps was crossing over bridges I rode down to Banks Ford to report to General Benham. Here I found the ballance of our Regiment, who had had a hard night of it in laying and relaying bridges for General Sedgwicks VI Corps, who were now alone on the other side of the river fighting gallantly and desperateley against superior numbers in the effort to turn the flank of the Enemy and make a junction with the main body of our Army at Chancellorsville. The fighting was going on desperateley and we were naturally very anxious lest they might be overpowered and completeley annihilated. This portion of our Regiment was ready to construct a bridge for them at any point on short notice should they signal us to the effect that a retreat across the river was necessary.

General Benham did not disappoint my expectations of him. He was riding about in a great state of excitement, his face red and bloated with a clot of blood upon one of his cheeks caused by a fall from his horse the night previous when he was too much intoxicated to retain his seat in the saddle. He had allowed the blood to dry upon his face, which gave him an extremeley savage and repulsive appearance.

The Officers informed me that he had been in a beastly state of intoxication all the previous night and had had a quarrel with General Wright, who commanded the First Division of the VI Corps, while the troops were crossing. I could not help congratulating myself that I was at least temporarily away from his immediate command.[3]

Having completed my duties there I started on my return, leaving the VI Corps fighting for the possession of Maries Hights. The afternoon was about half gone, no air was stirring and the sun poured down a schorching heat. I approached our position at U. S. Ford. and was mentally contrasting the quietness of things on our front and right with the noise and excitement of the battle which was raging behind me. Suddenly a most terrific cannonading opened from the right of our position. The explosions of a hundred or more pieces of artillery, the shreiking of the shells through the air, the rattle of thousands of muskets, the yells and screams of the Enemy and the defiant hurrahs of our men all intermingled in one huge *roar,* breaking forth so suddenly upon the stillness of a summer afternoon and so unexpectantly, that it almost chilled my blood.

My position at that moment overlooked the entire ground on which the battle was being fought, though the thick clumps of trees prevented me from seeing a very considerable portion of the combattants. This was the fearful onslaught of Jacksons Corps, so famous in the history of the war, when the XI and XII Corps were broken and driven pell mell to the rear. Our whole line was forced from its original position and for a few moments the entire Army of the Potomac was threatened with destruction. Fortunateley we soon had Troops in position to stop the flight of the two corps mentioned and check the onward rush of the enemy. Here and at this time the famous Stonewall Jackson met his fate.[4]

The object of his sudden attack was to turn our right and destroy our bridges and, had he not been killed just as he was, who can say that he might not have accomplished his object: our bridges destroyed and the Army taken in rear. Who can imagine what might have followed, perhaps annihilation, perhaps an overwhelming victory? Jacksons original plan, as I have since read, was to form his entire corps in column of companies and in this manner sweep down a narrow road by the rivers side to our bridges. Had this plan been adopted there is no doubt in my mind that his object would have been accomplished.

The firing did not cease that night until after 8 O'clock and we all dropped down to rest on the ground near the banks of the Stream, realizing that the battle had now been joined and a few more hours would perhaps decide the fate of our Army. It was evident that Lee was fighting us in detail, first throwing large bodies of men upon one and then upon another of our detatchments, this taking us at a disadvantage and showing the weakness of the *plan*, in spite of the excellent movements preparatory to the crossing. It was also evident that he had been largeley and suddenly reinforced.[5]

Sunday morning, May 3rd, the battle opened with the sunrise and continued with great fierceness until about noon. The day was bright, cloudless and beautiful but very warm and this was Gods Holy Sabbath day. While the good people of the land, in answer to the summons of the church bells were quietley and peacefully winding their way to their places of worship, two Armies were striving with the determination of demons, each for the destruction of the other. It was a sad, sad commentary upon the wickedness and weakness of human nature.

About 9 O'clock in the morning I rode over the bridge to the temporary hospitals and from thence to the front. The sights and scenes at the hospitals were of the usually horrifying nature and I will not dwell upon a description. Passing to the front, the battle seemed to be joined entireley upon our right and centre, while at the left all was comparativeley quiet. Our line extended from the river out, in the form of a horse shoe, and back to the river again.

General Hookers Head Quarters were situated at about the centre of the line and well to the front at the Chancellorsville House. As I passed to the front the wounded came back in streams, singly, by twoes and threes and sometimes a dozen at a time. Meanwhile, the roar of musketry and cannon was incessant and it seemed to me as though the whole army was falling back, so great was the crowd of wounded and stragglers. The latter were soon checked however by a detatchment of Cavalry who had been posted for the purpose of intercepting those who had not been wounded.

I went up to see the lines in position and when the balls began to drop uncomfortably near I remembered that I was not on duty then and returned to our bridges. Not, however, until I had seen enough to convince me that what we had so confidently counted upon as a *certain* victory was without doubt a repulse for us and possibly a defeat.

All that day we heard rumors of all kinds, one of which we hoped to be true, that General Heintzelman had arrived from Washington with 40,000 men. But like many others it proved false. That evening we moved our camp upon a bluff overlooking the river and country beyond. Just before sundown we witnessed a beautiful skirmish, if such a deadly contest may be called beautiful, which took place directly across the river in front of us in which we could see every man of both parties. It was one of the most skillfully conducted little fights I had ever seen and lasted until darkness put an end to the contest. Both sides lost quite a number of men but no permanent advantage was gained by either.[6]

That night two reporters of N. York papers called upon us and begged for hospitality. Spaulding took one and I annother. Mine was a Mr. [Thomas N.] Cook of the N. York Herald and a very agreeable companion I found him to be.

At daylight the next morning we were quietley sleeping and dreaming of our loved ones at home [when] we were suddenly called to our feet and senses by the screaching of shells falling among us at a fearful rate. In an instant our whole command was standing upright and dodging the shells as they approached us for it was not yet light enough to prevent us from seeing the fire of the fuses as they darted toward us, striking into the ground in our midst, crashing through the tree tops and every now and then disabling some of our mules and wagons. But fortunateley not one of our men were struck, though the pieces of shell as they burst above and around flew with terrifying noises in all directions. Our reporters, seizing their clothes in their arms with nothing on but their drawers, made rapid strides to rearward and were soon out of sight and range. A large number of Cavalry, who had collected about us during the night, with all the odds and ends of the trains in the Army amounting to several thousands of men and animals, were thrown into great confusion and disorder.

Many men were killed as well as large numbers of animals before a couple of our batteries could be brought to bear upon the single one which was doing all the mischief. When our pieces *did* get the range, the Rebel was soon quieted. Not, however, until much mischief was done and a good lesson was taught us not to encamp so *carelessly* in future, under the very eyes of a vigilant enemy.[7]

The battle continued during the day but with less vigor. General Hooker was evidently waiting for a junction with General Sedgwick and General Sedgwick, finding such a course impossible, was making good his escape back across the river.

On Tuesday, the 5th of May, the fighting commenced with redoubled energy. The escape of Sedgwick had disengaged the ballance of the enemys forces and a general assault was made.[8]

During the afternoon a Staff Officer came from the Chief Engineer directing me to repair and make ready all approaches to the fords [U.S. Ford], which were just above our upper bridge. We had three bridges laid and the 15th Regt., one. This order to get the approaches to the Ford in order looked like a falling back and I so remarked to the Staff Officer, who told me, in great confidence, that such was in fact the intention of the General but that the Troops must not know of it.

At 4 p.m. I had examined and placed in order all the approaches to the Ford. Soon after, it commenced to rain and it fell in torrents. In another hour the Stream began to rise and its velocity was very much increased. At dark the Stream had arisen nearly *six feet* and was running with fearful rapidity, the Ford as well as its approaches were perfectley useless and our bridges were threatened with destruction. It required *all* our skill and *all* our energy to take to pieces the upper one, using the material to piece out the lower ones as fast as the Stream arose and flooded the land at each end. It was a scientific operation and never before had we been placed where the occasion required so much coolness and *skill* in the manipulation of the bridges. A single false move of ours would have lost us the bridges and the Army, for even then the whole body was falling back rapidly in the thick darkness and concentrating on the flats opposite, ready to commence crossing the moment our bridges were completed.

At midnight we had succeeded in making two out of our three ponton bridges and, having covered them entireley with brush to deaden the sound, announced our readiness. We had but two small fires that night, one on each shore of the Stream so as not to attract the attention of the enemy. The rain had ceased falling but the clouds were black.

General Patrick stood upon the bank at the approach of the bridges, using all his authority to keep the men back and preserve order & regularity in the crossing. As we announced our bridges in readiness, he allowed a large fire to be built as a signal to the Army that all was now ready. The light from the fire exhibited to my astonished gaze a perfect sea of men, packed denseley upon the plain before me. So quietley had they been gathered there that I could not for some moments realize that I was not dreaming. Then they closed upon the bridges and for 7 hours the dense masses of men and Officers, Artillery and Ammunition Wagons poured over. Not a word was spoken and the noise of the crossing could hardly have been heard an eighth of a mile away. The last man and last piece of Artillery, the last straggler even was across and our bridges let loose at the south end had swung across to the north shore just as the fog lifted and revealed a line of Confederate sentinels silentley pacing up and down the shore opposite, without deigning so much as to speak a word or show any outward signs that they were even concious of the existance of an enemy.[9]

Thus ended the battle of Chancellorsville. The army of the Potomac, which one week before had broke camp and marched so confidently forth to give battle, had now returned and all, save the wounded and the dead, which it was necessary to leave behind, were again seeking their old camp grounds.[10]

For us it remained to get our material away, out of the water and up the hill, working all that long, rainy day, having had no sleep for two nights before and expecting momentarily to be fired upon by the enemy who were within easy range opposite. Some Artillery was brought down and planted in rear and over us and Barnes Brigade of Infantry was detailed to protect us in case we were fired upon, but the enemy allowed us to remove all our material without molestation and at about dark we had it all away.

We bivouacked that night in a cedar grove and, although the rain did not cease to fall, our tired bodies found the sweetest of repose upon the boughs which we cut for our mattresses. We remained here annother day and night gathering and arranging our vast material and, as we were alone on the outside of our lines, we used all the necessary precautions against surprise that the army regulations required. Then Gen. Hooker and Gen. Benham began to get anxious about us and an *order* to join the Brigade at Head Quarters was the result.

It was necessary to reach our new quarters and not be seen by the enemy, for the sharp-sighted fellows on the other side of the stream kept a keen look out for the Pontons, knowing full well that when they changed positions there was something about to happen. It was therefore necessary that the place of our encampment be concealed from them. During the day we brought our trains to within a few miles of our new position for camping and waited for darkness and then toiled all night through the ravines to get into our places. General Benham had issued the most stringent orders against being seen or heard by the enemy.

The place for parking the Trains was about a half mile from the river in a ravine. Captain Folley was in charge of one of the Trains. The night was exceedingly foggy and much perplexity ensued. All managed however to find their places except Capt. Folley. He, as did the others, parked his train and went to sleep. When morning broke and the fog lifted about 8 O'clock there stood Folleys Train upon a high plateau in plain sight of the enemy and, as the fates would have it, General Benham came riding along just at that moment. There then was a scene. Poor Folley was summoned from his sleep and for fifteen minuites stood and shivered under the most withering torrent of invective and profane abuse which this old West Pointer with his infinite rescources could command. Those who saw it pronounced it *awful*.

Poor Folley, that was the last of him. He might have rallied from the effects of the abuse but when he learned that General Benham had requested the Colonel "never to place Captain Folley again in a position requiring discretion or judgement," his heart was broken. All we who were his friends could say to cheer him up had no effect upon him and one night he came to my quarters to inform me that he had sent in his resignation. It was approved with much regret by Col. Stuart and signed most willingly by Genrl. Benham and a few days thereafter Captain George W. Folley, now citizen Folley, turned his back upon the Army and army life and his face homewards. Company C soon received its new commander in the person of Captain Van Brocklin.[11]

I must retrace my steps just far enough to allude to the cause which has very generally been given for the defeat at Chancellorsville. In reading the histories as they refer to that action you will notice that they all assert that one of the principal reasons why General Hooker fell back as he did was in consequence of the heavy rain that so raised the Stream as to endanger the bridges and that he, as a prudent General should do under such circumstances, fell back to the other side of the River. Now the truth of the matter is that the order for falling back, though possibly not *given* until after the rain, was fully decided some time before the rain commenced at all. As you will recollect, the order to get the fords in readiness was given to me early in the afternoon by a Staff Officer (Capt. Perkins) who informed me, as he also informed Maj. Spaulding, of General Hookers intention to fall back that night. The shower and subsequent rise of the Rappahannock River came opportuniley to his relief.[12]

With these few remarks I forbear further comment upon the Battle of Chancellorsville, leaving the question as to whether Gen. Hooker was knocked "out of time" and demoralized when the piece of rail shattered by a cannon ball struck and knocked him down when standing on the piazza of the Chancellor House, or whether King Alcohol managed to get the better of him or not, to be decided by those whose opportunities for knowing were better than mine.[13]

Scarceley had we settled down in our new camp ground than it became evident that General Benham was meditating mischief towards Colonel Stuart. I will not follow the whole of the correspondence which took place between them for it would not be interesting. Colonel Stuart made confidants of both Spaulding and myself and read us the letters that passed back and forth between them. So long as Gen. Benham confined himself to correspondence in the hope of

Stuarts committing himself in some way so that he might get the better of him, Stuart proved himself in every respect his equal. But the General brought things to a focus soon after by an order that the regiments should be drilled in Battallion movements every other day by the *Colonels Commanding*. This was too much for our Colonel. The General had discovered his weak point and intended to press his advantage unmercifully. Colonel Stuart was not as capable of *drilling* his Regiment as the poorest Private Soldier in the ranks. What was he to do? The order was imperative.

For once, twice, yes, three times he feigned sickness. Then came an order that *no* Officer would be excused from drill without the Surgeons certificate. This Dr. Hewitt positiveley refused to give him on the grounds that he was not sick. Then, the last proof was gone. In his extremity he called upon Spaulding and myself in secret council for advice. What could we do but conscientiously, as his best friends, advise him to *resign*? There was no other possible course to persue. The position in which he placed us was a delicate one because [his resignation] might be construed as desired on our part in order to accelerate our promotion. But, as our advice was asked, it was conscientiously given and his resignation was at once written and sent in.

That night I drew up a document that was signed at our request by every Officer to whom we presented it and within two or three days the 50th Regiment saw Colonel Stuart for the last time. The duplicity exhibited in this letter may seem to you as inexcusable and, as I was responsible for it, I may as well admit that the document was intended to sooth the proud spirit of his Wife more than on his account. She was a proud, high spirited Woman and felt keenly any implied disparagement of her husbands abilities, although no one was so well acquainted with his failings as herself. The letter did not fully reconcile her to his leaving the service for she knew full well that Spaulding & myself, who were his best friends in the Regiment, had drawn it up as a *plaster*. And yet it was so complimentary in its terms and so many Officers signed it that she was forced to accept the result as final and with the best grace possible. The Colonel, on his part, in order not to stultify the document which eased his passage home and saved his character and name from the charge of cowardice or incompetency, suddenly lost his voice. He could speak only in whispers for several months afterwards except occasionally when he forgot the part he was playing. [see Appendix 3, Item 8]

Gradually his name ceased to be spoken in the Regiment and he was absorbed back again into civil life, which he could better achieve. Not, however, until after he had written us several affectionate letters deploring in most heart rendering sentences the fate (bad health) which "forced him to leave his brave comrades in Arms before the war was successfully ended."

Lt. Col. Pettes now assumed command and Spaulding was soon after made Lt. Col., leaving me as Senior Major. We continued our drills and court martial which was composed of one Officer (myself). This duty brought about the first personal interview I ever had with General Benham. Something of my written proceedings did not strike his fancy [and] he sent for me to appear at his Quarters, hailing me with "Major, are you a Lawyer?"

"No sir," I replied.

"A Doctor?"

"No Sir."

"Any kind of professional man?"

"No Sir," I answered respectfully.

"Well, what are you then?" This in a very insulting manner.

"Well Sir," I replied indignantly. "I have always been *treated* as a Gentleman." I expected of course, that my head would soon come off, but he seemed to mellow down as gentle as a cooing dove and never afterwards spoke or acted towards me except with the utmost kindness.

The weather was now warm, extremeley so. Both Armies sat and looked at each other from the opposite sides of the Stream as if each was waiting for the other to move and then to spring at each others throats. Not a shot was fired or a word spoken by either picket line. The hot sun poured down its rays upon both armies, whose tents whitened the plains on the hights north and south of the river. For weeks all was silent as a summer Sabbath afternoon. But it was a delusive silence, like the calm before the storm.

CHAPTER FOURTEEN

Deep Run or Franklins Crossing June 1863

Scarceley had Colonel Stuart passed out of sight when the movement which had so long threatened culminated with an order from Head Quarters to hold ourselves in readiness for instant duty. For several days previous we had noticed unusual commotion on the other side of the river. Long lines of camps whose position had become so familiar to us by weeks of close observation faded away one after annother. Every morning [we] discovered the absence of one or more clusters that we had been watching at sunsett the night before until the hights which but a few days previous seemed almost covered as with snow, now presented but a scanty, scattering collection, a mere remnant of an Army.

Beyond question, the enemy had decided upon some bold movement and had taken the initiatory step. His hosts were in motion but whither? That was now the all absorbing question. Not until the change of position of the enemy had become unmistakeably apparant did our commander decide upon a course of action. Then one after annother of our Brigades, Divisions & Corps, following the example of our southern brethren, "Silentley folded their tents and stole away."[1]

At midnight on the third of June, orders came for us to be in readiness to move at daylight. At 3 O'clock we partook of breakfast and at half past four the whole regiment stood in line awaiting the order to march. We stood in line for two hours and were then dismissed. Nearly the whole of our Army had now gone. They took a north westerly direction and as no bridges were ordered it was evident that their line of advance was *not* to the Southward. All but the VI Corps of our Army had disappeared.[2]

The next day, June 5th, we were ordered to proceed with one Train of sufficient material to build a bridge 400 ft. long at Franklins Crossing. At 10 O'clock in the morning we moved our command of four companies down to within a quarter of a mile of the river bank (leaving the ballance of the Regiment a short distance back), and there we waited in the hot sun all that day for further orders.

There was a small fort opposite which was strongly garrisoned by the Rebels and in front of the fort was a peach orchard. During the day the Rebels would come out of the fort and lazily saunter around under the shade of the trees, making all sorts of insulting expressions towards us as we stood in the broiling sun, asking us to come over and sit awhile in the shade. We made our arrangements for crossing and, knowing that we were to meet with resistance, it was given to me to lead the command down to the water. I was to have my old Company C to unload the Pontons as they were brought down to the waters edge. Spaulding & Capt. Beers, now Major Beers, were to have immediate supervision of the construction of the bridge. The Regular Battallion of Engineers were to construct the rafts, while our men placed them in position.

General Benham decided to construct this bridge "by rafts," which is simply to complete 3 or 4 [abutments] at a time separateley and then connect them together in line. Our usual way being by "successive pontons," one immediateley in front of the other until the whole was laid.

At last, about 4 O'clock in the afternoon, my attention was called to the advance of our troops and looking behind I beheld a beautiful sight coming down from the hights in rear and about half a mile away from us. I saw a long line of steel, glistening and sparkling in the bright sun light, and as they came nearer the blue uniforms of our men presented a striking constrast with the somber brown of the earth behind them. The line waved and bristled through and down the ravine showing alternate lines of blue and silver. The bands played, flags waved and altogether it was a beautiful sight. The Troops were so skillfully handled that one could not miss the impression that a large Army was really coming whereas it was only the First Division of the VI Corps, General Wright in command.[3]

Four Batteries of Artillery came down almost to my company, unlimbered and sent their horses to the rear. The Troops of the First Division formed in rear of our Train. The Rebels came running down to their fort from the main line beyond and reinforced their comrades.

Still quite a large number of them stood looking at us difiantly from the parapet of the works and from the orchard outside of their lines as if they scorned its protection.

Then the signal was given. Our Batteries opened upon them and at the same moment I advanced with the head of the Train and old Company C. Quite a number of Rebels were killed before they could reach their earthworks, to which they betook themselves as soon as our Batteries opened and there they commenced a return fire upon us with musketry. As we arrived at the crest of the small hill near the stream with our ponton wagons, the animals were unhitched and allowed to scamper back over the plain while we pulled the wagons down the steep declivity by hand. Several of the men and many of the animals were placed "hors du combat" before we could get down to the stream.

Our experience at Fredricksburg had taught us a lesson: not to attempt to lay a bridge in the face of an opposing force until they were driven from their position. It was therefore the plan to unload our boats and cross over a sufficient force of infantry to dislodge the enemy and then construct the bridge.

Our Artillery poured forth their rounds of shell to keep the Rebels down but in spite of all their exertions the bullets came spitting and whizzing about in a manner that forcibly reminded me of the 11th of December. The noise of artillery and shrieking of shells above us drowned out the words of command so that the men knew the wishes of the Officers only by the signs they made.

We worked liveley and yet the time seemed very long. One after annother of my men dropped down and attempted to crawl away, some men hit in the arms, some in the body, others in the legs. I saw two hit in the knee within a minuiet of each other. There was a little depression in the ground, now partially filled with water, just back of the edge of the stream. Into this hole some of the wounded had found their way and some of those who were not wounded had shirked away among them. Observing this, I went to the place with the intention of driving out the well ones to duty when *up jumped* Lt. Col. Pettes, the Colonel of the Regiment. He commenced shouting lustily to the men to "get out of here," accompanying his words with violent jestures and oaths which sounded so strongly of sulphur as to put to shame the dense fumes of gunpowder. It was the first I had seen of him during the affray and, in spite of the awful surroundings of the moment, I could

Bridges laid at Deep Run (Franklin's Crossing) below Fredericksburg, Virginia, June 5, 1863. Note the water-filled depression in the lower right that was a part of Brainerd's narrative. Courtesy of the William Hencken Family.

not resist the inclination to indulge in a good laugh at the sight of his frantic antics.

We had succeeded in getting 4 or 5 Pontons into the water and had them manned. I turned around to see how the Regulars were getting on. They had a few boats in the water but were evidentley not making much progress. Captain Cross was in his shirt sleeves doing all in his power to urge his men up to their duty. I could see that this was a hard task to perform. Three of his men started to move away to the rear [but] he outflanked them and drove them back with vigorous blows with the flat side of his sword. A moment later his sword arm fell at his side, his head drooped and then his body seemed to *wilt* away. He fell dead, shot directley through the head, a portion of his body fell into the stream. Poor Captain Cross, he had the year before graduated at West Point and was considered one of the most promising as well as one of the ablest young Engineer Officers in the Army.[4] The Regulars did not stand up to the fight so well that day as did our volunteers. It was the first time we had ever fought side by side and ever after that they ceased to treat us as their inferiors.

At last we had launched our ten Pontons. It seemed like a long half hour, though that was all the time it took us to do it. We had lost two Officers and twenty two of our own men in killed and wounded. A bullet passed through the skirt of my coat but not a scratch did I obtain though the winged messengers of death came so near at times as to make my ears tingle.[5]

When we had our ten boats launched and manned, the signal was given and down poured the Infantry. The boats were soon filled and quickly rowed across and then it was all over. We captured about 280 men in the little work opposite, many of them professing to be very glad that they were taken and we treated them all as though we were the best of friends.[6]

The Troops of the 1st Division, VI Corps crossed and at once fortified themselves in a semi-circular position, the right and left resting on the river [and] the centre only about a quarter of a mile away. Now that we had established ourselves on the Southern bank of the river, we began to comprehend the object of the movement which at first seemed to us so foolish and unreasonable. The whole object was to make a demonstration with a large force as if it was really the intention to advance towards Richmond and thus withdraw a considerable portion of Lees Army to check our advance and thus give time for our main body to get in the advance in the race for Washington. So far as I have now been able to learn the movement was quite successful.

The Rebels, amazed at the audacity of the movement, checked for a time the rapidity of their march and hesitated whether to advance on Washington or return to the protection of their own Capitol. They finally decided to detatch an entire Army Corps to watch our movements. It was not until after we had withdrawn across the river that they discovered the movement was made with less than one Division. When they again got in motion, whether for Washington or further north, our Army was abreast with them and in condition either to race or fight as they might elect. The present movement had been commenced by the Rebel commander with a view of causing our people of the North to taste some of the bitterness of war waged on ones own soil.[7]

The Soldiers of the VI Corps were soon strongly intrenched in their semi-circular position and, having accomplished the object of the movement, made no further efforts towards an advance. The Enemy, alarmed at the foothold thus gained, endeavored to drive us out.

Detail of the bridges at Deep Run or Franklin's Crossing below Fredericksburg, Virginia, June 5, 1863. Courtesy of the William Hencken Family.

No assaults in large masses were made but the constant picket firing which was kept up for several days became very bitter and deadly. From 50 to 60 per day was our average loss and we thought that of the enemy much greater. We completed two bridges, so that our facilities for keeping up supplies or for a sudden retreat were ample.[8]

During all this time I was suffering indescribable torment from a tremendous boil or carbuncle under one of my arms. The Surgeon finally froze and then lanced the huge tormentor.

Just inside our lines was the ruins of a once beautiful residence now a total wreck. I heard it said that this was once one of the most substantial and aristocratic residences in Virginia. It was a sad sight. Nothing remained of the mansion but a portion of the walls and our lines ran directly across what had been a beautiful lawn. The sufferings of these people of the South, their privations, loss of property and everything that goes to make life sweet can not be immagined. Yet who but themselves could be blamed. They had sowed the wind, they were now reaping the whirlwind.[9]

While I was examining the ruins mentioned, who should approach and put out his hand but Mr. Gifford, now Captain Gifford of the

49th N. York. Formerley a private in the Gansevort Light Guard of Rome, his Army life had been marked by some wonderful experiences and hair breadth escapes.[10]

A whole week had passed since our crossing and nothing more than a sort of rambling fight had been kept up. On Saturday morning, June 13th, we knew by the tickling of our fingers that a change was on the tapis.[11] Our baggage was all sent over towards the Potomac and we [were] in light marching order. That night at 11 O'clock and while the rain was really pouring down in torrents, we were ordered to cover the bridges, which was done by laying brush thickly over them. This duty was soon performed and then backward came our Artillery. The brush on the bridges so completeley muffled the noise that a person standing on either shore could hear nothing but the jingling of the harness as Batterey after Batterey passed over. Then came the Troops in solid column four deep, well closed up and no talking in the ranks.[12]

The rain did not cease falling until daylight and then our last man was safeley over, our bridges up and loaded upon the wagons. Daylight revealed the Rebel sentinels on the other shore and it all seemed like a repetition of Fredricksburg and Chancellorsville. It was Sunday morning but there was no rest for us. We had to reach Acquia Creek 12 miles away as soon as possible. By 12 O'clock on Sunday night, [with] no sleep the night before and by hard work, we had unloaded all our Pontons from the wagons, made rafts of the boats on which were loaded the wagons, placed our animals on transports and laid down at the end of the long pier which was here built out into the Potomac river and took a sweet sleep lasting for three hours.

The Army had passed on far beyond us towards the upper Potomac. Besides our own, there was now but one Regiment here who were acting as rear guard, [with] the enemy pressing us with a small force of cavalry. We did not succeed in getting away from Acquia Creek until midnight of Monday. It was necessary for our Gun boats to preceed us up the Potomac to prevent the enemy from building batteries with which to destroy us and our convoy. We were all stowed away on the steamer ROBERT MORRIS and arrived in safety at Alexandria on Tuesday morning, June 16th, at 7 O'clock.

We found the people of the City in a state of great excitement. A *raid* was anticipated that night. Some of the inhabitants who had the means of knowing (Rebel sympathisers) had started the report which circulated rapidly among the people and great was the consternation, especially among the fine Union people who still remained in the place. Subsequent events proved that the raid was really planned and only failed of execution by the treachery of some one who gave the information to our authorities who stationed enough Troops in and about the Town to render the attempt too hazardous.[13]

Meanwhile, we were as busy as bees getting our Trains in order. Hurry was the order of the day. A Train had to be sent forward immediately to the upper Potomac where the main body of the Army now was. That day I had one meal and that night slept but three hours.

Next morning I rode into Washington to execute an order from Head Quarters. On my way down Pennsylvania Avenue, I called a moment to inquire after my friends, the Purdys. My old friend Mrs. Purdy was dead, her husband travelling in Europe, Mrs. Thompson in N. York & Mrs. Wallace with her Father abroad. I then met Capt. Jewell of the 146th N. York. He was dressed in citizens clothing and had *abandoned* his Regiment. I speak of him because I had known him in Rome as a stylish young man. [He was] a fine Soldier when there was nothing to do but deserted his colors the first time he was called upon to undergo any real hardship. He was soon after dishonorably dismissed from the service for desertion.[14]

Spaulding and the command had gone on up the canal with the bridge Train. I made up my escort during the day. It consisted of 4 Trains and wagons, all the Officers and 12 men of my old company. I was to go *overland* and meet the main command at Nolands Ford. I left Washington with my little command at 12 O'clock that night. It was a lonesome journey, especially after passing through Georgetown and beyond our picket lines. I emerged into the open country [where] there was the inner line of Pickets, then annother line and then an outer line. Beyond this, the Cavalry line and finally the outer line of Cavalry, at all of which I had to halt and have my papers and orders examined by the proper Officers before I was allowed to proceed.

We reached Great Falls about noon and took a thorough soaking, having to march through a very heavy thunderstorm. At dark my little command reached Seneca and then we bivouacked for the night. We had marched about 25 miles that day and it did not take a long time to rock us to sleep. At 5 O'clock the next morning we were again on the march, passing through Poolsville, a thorough Secession hole. I saw the Signal Officer who said that there was a Cavalry fight going on on the opposite side of the river and that my Regiment was at Monocacy

[Church]. We arrived there (Monocacy) at 11 O'clock, having marched 12 miles that morning.[15]

Here I found Spaulding with the regiment. He had received orders not to go further, as Nolands Ford was then in possession of the enemy. Had they gone on a few miles they would all have been captured.[16] All we could do was to await orders, which finally came at dark, to return back to Edwards Ferry. We started at once, Spaulding down the canal with the main body and I with the wagons and transportation by land.

The night was pitchy dark and the rain poured in torrents. To make matters worse, I missed the road and wandered about in the mud and rain until 1 O'clock, when I came into Poolsville by annother road than the one on which I had passed but the day before. Then when I found where we were, I halted and laid down in the wagons until daylight, when the march was resumed, reaching Edwards Ferry about 8 O'clock. Some warm coffee tasted good that morning and I hoped to get some rest. I had disposed of myself on a bale of hay for that purpose when an order came for Spaulding and myself to return immediateley to Washington, leaving the Trains and command to build a bridge at Edwards Ferry and await further orders.

We were 35 miles from the City of Washington, had had no sleep and but little to eat for several days [but] at 11 O'clock we mounted our horses and started down the tow path for Washington, a hard ride that afternoon under the circumstances. We arrived in the City at about 8 O'clock, rode up to a restaurant and ordered a dish of deviled crabbs, which cost us *four dollars*, and then we rode up to our old Quarters at the Navy Yard where we at last obtained one good, sound, uninterrupted nights sleep. Strange as it may appear, the deviled crabbs did not desturb our repose, which was made all the sweeter by a batch of nice letters from the loved ones at home.[17]

Meanwhile, our Army, keeping on the inside of the circle, crowded the enemy away from the Capitol around to the Potomac. No general engagement was brought about but a series of small skirmishes was fought chiefly by the Cavalry. At last a crossing was effected by both Armies and a race commenced in the direction of the North to meet at last in a death grapple at Gettysburg.[18]

My stay in Washington was short, amounting to but 4 days including Sunday. The time was all occupied in getting other Ponton Trains ready for the field [and] perfecting the outfit and details of their organization. General Benham had established his Head Quarters on the old ground formerley occupied by Col. Stuart. The people at the Capitol, as well as throughout the Country, were very much excited and alarmed, for it was rumored that Lee, with an immense Army, was across the Potomac and his advance had already passed through Maryland and was now invading Pennsylvania with the intention evidentley of carrying the war into the North and perhaps to destroying the Cities of Philadelphia and N. York.[19]

On the 25th day of June, 1863, we recieved orders to march and at 4 O'clock p.m., with the whole of our Trains and the ballance of the two Regiments, General Head Quarters and all, we marched down through Pennsylvania Avenue towards the setting sun. My command led the Brigade. I was riding quietly at the head when a street car passed and on the hind platform stood my old friend whom I had not seen for many months, Major Ludlam of the 8th Ill. Cavalry. The Major saw me at the same moment and in his anxiety to meet me, forgetting that the car was in motion, jumped from the car towards me. In a moment, the Major, who was a tall man, reversed his position completeley, his long legs raised high up in the air until a perpendicular was assumed, his head resting on the pavement. There he poised himself for several moments and then came down flat in the dusty street. We were glad to see each other notwithstanding the disagreeable mishap and exchanged a rapid conversation for the few moments allowed us.

At dark our Train, which was more than two miles in length, was well out of the City and at 10 O'clock we halted at Rockville 10 miles away and bivouacked by the roadside amid a driving rain storm.

Chapter Fifteen

Gettysburgh 1863

The end of the last chapter, which was written some months ago, left me in bivouac with my Regiment at Rockville on the road from Washington to, we knew not where at that time. The next day we moved on through the mud and rain, passing through Darnstown and again, just before nightfall, bivouaced near Poolesville, having marched 18 miles that day. When the head of our column reached Poolesville, the rear was more than three miles behind. I well remember what a tired and disgusted crowd we were that night. Marching 18 miles through a drizzling rain is not conducive to amiability.

That night and the next day we remained at Poolesville. One week previous this day I had left this place for Washington and here we were back again, only two weeks since we had taken up our bridges and left the Rappahannock and yet it seemed a month. Everything and everybody was on the move and no one seemed to know anything. Heretofore we had been able to derive some cause for certain actions and conclude something about results but now everything seemed beclouded in an impenetrable vail of mystery. All we knew or surmised was that Lee had advanced into Pennsylvania and that unless something was done, and that speedily, the results of his *raid* would prove disastrous, perhaps fatal, to our cause.[1]

Sunday morning, June 28th, the sun rose bright and beautiful and with his rising we rose also. As the light came up from the east we saw a long brown column of dust and smoke between us and its brightness and you may rest assured we did not feel very comfortable when our outpost sent in the message that Wade Hampton, with a column of Cavalry 7000 strong, was then crossing the Potomac and making his way to join the main Rebel Army under General Lee.[2]

Here was our helpless little command, the very "right arm" of the

Army of the Potomac, with all its invaluable Ponton Trains and material within less than three miles of him and he moving on in his resistless sweep towards the North, unconcious of our presence. What a magnificent haul we would have made for him had he but known of our whereabouts. The smoke was caused by the burning of the canal boats, locks &c which they destroyed as they passed through.[3]

That night our command reached a little hamlet called Buckleyville.[4] We arrived here after dark having again marched 18 miles. My wagon stopped in front of a cosy looking house and I was about to retire inside of it (the wagon) when a Lady stepped out from the house and offered me the priviledge of sleeping upon their Piazza, their beds being all full. I accepted the offer and next morning found my hostess to be a Miss Duval, the eldest of three maiden sisters who kept the house. They gave us (a few Officers) a nice warm breakfast and treated us as *ladies* do who know how to deal with Gentlemen. It was here while at breakfast that I learned of our change of commanders. General Hooker had been relieved and General Meade was now in command of the Army. The announcement was received with mingled feelings of approbation and uncertainty or distrust. This feeling soon gave way to one of entire confidence and cheerfulness.[5]

That day we passed through Fredrick City at noon. The streets were crowded with many stragglers who were intoxicated and two were fatally shot in a quarrel as we were passing through. They were cavalrymen. I saw one chasing the other on horseback and when he came within a few feet, he fired his revolver. The other returned the fire and both fell from their horses, mortally wounded. We passed through the Town and went into a clover field beyond, where both men and animals enjoyed the most delightful repose for *two long hours*. At 2 a.m., June 30th, we were again on the march.

The roads were filled with Troops and trains all rushing forward but in good order towards the north. One column filled the road, annother on the right and annother on the left in the fields beside the roads, making three parrallell columns all moving as rapidly as possible. It was an irresistible current and presented a grand appearance.

The country through which we were passing was fairley diversified with hills and valleys, well cultivated meadows and clear running streams. The people were evidentley *Loyal*. They stood at their doorways or by the roadside, men, women and children, greeting us as we passed with pleasant words and every demonstration of friendship. A Lady came out from one of the farm houses and presented me with a boquet of which I was very proud I assure you. Mine was not a solitary instance of their favor however, many Officers and Soldiers were delighted by the same attentions. This treatment contrasted strangeley with the manner we had been accustomed to receive from the people of Virginia.

During the day we were ordered to allow a Brigade of the Pennsylvania Reserves to pass us. What was my astonishment when I recognized the Colonel in command of the Brigade as my old friend and shopmate of twelve years before, while we were apprentices together at Norris Locomotive Works in Philadelphia, William McCandless. We had not seen each other for more than 12 years and had no knowledge each of the other during that time. Changed as we were, in uniform, hurriedly passing, a glance from among a multitude of faces—it was "William," "Wesley," and two extended hands. No halting, but we rode beside each other for several miles and told each other the story of our lives, then good bye again and we have never seen each other since.[6]

We passed through Liberty and bivouaced at Johnsville with all our immense Ponton & Siege Trains away from the Army, for aught we knew in the direct path of the enemy. The question naturally arose, what was the use of all this valuable material *here*? There were no streams in this direction to cross and no towns to besiege. Then why expose all these Trains to capture by the enemy off here in the country where they were *not needed*? It was evident that in the confusion incident to a change of commanders we had been allowed to drift along with the Army for which no one in particular was responsible until we had marched far inland where neither we nor our Trains were wanted and in imminent danger of capture by the enemy.

The next day we were ordered to return to Washington with our Trains as best we could. The country was filled with stragglers from both Armies, chiefly our own. We returned through Liberty and New London and bivouaced beyond New Market. Captain Chester & myself went into the town about midnight [and] found the inhabitants much excited and alarmed by the action of a dozen or so of our stragglers. The Captain & I, with our Orderlies, captured six of the desperadoes and marched them to camp with us. We had but four hours sleep that night.

At daylight, July 2nd, we were again on our way. The day was awfully hot. We could distinctly hear the cannon and see the rising smoke from the direction north and west of us. We knew a battle was in progress but we were not at all satisfied to be moving in an opposite direction. The cannonading was terrific all this day and the day before. We felt that a great battle was in progress and we were not permitted to participate. This was the first and second day of that great Battle of Gettysburgh, which has been termed the "turning point" of the war.[7]

That day we made a *forced march*. About 9 in the evening, when within about three miles of Colesville, a Negro met us and informed us that a large force of Rebel cavalry was in the Town, probably Imbodens Cavalry. He told us of a bye road which would lead us around the Villiage and we gladly accepted his information. We were in no condition just then to meet a superior force of the Rebels and our orders were to *dodge* them if possible, should we meet any.

It was a bright moonlight night and we could see the lights from the Town as we passed around it. Every man in our little command felt the importance of keeping the most perfect silence. Not a word was spoken, we almost held our breaths and trembled lest the jingling of the harness or the braying of our mules might betray us. Fortunateley the mules seemed to comprehend the situation and not a bray was uttered by them. At 12 O'clock that night we reached a Quaker settlement 19 miles from Washington. We had not had a regular ration since the morning of the 1st, nearly 40 hours.[8]

As we dropped down on the roadside exhausted and *almost* too tired to sleep, who should we see but two Ladies who came out to us from a house which set well back from the road. Following the Ladies were two stout negros, each with a pail of milk, nice fresh pure milk, with cup and glasses. The Ladies had baskets filled with sandwiches and in the sweetest and most Ladylike manner, invited us to eat and drink our fill. Did ever you see two Angels? I never did before.

These were Ladies of wealth and refinement. Their names were *Lea*. O, how our men did bless those women that night. The memory of that kind action will never grow dim. Annother of the families was named Byron, I think. The next morning the Misses Lea invited the Officers into the house where an elegant breakfast awaited us. There [was] a boquet of sweet flowers for each one of us and after a pleasant chat we rode away, as you may well imagine, our hearts filled with gratitude.

The Misses Lea had completeley emptied their cellar for our benefit. When we remonstrated with them, they politely informed us that they knew where more was to be found. One of the Ladies informed me that the day before a passing squadron of Rebel Cavalry had stopped there to inquire for provisions. They had none, she said. So complete were their arrangements for hiding their good things that not even the hungry Rebels could find them. But when *we* came along they had enough to feed half a Regiment. That kind of *Union* people were good for something, *their* loyality amounted to more than words.[9]

All that day, July 3rd, we passed through a country filled with friends. I was in command of the rear guard. We came into the City of Washington about midnight. Already the people had commenced to celebrate the Fourth of July and with it the news came of the great victory at Gettysburgh. These were *anxious* times, my son. We all felt that the fight then impending was the most desperate of any that had as yet taken place.

We had had no positive news from any source but as we approached the City and saw the rockets darting into the sky and as we looked towards the Capitol [where] the immence dome was illuminated, then annother and annother of the prominent buildings shone out brilliantly into the night, we all felt that a victory had been won and in spite of our sore feet and weary bodies we sent up an involutary shout, for our hearts were full of that peculiar sensation which none but those who have passed through great perils and suddenly find themselves in a place of peace and rest and security can appreciate.

We marched to our old Quarters near the Navy Yard on the Anacosta and soon all except the guard were sleeping as peacefully as though we were once more at our own beloved houses.

I had no idea but that we should remain in Washington for several days, perhaps weeks. The day after our arrival being the 4th of July, the whole country had a double cause for rejoicing. The National Holliday was made wonderfully welcome by the intelligence of the great and decisive victory of Gettysburg. Of course the Capitol of the nation celebrated the event in a worthy and becomeing manner.

As for us, who had arrived during the night, the sweetest thing was sleep and we passed the most of the day in laying in a good stock of the "balmy." Towards evening I mounted and rode over to see my old friends the Naylors, who as usual were happy to see me and a pleasant visit was the consequence.

The next day, July 5th, was a bright and beautiful Sunday which I enjoyed after the usual Sunday morning Military Inspection, Guard

Mounting, &c by attending the Episcopal Church not far from our camp. The day passed so tranquilly and peacefully that I really flattered myself that now we were to have a season of rest. Alas, for the vanity of all human expectations. On Monday morning early came the order for me to take two Companies, J & H, and move towards Harpers Ferry.

We passed through Washington, Georgetown, and Big Falls and arrived at Seneca late at night, having marched all day through a drenching rain storm. It seemed to be our fortune to have our fine weather when it made but little difference to us and our disagreeable weather generally commencing at the same time our marching orders arrived. Having established ourselves in a cedar grove, we enjoyed the ballance of the night in drying ourselves before huge fires and getting what little sleep we could in our soaked condition.

The next morning we resumed our march [and] the rain still continued to pour. At Poolesville and at Edwards Ferry I halted and sent messengers ahead to each of the places for orders but there were no orders from General Meade and I continued on my journey.

We marched rapidly by our old camping grounds at Berlin where a messenger overtook me with orders to proceed to Harpers Ferry and report to General Kenley on Maryland Hights. At eleven O'clock that night in the darkness and rain we were halted by the Pickets of General Kenleys command and we lay down by the road side in the *intense* darkness, the rain pouring upon us, and waited for the daylight. At the first *streak* of dawn I was in my saddle. My Orderley being too much fatigued to follow me, I allowed him to remain with the command and started off alone up the mountain to find General Kenleys Head Quarters.

It had now ceased raining and a dense fog enveloped the mountain side. For several hundred yards my horse and I did very well together, but soon it became so steep that I could not hold my seat in the saddle and dismounted for fear that my faithful animal might lose his foothold and both of us go tumbling backwards down the steep. It was evident that what I at first thought a bridle path was nothing more than a short cut used by the Soldiers in climbing up or descending the mountain. Certainly no horse had ever been up that way before and how *Bill* and I succeeded in getting up there will ever remain a mystery to me. I suppose the horse must have felt as I did: after having gone so far it was more dangerous to attempt to return than to keep on in our course. Go back we could not without falling, go on we *must* and we did. We were both *desperate*, for now it had become a very serious matter with us. At last we reached the top and when we did so we both lay down in complete exhaustion. Looking back down that mountain it did not seem possible that we could have ascended the way we did.

At the summit I fell in with a small force of Infantry and learned that Head Quarters were over to the west about a mile away. Down the steep on the other side and up annother, smaller mountain, I at last reached a little old dilapidated house which the Sentinel on duty informed me was General Kenleys Head Quarters but the General was still asleep. It was but a minute however before I was ushered into his presence and received my orders, which were that he had nothing in particular to communicate until General Meade could be heard from. The Rebels were still in possession of the Town of Harpers Ferry and I could return to my command and await developments.[10]

Now in order to return I had my choice of two ways: either go back by the way I had come up, viz. slide down the mountain side; or go around by the side of the river, following the canal and mountain which came around in the shape of a horse shoe with about a mile which was exposed to the full view and easy range of the Enemy on the opposite side of the river. [It was] in fact, a regular gauntlet to run. But rather than attempt a descent of the mountain I chose the latter route.

As I came near the point where [there was] direct exposure to the fire of the enemy command, I saw an Officer about a quarter of a mile in advance, rushing along at full speed, his head bent down close to the neck of his horse and the enemy on the opposite side of the river were firing away in a very liveley manner, evidentley intending to stop him. But he sped onward and was soon out of my sight around the point of the mountain.

Now to think of running this gauntlet myself was no very delightful undertaking but, having come thus far and having seen annother man pass through before me, I could not think of turning back. So, putting spurs to Old Bill, we dashed forward. Soon the bullets began that dreadful whispering around and about us, overhead, behind, before us, spattering against the solid rock beside us, striking full against some boulder in front of us and then going whirring off in annother direction. The horse seemed to comprehend the situation as fully as I did myself [and] he flew at his utmost speed. At length, after what seemed to me a *very* long ride but which in reality was of but a few minuites duration, we rounded the corner, the horse fairly covered with foam and we were out of danger.

That afternoon General Benham with his Head Quarters arrived and ordered me to remove my Train & men a half mile to the rear for more safe quarters. During the day the young & gallant Lt. Meiggs came up with a small Battery of mountain howitzers and ran one of them out to the bend in the road commanding a range of Harpers Ferry. I saw him execute some of the handsomest firing it had ever been my fortune to witness. Sighting the pieces himself, the shells seemed to strike every time just where he wanted to have them. Young Meiggs was a gallant Officer, just from West Point, daring and ambitious. He too met with an untimely fate. A few days after the events which I have described, he was following the retreating enemy as they fled from Harpers Ferry, fell into an ambush and was cruelly murdered.[11]

We remained here from the 8th to the 14th of July, six long days, hourly expecting the order to advance and drive the small force from the Town, then construct our bridges. We knew that Lees Army was retreating from Gettysburgh and wondered why some Corps did not come, take possession of Harpers Ferry, cross up the right bank of the Potomac and cut off his escape. Was Antietam to be repeated? We waited and wondered. Occasionally we heard the deep booming of cannon as of distant thunder and we thought sureley a battle was in progress and Lees Army of invasion would be entireley annihilated. These sounds were repeated daily and still no Troops arrived.[12]

On the seventh day after our arrival, we received our orders to prepare to lay a Ponton bridge across the river. The order came at 11 O'clock at night during one of the darkest of egyptian nights. The rain [was] pouring in torrents as was usually the case when our orders came to do anything. At 2 in the morning we were up and moving [and] the rain continued to fall. At daybreak we were at our position awaiting for a Division of General Negleys to join us and be taken across. Our boats were in the water an hour before the Infantry arrived. When they did come we soon had a Regiment on the river. The Rebels could not stand this sight without firing a shot. So, they simply emptied their pieces and they fled. We could see them scampering up the hill and away as fast as horses could carry them for it proved to be but a Cavalry outpost.

At 10:30 that morning we commenced laying a bridge and at 1:30 that afternoon all was completed and General Negleys Division poured over it. We rested the ballance of that day. Our forces meanwhile crossed over and again established themselves on Bolivar & Loudon hights. Army Head Quarters were established at Berlin, six miles below.

It seemed as though the whole campaign of the year before was being enacted over again. Annother bridge Train was sent up from Washington by General Benham, who, by the way, had returned there. Spaulding, with a portion of our command, left for Berlin to lay that bridge, leaving me in command at Harpers Ferry. Then more material from Washington and two bridges, each 1500 ft. long, were completed there. I sent annother company down, so that my command now consisted only of Cos A and F.

The II, III & XII Corps with all their Trains crossed at Harpers Ferry. The ballance of the Army crossed at Berlin. Lee with his Army was well down the Shenandoah Valley. He had made his escape in spite of the high state of water in the river which at one time threatened his destruction. Antietam, for which McClellan was so severeley censured, was again repeated under General Meade.[13] [See Appendix 2, Item 9.]

CHAPTER SIXTEEN

After Gettysburgh 1863

The waters of the Potomac, which for several weeks had been very high, now began to fall and it required much care and labor to prevent our boats from being broken to pieces by the rocks which began to show themselves in unpleasant proximity to their bottoms. Our men were constantly on duty. One company was detailed to gather the debris as it came floating down from above. An entire Ponton bridge, or rather the wreck of one, was among the collection. This bridge had evidentley belonged to the enemy and had been abandoned by them when their crossing had been completed.[1]

My camp was now situated at Sandy Hook on the East side of Maryland Hights. One day I took occasion to visit the Hights for pleasure. The view from this position is one of the very finest in America. I will not attempt to describe it but hope you may live to visit this and many other places of which I speak and I am sure if you can remember even a little of what I am telling you, the visit will be a very interesting one to you.

At the time I now speak of I was introduced to the gallant Colonel Porter of the 8th N.Y. Artillery, who was then in command of the Hights. Colonel Porter was an educated gentleman of wealth and refinement. He had left a home such as few men possess at Niagara Falls, where he owned great amounts of property, to take command of this Regiment. He recieved me in a very cordial manner, offered wine and cigars and I soon felt well acquainted with him. His tent on the mountain side opened towards the West and the view up the Potomac from its entrance was wonderfully beautiful. We chatted an hour or two. The Colonel entertained me with stories of his experiences in foreign travels and I left him much delighted with the visit. The next time I saw Colonel Porter was about a year after. Then his dead body lay between our lines at Cold Harbor. I shall never forget the first and the last time my eyes beheld the gallant Colonel Porter. That beautiful summer afternoon on Maryland Hights and that dreadful summer afternoon after the bloody repulse at Cold Harbor.[2]

On the 27th of July, 1863, about twenty days after our arrival, orders came to leave one Company in charge of the bridge at Harpers Ferry and to proceed with the ballance of the command and all the bridge material to Berlin, thence to Washington. Accordingly I left Co. A at Harpers Ferry and proceeded with Co. F to Berlin.

Riding some distance in advance of my command, I was passing a house near the road side when several women called to me to protect them as two cavalrymen were rifling their house. The two villains came out as I rode up to the house and I ordered them to mount and follow me, at the same time informing them that they were in arrest. The two mounted and came on with me a little distance when I discovered mischief in their eyes and loosened my pistol from the holster to be prepared for emergencies. Observing my movement, one of them wheeled and as quick as a flash, putting the spurs into his horse, was bounding away from me. I ordered him to halt, to which he paid no attention. Then I fired, and again, and again, but by this time he was out of range.

The other one now showed his intention to quit me in the same manner but my pistol brought to bear within three feet of him with three shots to spare had the effect to keep him quiet, though I earnestley invited him to try the experiment. However, the first one did not succeed in making his escape, for just as he thought himself out of my range he came in fair shooting distance of the command who were following me. A shout from me to "Stop that fellow!" and fifty muskets leveled at him brought him to a stop.

The two were conveyed to Berlin and placed in custody of the Officers in command at that post with a statement of their conduct on which to base charges for a court martial. That was the last I ever saw or heard of them but I can now tell you how thankful I was that I did not shoot that man though I fired three times intending to. When it was over I felt so grateful that I could hardly keep from getting down from my saddle and on my knees thanking God that my hands were not stained by the blood of one of our own Soldiers. An act which I did that night over and over again and have done many times since and now do every time I think of the circumstance.

We travelled all that day and all night, arriving at Washington the next day, July 28th, towards evening. Leaving the bridge material at Georgetown, I marched the command to our old ground near the Navy Yard and we all enjoyed a good rest and nights sleep. Two letters from home helped very much to quiet my nerves. The next day we brought our boats around by water and again commenced guard mounting, drilling &c.

It now occurred to me that a Leave of Absence for 30 days would be a pleasant thing to have, particularly as the campaign, which had been a very active one, now seemed to be brought to at least a temporary stand still. Accordingly my application was made out in due form and fowarded through the proper channels. The next day it was "respectfully returned, disapproved." Well, there is nothing like getting accustomed to disappointments in this life and so I concluded to make the best of it.

We soon settled down into our old ways of drill, drill, drill, guard mounting, dress parade and the making of reports. However, the monotony of the life was very much modified in my case for I had one hundred and eighty men under my special instructions in the art of rowing, an exercise which was my special delight. Then I had the Naylors to visit on every occasion when I could get away. As we were likeley to remain here some little time and as I could not go to see your Mother, I concluded to write her to come and see me, as Mohamed did to the mountain, only reversed. Accordingly, I wrote her to that effect and she, like a good woman, lost no time in answering my request.

The 6th day of August was designated by the President as a day of Thanksgiving to God for the victories and mercies so recentley granted us as a Nation. We therefore had a rest that day which we all appreciated for the weather was schorching hot and debilitating and General Benhams freaks seemed to increase as the heat became more intense. Every day he added some new duty or issued some new order, until the multiplicity of his instructions became so cumbersome that it required a good portion of the time of the Officers to study the order book lest he should do something that might conflict with some previous order. This *extract* will show to a very limited extent how this old Martinette afflicted us.

GENERAL BENHAMS PLAN OF CONSTRUCTION BY CONVERSION

The details of the execution, or the drill for this construction is very Simple, being as follows. Where the shore will allow, the train of ponton trucks is drawn along the edge of the shore, closed and wheeled to distances of about 20 feet with the sterns of the boats ready to be run at once into the water and the balks slid off opposite to them. The chess wagons (as chess for two bays are carried in each) are brought at the same time in front of the space between *each pair* of pontons and the chess there run off, the space between each odd ponton and the next higher numbered boat being left free for communication. If the immediate shore or river edge is a high steep bank (the only case where this method may not be available) and unapproachable by wheels, the pontons can be placed in the water at the most convenient points and floated to their proper positions and the balks and chess can be very speedily carried and arranged as above proposed with the large force of troops usually assisting on such occasions, by whom it can, in almost all cases, [be] easily passed down to the boats.

The material being thus placed and ready and the pontoniers arranged in squads of a non-commissioned officer and 6 to 12 men each, according to the number available, and the boats being numbered from the near shore abutment or pivot flank of the raft, and the squads assigned and in position at the respective sets of balks (the two outer balks first) between *each* or any two of the boats. By the easy movement of the partial rafts in the water it has been found that even the laying of these in succession is not requisite. The remaining balks are then placed and lashed and, as soon as the lashings are completed over any even numbered boat, the placing of the chess commences, being laid both ways from the centerline of the ponton over the adjacent bays, the lashing rails being placed on immediately that any bay is completed. The rails over each pair of boats, as between Nos. 1 & 2, Nos. 3 & 4, &c., should be laid as inner rails and *between* the pairs as outer rails to prevent irregularities. Each squad will lash the rafts upon the bay above its boat or between that and the next higher numbered boat.

As soon as any squad has lashed the side rails, it takes position at once in the boat concealed below the gunwale except the *one* or *two* oarsman previously designated, who raise the oars ready to let fall at the word to swing, each oarsman having a relief under cover near him, ready to seize the oar in case of any casualty. While the balks are being lashed a minimum number of anchor-boats on the outer side are held ready with good crews with anchors from say two not adjacent boats and are kept head outward on the gentlest strain of their cables. At the word to move the strain of the oars should be strong on the chord of the arc of movement, as the carelessness which permits these boats to come athwart this line very greatly obstructs and delays the swinging of the bridge. In the rapid passage of the moving end of a long bridge it is best that the outer or farther ancher boats should be simply towed alongside the end boat of the raft.

While this work is being executed a special squad can prepare the thither abutment and if no opposition is expected another such squad can cross with two or three pontons to complete the opposite or farther abutment.

As soon as the squads on the raft shall have taken their position and the officers are all *in* the boats, if opposition is expected, the oarsman at the word should drop and fly their oars together at the wheeling flank strongly, near the middle gently and at the pivot, very slightly. In these last cases a constant watch must be kept by the officers and men not to advance their portions too much. With care at these parts the bridge may be carried round and kept very nearly straight and of course with the least strain or injury to it. I was surprised to find no damage of consequence resulted from a cramping of the pivot end of the bridge by which it was curved in to about a quarter circle of a radius of not more than 150 to 200 feet. To avoid this the pivot end should always move on a small circle well clear of the abutment.

As the bridge nears proper position, the up stream anchors direct from the pontons of the bridge should be thrown, first from near the pivot end, to aid in judging of the position for which range poles on the proper line or distance above the intended position of the bridge should be placed upon the shore. As the bridge comes between the abutments previously placed, the connection for vehicles is made at once, while the anchor-boats with the down stream or steadying anchors move off to drop them in position. If opposition is expected or offered, the farther abutment of course not being placed, the bridge is held by its upper anchors as a raft until the storming columns of these concealed in the boats and as many others as shall be required from one shore shall have passed over it.

[handwritten copy of circular order, undated]

Half a dozen such orders as above a day and that in the "dog days" was pretty hard for poor human nature to bear.

At last, on the 7th of August at 9 p.m., your Mother arrived. I met her at the depot and we preceeded to our boarding place in the city where I had previously obtained rooms and had also suceeded in getting permission to be absent from camp during the night for a limited period. We camped at the house of a Mrs. Taylor on 4th St. not far from Pennsylvania Avenue. Our quarters were not of the best but it was the very best I could do. The weather during her entire visit was exceedingly warm and we really suffered very much from the heat.

The day after her arrival your Mother went up to camp with me. This was her first visit to a camp and of course everything was new and strange to her. She enjoyed the novelty of the sights she saw and I enjoyed her visit beyond expression. Fortunateley for me I had been detailed as a member of a Court Martial and this enabled me to take more time from camp than I could have done had I been on ordinary duty. In fact, I was placed by that order beyond the orders of our Lt. Col. Pettes, who I verily believe would have deprived me the pleasure of a visit with my Wife had it been in his power to have done so.

Lt. Col. Spaulding having obtained leave of absence, the command of the Battalion devolved upon me. If you think I had nothing to do those dog days, consult the following, which is part of a leaf from my diary:

"Aug 18; Left Janes (that means Jane Taylors, our boarding house mistress. Your Mother & I called her Jane for the fun of the thing) at five a.m. (it was about two miles from Janes to the camp, a nice little walk). Drilled in Pontoneering from 7 to 9. Attended Court Martial from 10 a.m. to 3 p.m. Battallion drill (in which I also drilled the Battallion in the bayonet exercise) from 4 to 6 p.m. Dress Parade at 6 1/2 p.m. My Wife passed most of the day in camp.

This evening we 'changed our base' from Janes to Mrs. Moores, arriving there at 9 in the evening. Mr. & Mrs. Moore lived on 4th St. They were old acquaintances of ours in Rome, took compassion on us and consented to accommodate us for the time of your Mothers visit. This added very much to our pleasure as it added to your Mothers comfort."

Sometimes I would get away for half a day or so and visit some of the places of public interest in and about Washington with your

Mother. When I could not do so she generally went out with Mrs. Moore or a young Lady who was then visiting there. So the time passed rapidly by but these pleasures were not for the Soldier. Knowing that the court would adjourn on the 27th, I arranged to have your Mothers visit close on the 26th, as after that time I could not obtain permission to leave camp at night and I had no accomodations for a lady inside our lines.

At 6 1/2 O'clock on the morning of August 26th your Mother left me standing in the Depot looking sorrowfully towards the train that conveyed her away from me. The day was chilly, windy, dusty and cheerless. Your Mother went to visit our friends in Byron, NY, and I returned to camp with its routine of duties and deprivations with a heavy though not a desponding heart. In fact, I returned to my duties again with renewed energy and vigor as the only way to avoid being miserable as was natural when the mind dwelled too much on what it would like to be.

Distinguished parties from all parts of the country and the world often visited us to witness the [ponton] drill. President Lincoln and the Cabinet watched on more than one occasion.[3] Officers of the French Army admitted that we excelled even the French Engineers in the rapidity and correctness of our Pontonering. I dont remember that any *English* Officer ever admitted that we were quite as quick as *their* Engineers.

> A Pontoon Bridge,—On Monday last the experiment of throwing a pontoon bridge over the Eastern branch was repeated by the corps under the command of General Benham. Among the distinguished persons present were Postmaster General Blair, Secretary Fix, General Heintzelman, General Stoneman, General Frank Blair, Captain Wise, of the Navy Department, and quite a number of army and navy officers. The occasion was also graced by the presence of ladies.
>
> In eight minutes from the time the signal flag was raised the bridge was laid; in eight minutes more it was swung across the river to the opposite shore, and in twenty-five minutes from the commencement of work, parties in carriages and on foot were crossing on the bridge at once firm and substantial. Everything went off successfully, as might have been predicted of any operations under the direction of General Benham, to whom and his admirable corps of officers and men the coun-

try is indebted for most valuable and efficient services on occasions when a single mistake might have hazarded the safety of an army.

[newspaper and date unknown]

Sometimes we used to fill the boats with armed men concealed under the covering of the roadway. As soon as the end of the bridge touched the opposite shore these men sprang from their concealed positions, ran ashore & went through the motions of driving away an enemy supposed to be posted there to prevent our landing.

In order that you may know how a detail for Guard is made I have pasted in the [following]. Those orders detailing the guard for the next day were always read out by the Adjutant on Dress Parade. On this occasion you will see that Capt. Folwell was on duty with me.

Head Quarters Engineer Brigade
Washington D.C. Sept 15th 1863
Announcement

For Field officer of the day	Major W. Brainerd	50th N.Y.V.E.
For Captain of the Guard	Captain Wm W. Folwell	50th N.Y.V.E.
For Officer of the Guard		15th N.Y.V.E.
For Guard		
1 Sergt 4 Corpls—78 men		50th N.Y.V.E.
1 " 3 " 21 "		15th " " " "

The Adjutant of the 15th will officiate at guard mounting

By order of Genl Benham
Channing Clapp
Asst. Adjt. Genl.

[handwritten order]

The great proportion of our Regiment was kept at Washington while the Army was now and then making a forward movement into Virginia to return again as soon as the City of Washington was threatened. Capt. McDonald with Co. K was at Culpepper, Co. D at Rappahannock Station, Co. G at Alexandria and Co. C was also sent there with annother Ponton Train of 10 boats.

September had come. The nights began to get cool and the days short. Still there had been no decisive movement by either Army. The indications were that the Winter would not close in before we should have annother campaign, though it must of necessity be of short duration. Swinton in his *Army of the Potomac* calls this a "Campaign of Manoevers" and includes within it the time from July, 1863, to March, 1864.

During this stay in Washington it was my good fortune to witness the crowning of the dome of the Capitol with the statue which now stands there. A large concourse of people were present and the event was one long to be remembered by all who witnessed it.[4]

My time as usual was fully occupied. To my other duties was now added that of drilling the Officers in the bayonet exercise and the evenings were employed in recitations on field fortifications, sapping & mining, &c. Sunday was a great day, for the people of that entire section of the City, men, women & children, people in carriages and on foot, flocked to our camp to witness our Dress Parade. Of course, every man in the Regiment did his best on those days and to command the Regiment on Sunday afternoon was to be made quite a Lion. This, Spaulding generally monopolized, but some times it fell on me, and whenever it did I was sure to present something new for his especial benefit.

Considerable sickness prevailed during this season. One day Major Arden and myself had ridden down the Potomac some 7 or 8 miles to a fort where Col. Seward was in command. It happened that he was very sick at the time with dysentery and while we were there his venerable Father, then Secretary of State, came down to visit him. A tug boat brought him almost to his Sons tent. I shall never forget how weary and utterley care worn the old Secretary looked that day. The cares of State rested entireley upon him. The complication of our Nations affairs abroad, then in such a delicate position, were fully comprehended by him alone. His son was not expected to recover and he had hastened to see him. It was the last time my eyes ever beheld him.[5]

The next day, Sept 23rd, came our marching orders and what a fearful scattering of wives and sweethearts then took place. Many of the Officers and men had taken advantage of our long quiet spell to receive visits from their wives and friends & the order *to march* fell upon them like a thunderbolt.

We left our camp at 4 O'clock p.m., crossed into Virginia and bivouacked at 9 O'clock that clear, cool, starlight night several miles into Virginia towards Fairfax. We marched the next day but a short distance, being much hindered by the teams, which could not be got into the road. We kept on the road until 8 1/2 p.m., the moon coming up and making the march a beautiful one. The next evening we bivouacked on the old battle field of Bull Run.

During the evening of Sept. 25th, while we were marching *south*, we met Battery after Battery moving *north*. The roads were dusty and it was a stirring sight to see. As they passed us the drivers seemed to be urging their horses at their utmost speed. We could not imagine what this mysterious movement meant. The Artillery proved to be all the Light Batteries of the XI and XII Corps, who had been ordered from the Army of the Potomac to reinforce Burnside at Knoxville, Tenn.

The rapidity of the transfer of the XI & XII Army Corps with all their Artillery, Baggage Wagons, Ambulances &c from the Army of the Potomac to the West was one of the most astonishing feats of the War. The XI & XII Corps never afterward operated with the Army of the Potomac.[6]

Chapter Seventeen

The Campaign of Manoeuvres, Court Martial and Winter Quarters 1863–1864

We marched past Manassas Junction [and] Warrenton and arrived at Rappahannock Station, which seemed to be our destination, at noon on the 27th of September. Rappahannock Station was about 12 miles from Culpepper where the main body of our Army then was. We found Company D already there in charge of a bridge across the Rappahannock River. We were soon encamped and our regular routine again commenced. No sooner had we become comfortably fixed, than as usual, orders came to be in readiness to move again at a moments notice. The enemy were threatening our right flank.

On the 2nd of October we had a fearful rain storm, the water literally *poured* down. In the midst of it all a Division of the VI Corps marched up near us and halted. For three long hours the poor fellows stood in the drenching rain awaiting orders. Then they bivouaced in the mud and soon the blazing fires showed the true soldierley scene which had now become so familiar, but always was and always will be picturesque. Then the weather cleared away again and with it, our alarm. Somehow it always did seem strange but in nine out of ten cases true, that marching orders always accompanied stormy weather. Something very peculiar about that, was there not?

We remained at Rappahannock Station until the 9th of October, doing guard duty, taking care of the bridges and drilling &c. We would occasionally ride over to Culpepper 12 miles away when we had any business at Head Quarters. The main body of the Army was there, although a Division was near us and still other considerable bodies of Troops in our rear, posted at the most exposed positions where our communications with Washington were likeley to be attacked. Both Armies sat and snarled at each other, neither caring to bring on a general engagement. Small skirmishing was of daily occurance but both Commanders manoevered principally for the most advantageous position in case he should be attacked by the other.

Suddenley, on the night of Oct 9th, came the order to move. There was no mistake this time. The night was spent in cooking rations and at 3 O'clock on the morning of the 10th we were once more on the road. General Benham and staff went ahead to the RR Station and took cars for Washington, turning over the command to me, or rather the command of the march, for Spaulding with two Companies, H & E, was to remain a while until all the Troops had crossed the bridges, then take them up and follow. The main Army was moving northward to the right of us keeping closeley by Lee, who was now on the war path directley on a course for Washington. It was to be a race as to who could first reach Centreville, for that was the key to the position.

Soon after we had taken the road Lt. Col. Pettes was taken sick and obliged to ride in an Ambulance. His orders from General Benham were *imperative:* to march the command that day to Bull Run Creek and not to halt or go into camp until that stream had been crossed at Blackburn Ford. These orders he in turn communicated to me. We marched rapidly and at 11 O'clock that night reached Blackburn Ford, 27 miles from our starting point. The night was raw and chilly, Bull Run Creek 100 ft. wide and water up to our waists was before us. We had marched 27 miles that day but our orders admitted of no discretion. Bull Run Creek must be crossed that night. In it and through it we went. Then [we] built fires and stood around them until 3 O'clock in the morning to warm and dry ourselves. The compliments paid our General Benham that night, if not *loud*, were *deep*, and I would not care to repeat them.[1]

The spot where we bivouacked was on the old, twice fought over battle field of Bull Run and human skulls and human bones were scattered around quite plentifully. These were the bones of our comrades who had left their homes a little sooner than we in the fond anticipation of returning within 90 days time to a grateful people who would be only too glad to crown them with victorys laurel wreaths.

The next morning at 7 we were again on the road, passed through Centreville and reached Fairfax about noon. At 5 O'clock that afternoon we made ourselves comfortable for the night at Baileys Cross Roads. Not so Spaulding with his detatchment. He came very near being captured with all his train. The enemy crowded him so closeley after his bridges were taken up that it was at one time feared by Gen. Buford, who, with his Cavalry, was attempting to protect him, that he would be lost. Finally Warren with the whole of the 5th Corps halted and fought the Battle of Bristoe Station. This gave Spaulding an opportunity to escape by cutting a new road through the woods and by most excellent Engineering combined with almost superhuman exertions of his men, he finally succeeded in getting safeley away with his entire Train and equipage. On the 12th day of October we again marched into Washington and occupied our old camp ground.[2]

The Army soon returned to its old position at Rappahanock and Brandy Station, Spaulding with a Detachment of the Regiment accompanying them, leaving me in the general charge of the Engineering material at the Depot. Gradually the Army went into winter quarters, several other companies were sent down to Spaulding and our camp began once more to assume the appearance of winter quarters also. Soon after our arrival I was placed in charge of all the Ponton material at the Depot and being now the Senior Officer next to the Colonel, the command of the camp devolved chiefly upon me. One company after annother was sent down to Rappahannock Station where Spaulding, who was in command of the detachment which comprised the greater portion of the Regiment, was in command.

Active operations soon ceased altogether and it was now a question of subsistance and comfort for the Winter. We kept up our usual routine of drill and guard mounting, Dress Parade &c so that the time passed rapidly enough. The air was also full of rumors of what the Armies were doing or about to do down at the front, but these rumors generally amounted to nothing.

My "Gillotine," as it was called, was once more put in operation and I had now become so proficient as a Court Martial that I could dispose of half a dozen cases in the morning and attend to drilling the Battallion in the afternoon. My Court Martial record will show you how these cases were disposed of in those days. The 15th Regiment, or what was left of it, were encamped near us and I used to drill the two regiments consolidated every afternoon. So passed the months of October and November without any special event of importance to change the monotony. We crossed the Anacosta occasionally to make Gabions for practice and this gave me an opportunity to renew my acquaintance with my old friends the Naylors where I passed many pleasant evenings with the old Colonel and his daughters.

Lt. Hoyt was dishonorably dismissed [from] the service for dishonesty in the Quartermasters Department of which he had control. As he was formerley a Lieutenant in my Company, his dishonorable dismissal, while he richly deserved it, made me feel very sad.

On the first of November we had orders to fit up a bridge train 1250 ft. in length which was done immediateley. We worked all of two nights to complete it. Then came the order to break camp and be ready to move at a moments notice. So we tore up everything once more and stood in line all of a raw, windy, November day awaiting *orders*.

It was known that something was going on down at the front but what it was we knew not. All that day, all night, all the next day we waited, and then the order was countermanded and we were allowed to settle again. It turned out that the Battle of Rappahannock Station was being fought while we were waiting under Arms. We were held in readiness in case the fight developed into a general engagement, but as it did not, we were again allowed to go into camp.[3]

On the 27th of November the same thing was repeated. Orders to move and we remained in readiness *five* days until the 5th day of December, when, as we had again broken our camp all to pieces, we were allowed to rebuild it. This tear up was [the] consequence of the Mine Run move, of which you can read in the histories. After this both Armies finally settled down for the winter.[4]

I must not neglect to tell you of quite a romantic affair which occured about this time. We had a Sergeant in the Regiment by the name of Lang, an old German Soldier and an excellent swordsman. The Officers took lessons from him in the broadsword exercise. On Sunday night, Dec. 20th, a cold, wintry evening, he came to my tent and asked permission to be absent from camp for a few hours. I judged from his manner that something was wrong but gave him the permit. My curiosity being aroused, the next morning I made some inquiries and found that the night before Lang had fought a duel with an old enemy of his, now a Lt. Col. in annother Regiment. They went out that cold night and met in the old Congressional burying ground at midnight where they fought with sabres. Lang cut the Lt. Col. seriously in the abdomen and I presume honor was satisfied.

Christmas Day was passed very pleasantly at Colonel Naylors where I was invited to dine. We had a dance in the evening. The Armies had both gone into winter quarters so that many of the Officers & men could now visit Washington. All endeavored to enjoy themselves as well as possible and in this way ended the year 1863.

New Years day, 1864, will long be remembered as "The Cold New Years." Never before or since in my reccollection did such an intense and stinging cold day fall upon us. The news from all parts of the land was chiefly of the cold. People perished in all directions. Trains were delayed on almost all the principle [*sic*] railroad lines of the country. The papers for weeks following were filled with the events of that cold New Years day.

For a few days previous the weather had been comparativeley mild and pleasant. Then suddenley the wind chopped around to the Northwest and came upon us in a fierce, biting, penetrating, fridged hurricane. The morning was quite mild and, according to time honored custom, the Officers of the Army and Navy who were then stopping in Washington called in full uniform upon the President. This ceremony was always opened promptly at 12 O'clock noon. A large number of Officers availed themselves of the priviledge but I did not feel equal to the task that morning and so passed the time in camp.[5]

I made out however to call on my old friends Mrs. Moon and Mrs. Wallace and that evening went over to call at the place where I always found the most sincere enjoyment and pleasure, my friends the Naylors.

As usual the evening was passed in the most pleasant manner and 12 O'clock came before I was aware of the fact and I started for home. Heaton, my Orderley, and I mounted on Old Bill and [the] Gray (Frank). We had something near two miles to ride before reaching the Anacesta Bridge and were almost perishing when we reached there. As we came upon the long bridge and faced the piercing wind, the horses actually stopped and had to be spurred forward. I thought we would never get across. About the center of the bridge we found the draw was raised, as was customary after 12 O'clock, and we halloed a long while before we could make the guard on the other side hear us. Finally, as all things have to have an ending, the Guard came, let down the draw, examined my papers and let us pass. When I reached camp that night I was about as near dead as alive. Certainly I never suffered so much in the same length of time before and I shall never forget that ride across the Anacesta Bridge on cold New Years night, 1864.[6]

Soon after the 1st of January a General Court Martial was ordered.

John W. Heaton, Wesley Brainerd's orderly. Courtesy of the William Hencken Family.

The detail for the Court consisted of Capt. Personius, Capt. Schenk, Lts. Rider & Burke, with Capt. Sergeant as Judge Advocate and myself as President of the Court.

Courts Martial are governed by very simple but very effective laws. The main object being "here to do Justice" and not as is too often the case in civil law "how not to do it." The principle [*sic*] authorities which govern Courts Martial are De Hart and Benet.[7]

It was customary to *fine* any member of the court who came in late. The *fine* consisted of cigars for the entire court. The best patronized member of the court during the winter was the *Stove*, of course.

The hardest case the court had for trial was that of *Doxy* for desertion. It was an aggravated case and occupied the time of the Court for two, long weeks. Sherman, his council, was a criminal lawyer who made a business of attending Court Martial as council for Soldiers who could pay him a good fee and who were accused of all manner of crime. He was thoroughly acquainted with all the kinks of the law and generally gave a court great trouble. He boasted that in every case intrusted to his charge he had either *acquited* his man or *broken* up the court. My court heard of this boast and we concluded that he would certainly not do the latter and if in this case he did the former, it would be pureley upon its merits.

Poor Doxy was sentenced to be dishonorably discharged, his body branded with a letter D on each thigh, his head shaved and he drummed out of camp. The sentence was rigidly executed in the presence of the Regiment and thus ended the Doxy case.

Mr. Sherman, council for the accused, was bitterley sarcastic in his language and manner as he took his final leave of the court, which he did not succeed in breaking up nor did he acquit his client.

The records of the case, as every question and answer was reduced to writing, were very voluminous. The President was pretty well tired out after the second reading of the record. About these days Jefferson Davis issued his famous proclamation to the Southern people. Desertions were becomeing alarmingly frequent and numerous. Their cause was at a low ebb hence the sublime imprudence of the proclamation worthy of Jeff Davis.[8]

Chapter Eighteen

Commencement of Campaign of 1864

Annother year has passed since writing that which you have now read and yet it seems but a few nights since I sat in this same place by the winter fire. Your Mother is at her work and I am scratching away on these reccollections which perhaps neither you nor any one else may ever read. You will discover, should you live to my age, that the years follow each other so rapidly one is *startled* at the thought of *annother* anniversary of an event following so soon upon the heels of the one just passed. As I look back upon the page preceding this [one], it seems that it was only last evening and yet 1873, a *year* of more than ordinary interest and importance, has actually passed away.

Ten years have passed since the occurrence of events of which I am now about to write. The events of 1864 stand out prominently as one of the most wonderful campaigns of history, the one campaign of all the war in which was enacted the most desperate and decisive battles, the greatest loss of life, the most brilliant display of generalship, of endurance, of tenacity, of suffering and heroism [and] resulted finally in deciding favorably for us the issues of the contest.

When I commenced this chapter night before last, it was on the eleventh anniversary of the night before the Battle of Fredricksburg, Dec 10th, 1862, and something, I know not what, impelled me to commence writing these sketches again, a task that had been laid over for nearly a year. Annother rather singular coincidence I will mention. Yesterday, the 11th, eleven years ago, I was laying in the Hospital among the wounded and the battle was in progress. Alexander H. Stevens, then Vice President of the Southern Confederacy was then there also, said to be on a visit to the Southern Army, and witnessed the battle. Yesterday he again stood up in the United States House of Representatives and made his first speech, 14 years having elapsed since he left that body and joined the enemies of his country. Was ever a Government or a people so magnanimous before? Eleven years ago to day, in company with Capt. McDonald, I crawled out of the hospital against the Surgeons orders and ascending a high ground near by and witnessed the battle then going on, the long lines of blue charging and again falling back leaving the ground covered with our dead.[1]

Now to return to my narrative again. Your Mother had written me to expect her in Washington on the 7th of January and so I went to the Depot at 10 1/2 in the morning to meet her. The train came in but no Wife was there. Train after train came into the Depot but the one I longed to see came not and I waited all that long cold, dismal day and returned to camp at half past ten a disappointed man. The next morning I was on hand again and waited until half past one in the afternoon but no Wife came. As I was then obliged to return to camp on business, I left my Orderley at the Depot and sure enough, at half past three *in she came*. Heaton escorted her to camp and soon we were ensconced near the camp ground at the house of Mrs. Scarff, where I had made arrangements for our accommodation. Your Mother remained with me from the 8th day of January until the 4th day of April 1864.[2]

The family with whom we boarded, Mrs. and Mr. Scarff (I place the Ladies name first as she was the *man* of the house) were ordinary, good, kind hearted people of the working class, the husband being one of the Foremen in some department of the Navy Yard. They were honest, quiet people who worked hard for a living and attended strickly to their own business. They possessed a fortune in quite a large family of healthy children who took to *grease*, with which the table was generally and plentifully supplied, as naturally as ducks to water.

To me the change to a genuine bed, out of camp with a real roof to cover me and a family table to go to was quite a luxury. The camp fare to which I was accostomed had prepared my stomach for anything cooked in a regular manner. But to your Mother, acostomed to the delacacies and cleanliness of a well ordered house, the change was not so agreeable and, had it not been for the stimulating influence of my society, she would hardly have been able to accostom herself to the difference. Taken altogether, our temporary home was about as good as could have been expected under the circumstances, especially as we knew it could be but temporary. Almost any place *together*, was to us better than any amount of luxuries *separated*.

My old friend, Col. Naylor, had offered me one of his family carriages to use during the visit of your Mother. His kind offer was accepted and I had it rigged up for two horses at a trifling expense. The Col. also lent me a double harness and so old Bay & Grey were soon transformed from war steeds to civilian horses and taken altogether made quite a stylish rig. The horses seemed to enjoy the change quite as well as Heaton and Johnny, and the pleasure of having a carriage in which to ride at any time added very materially to the enjoyment of your Mothers visit.

Whenever so disposed she could take Heaton for a drive, in case I could not go, and visit our friends the Moores. In this manner during her visit she visited Halls Hill, Arlington, the Soldiers home, the Naylors, the fortifications about the city, Chain Bridge, Georgetown, Alexandria, Meridian Hill and other places of interest. Occasionally she and I rode out together, but not as often as we would have liked or could have done, had Colonel Pettes been disposed to act less of the martinette and be a little more accommodating.

On the occasion of one of the regular weekly Presidential levies your Mother and I called upon the President. The crowd was so great, as it always was on those occasions, that we could do nothing more than simply shake hands and receive his "*Howdy Doo*," the same as all the rest, but the sight was well worth seeing. Rich and poor, high and low, the dirty and the richly clad, shoddy and modest worth were all treated alike.

On annother occasion we visited Grovers Theatre. It was on the evening of Friday, March 4th, and [Edwin] Booth played the character of "Richelieu." Opposite the President sat Lord Lyons, the British Minister. The boxes were gaily decorated with flags and the house was crowded. Your Mother and I both noticed the sad, woe-begone, absent look of the President. Our seat was placed where we obtained a good view of him. Little did we then think that in thirteen months from that night, on the next Good Friday night, that great and good man was to meet his death by the hand of an assassin who proved to be the brother of the man who was then playing before us.

Meanwhile both Armies remained in their winter quarters on the Rappahannock river with the Head Quarters of General Meade near Brandy Station and the major portion of our Regiment encamped near

Rappahannock Station. Spaulding, now Lt. Col., was in command of the Detatchment and it was a standing joke among the Officers that Col. Pettes remained at Washington with the *Regiment* of two companies while Spaulding commanded only a *Detatchment* of ten companies.

The camp at Rappahannock Station was one of the finest in the Army. Situated upon a rising ground about half a mile from the line and bridges which it overlooked, it commanded a fine view of the country for miles around. Large log houses with real windows and doors, wooden chimneys and immense fire places were constructed for the Officers along the crest of the ridge, affording from the doors of each a view down the company streets, which were located on the slope of the hill containing the little, comfortable huts of the men laid out in systematic lines, all of the same size and shape and covered with their shelter tents for roofs.

In front and facing to the South was an open space for parade ground and color line. To the right and Westward stood the wagons containing Ponton boats and material, drawn up in regular lines by columns. On the left to the Eastward were the corrals for the animals, provender, Hospital &c., while to the rear or North, and back of the Officers Quarters, were stationed a section of two pieces of Artillery detailed by General Meade from a regular Battery for the better protection of the camp which, like all the larger camps, were that winter considerably isolated and separate from each other.

Surrounding the whole camp was constructed an immense abbatis twelve or fifteen feet high, made of fallen trees placed in position with the branches cut off, sharpened and pointing outward, with regular gates or openings for entrance and exit. This abbatis formed an additional protection against Cavalry raids which were quite frequent and indulged in during the Winter by the Cavalry of both Armies. Such was the camp at Rappahannock Station and here the Officers and men passed the winter enjoying the *romance* of war with just enough hazard to make the situation interesting.

Our duty was chiefly to drill and take care of the two bridges which spanned the river. I can fancy no position in life which presents so many charms to the mind of one martially inclined as a Winter camp of this description. The long evenings spent in pleasant, agreeable intercourse around the huge blazing fires of the Officers Quarters after the duties & fatigues of the day. The readings, the smokings, the attentive listening to narratives of personal experience and exploits,

Winter camp, 50th New York Volunteer Engineer Regiment, Rappahannock Station, Virginia, March 1864. Library of Congress.

recountings of the past and speculations of the future, with occasional rumors of an intended raid upon the camp by detatchments of the enemys Cavalry and sometimes a midnight alarm, which aroused the whole command and placed them in position to repel an anticipated attack. All this and more than I can enumerate served to pass the time more rapidly than one who has never experienced the like can readily imagine.

During the month of March General Kilpatrick made his famous raid to Richmond with a view of cuting off the communications leading to that City. Young Dalgren accompanied the expedition in command of one of its Divisions. By some means he became separated from the main body when near Richmond and was cut off and captured. The inhuman treatment to which his dead and mutilated remains were subjected is now a matter of history. The expedition resulted disastrously but served to dispel the lethargy with which we were all afflicted. The ritcheous indignation aroused in consequence of the inhuman treatment to young Dalgren, made an impression which has not yet and never can be forgotten.[3]

On the 21st of March it was announced that *General Grant* would assume command of all the Armies and make his individual Head Quarters with the Army of the Potomac. About this time the question arose about allowing Soldiers to vote. My Friend, Mr. Stevens of

Officers' quarters, 50th New York Volunteer Engineer Regiment, Rappahannock Station, Virginia, March 1864. Courtesy of the Library of Congress.

Rome, an opponent of the war or what was more familliarly known as a "Copperhead," voted against allowing the Soldiers to vote.[4]

Appearances now began to indicate the early commencement of active operations. Regiments and Batteries were ordered from Washington to the front. Immense supplies followed and we knew by the "pricking of our thumbs," as Shakespeare says, that soon we were to be up and at it again and in earnest.[5] March 30th was the day appointed for the 15th and the ballance of our Regiment to leave Washington for the Rappahannock. At 7 O'clock in the morning the Regiment marched. The day was ushered in by a heavy, disagreeable snow storm and the ground was covered to a considerable depth when the Regiment started. Soon the snow turned to sleet and the marching was about as disagreeable as could be imagined.

As my court martial duties were not yet completed I could not go with the Regiment. Your Mother stood with me at the door of our boarding house and witnessed the Regiment passing by. It was the first time she had seen any save the pleasant side of a Soldiers life and her sympathys for the poor fellows trudging along through the storm with all their worldly effects strapped upon their backs, trudging towards they knew not what, were deeply effected. The effect of the melancholy spectacle was intensified by the reflection that we too must soon part company, perhaps never again to meet on this side of the mysterious river.

The time for separation soon came. On Sunday we attended church together and then we rode down to see the Moores and bade them farewell. We spent the evening together and the next morning, Monday, April 4th, we paid a visit together to the now deserted camp and that evening at 7:30 your Mother took the train for the North and her friends while I returned once more to my old tent and resumed the life of a Soldier.

The court martial of which I was the President continued its sittings until the 8th day of April when it adjourned "sine die." Our aim was to render equal and exact Justice to all, always tempering Justice with Mercy so far as was consistant with our duty to the service and the oath we had taken. A few of our accused ones were sentenced to be shot to death with musketry, but President Lincoln, in the magnamity of his great heart, either pardoned or commuted their sentences, so that I am not aware that a single Soldier died under our sentence, though strictly speaking, according to the Rules of War, they deserved death.

I was not sorry when we had closed our last case, though the experience of the months devoted to that duty afforded me a change of scene and much information and I was more than ever impressed of the general fairness of court martial as instruments of the Law and sincereley wish that many of its most prominent features might be introduced into civil courts of Justice.

As soon as the Court Martial finished up its business and, according to orders previously received, I prepared to rejoin my Regiment. Bidding my friends and Washington farewell once more, I left on Sunday morning in company with Capts. Schenk & Personious, also members of the court, with 23 men by steamboat to Alexandria. We were obliged to remain there one day in consequence of a bridge having been carried away at Bull Run and that evening attended Church and returned to the Marshall House, now famous as the place where Col. Ellsworth was killed. The road was not repaired until Tuesday, when we left Alexandria and arrived at the camp at Rappahannock Station at about noon.

Taking up my quarters with Spaulding, here I found everything and everybody alive in preparation for the opening of the campaign. Our Regiment was no longer to move as one body but each of the

Ponton wagon and wooden ponton, 50th New York Volunteer Engineer Regiment, Rappahannock Station, Virginia March 1864. Courtesy of the Library of Congress.

Majors was to have command of a separate Battallion and complete Train & assigned to each of the three principal Corps of the Army, which was then divided and commanded as follows: the II Corps commanded by General Hancock, the V Corps commanded by General Warren and the VI Corps by General Sedgwick. The Cavalry was a separate Corps and Batteries of Artillery were assigned to each.

The question among the Majors was which Corps to join as it was a matter left for us to decide. It was a general impression based upon previous experience that the Corps of General Hancock would have to do fully, if not more, than its share of the fighting. This opinion prevailed extensiveley among Officers who were acquainted with the character and disposition of that General and subsequent events proved the correctness of their judgement. Major Ford preferred to go with General Warren as he had some personal acquaintance with that Officer and for the same reason Major Beers preferred to be assigned with General Sedgewick. As they had so arranged matters between themselves before my arrival, there was nothing left for me but to consent to be assigned to the II Corps under Gen. Hancock.

I had no objections to being assigned to General Hancocks Corps, especially as a rumor was afloat that he was to take his command on a distant and separate expedition, one that involved much hard fighting,

Canvas ponton, 50th New York Volunteer Engineer Regiment, Rappahannock Station, Virginia, March 1864. Courtesy of the Library of Congress.

great hazard, and a corresponding amount of *glory*. So all things were satisfactorily arranged and the work of organization and preparation was vigorously commenced.

The companies assigned to my Battallion were B, F & G, commanded respectiveley by Captains Schenk, McGrath and Personious. The train to be handled by my Battalion was already made up under Spauldings directions and consisted of 54 wagons and two ambulances. The Train was put in direct command of Captain Personious and the companies divided into sections, each with appropriate duties to perform.

On the 14th of April, accompanied by Spaulding, I rode over to General Hancocks Head Quarters and reported to the Adjutant General, Col. Francis A. Walker, for duty. I found him a very pleasant, genial gentleman and an acquaintance was commenced which afterwards developed into quite a close friendship.[6] I then rode to General Head Quarters, had an interview with Maj. Duane, the Chief Engineer, and returned to camp after a ride of 16 miles.

Colonel Walker had informed me that General Hancock desired my command to be moved nearer his Head Quarters as soon as practicable so the next day we broke up the camp at Rappahannock Station with all its pleasant associations. [See Appendix 4, Item 6.] I marched my command through Brandy Station and selected a camp ground near Stevensburgh on a little stream called Mountain Run, about a mile from the Head Quarters of General Hancock. Having selected my ground I rode over and reported myself and command ready for duty.

About the same time that I reported to General Hancock, each of the other Battallion commanders with their bridge trains reported also to their respective Corps commanders. Spaulding had a separate command of two companies with a canvass or flying train and was attached directley to General Head Quarters of the Army.[7] Colonel Pettes remained in Washington with one company at the general Depot on our old camp ground near the Navy Yard. This was the disposition of our Regiment at the commencement of the campaign of 1864.

The spot that I had selected for a camp ground was in the peach orchard of an abandoned estate with the ruins of the house situated at about the centre. The view looking to the South was charming. The weather, with occasional rains, was getting quite settled and in two days my little command were comfortably fixed in as neat a camp as could be found anywhere. Officers and men were delighted with the change from winter quarters to a regular Summer camp again and each vied with the other in making everything pleasant. The "Esprit du Corps" of my Battallion was perfect and all aspired to make it the *first* Battalion, a model for the Regiment and so it was both in dicipline and morale. You may rest assured I was proud of my little command and happy in the conciousness that they in time were satisfied with their commander.

"Camp Peach Orchard," as we named it, was oftentimes referred to as an auspicious opening of the campaign. We built a few bridges across Mountain Run and spent the ballance of the time in drill and preparation for what we knew was soon to follow. Taken altogether we were as busy as bees.

Confidential

This order will not be published to the troops.

Head-Quarters, Army of the Potomac,

April 20th, 1864

Special Orders, No. 111.

The following instructions respecting the supplies to be provided for the approaching campaign are published for the guidance of Corps and other independent commanders and the Chiefs of the Staff Departments concerned:

1. Paragraph 2 of General Order No. 13 of March 30, 1864, from these Head-Quarters is so far modified as to direct that one hundred and fifty rounds of small arm ammunition per man be kept constantly on hand. Fifty rounds will be carried on the person, forty rounds of which will be in the boxes. The remaining ten rounds will not be issued until marching orders are received. One hundred rounds per man will be carried in the supply trains.

2. Three days full rations will be kept in the haversacks and three days rations of bread, coffee, sugar and salt in the knapsacks. Ten days small rations and one days salt meat will be carried in the supply trains. Thirteen days beef on the hoof will be taken. The three days small rations to be carried in the knapsacks will not be issued until marching orders are received. They will, however, be kept on hand ready for issue.

3. Ten days forage will be taken in the supply trains and other wagons.

4. All Commanders are notified that the rations must be made to last the full time for which they are issued. No new issues will be authorized under any pretext whatever until the expiration of the time for which the original issues were made.

5. Corps and other independent Commanders will issue the most stringent instructions to prevent this order being made public or placed where newspaper correspondents can have access to it. Any officer convicted of allowing it to pass out of his hands, will be severely punished.

By command of Major General Meade:

S. Williams,
Assistant Adjutant General.

Official:

H. P. Wilson
Actg Assistant Adjutant General.

Major Brainard
Comdg Batt Engineers

[official printed circular order]

During the night of the 20th an Orderley rode up to my quarters with the information that on the morrow General Hancock would review my Battallion and transportation in person. On the following day promptly at 12 O'clock my command was drawn up in line with white gloves and blackened shoes, their burnished arms and polished accouterments glistening in the bright sunlight. The animals were hitched to the wagons and everything in apple pie order. Mounted upon my bay, who that day appeared to excellent advantage, I stood awaiting the arrival of the General. Soon, at a brisk trot over the crest of rising ground to the eastward, came the commander accompanied by a large and brilliant array of General and Staff Officers.

I brought my men to Shoulder Arms and we stood like so many statues until the General and all who accompanied him had ridden down our front, back by the rear and taken their position in front of me. Then, wheeling my companies into columns with the drummers on the right, we marched and wheeled and passed in front of the reviewing Officers in a manner worthy of the oldest veterans. Arriving back to the first position, the column was halted and again brought into line. At a signal from the General, I then accompanied him to the inspection of the Quartermasters Department, all of which was in the most perfect trim, as perfect as a train of that kind could be, not a line, not a buckle missing.

The inspection completed, General Hancock turned to me and said, "Major, I am very much pleased with the creditable appearance of your command. Tomorrow General Grant is to review my corps and your Battallion will take its position *on the right of the line*." Saluting me with a, "Good Day, Major," he rode rapidly away.

This was my first interview with General Hancock and my admiration for him as a Soldier from that day to the present has been simply unbounded. When the Generals words were communicated to the command they were perfectley delighted and all determined if possible to make a more creditable appearance at the Grand Review which was to take place on the morrow, when we were to occupy the position of honor on the extreme right of the entire Corps.[8]

Chapter Nineteen

Review of Second Corps, Crossing of the Rapidan and Commencement of the Battle of the Wilderness from April 22nd to May 7th, 1864

Friday, April 22nd, the sun rose bright and beautiful. Just the kind of weather so ardentley wished for by those who were to participate in the Grand Review of the II Corps and who were that day to pass for the first time in front of Lieutenant General Grant.

Early in the morning the different Batteries, Regiments, Brigades and Divisions left their camps and filled the roads leading towards the field selected for the review near Stevensburgh. Each division approached the ground by a different road and by 11 O'clock all the Troops were in position, drawn up in three parallell lines, each nearly a mile in length and facing to the South.

At early dawn my little command was busy brushing up their clothing and burnishing their Arms and accoutrements. When we reached the ground the Troops were nearly all in line. The old veterans gazed in astonishment at our insignificant little Battallion with three drummers and a fifer at our head as we marched past one after annother of the Regiments towards our position on the right. There were posted, as a mark of distinction, the remnants of the III & I Corps, consolidated and known in the Army as "The Old Red Patch," all of whom were veterans who had participated in every battle fought by the Army of the Potomac. The "Red Patch," a square piece of red flannel fastened to the cap, was bestoued upon them by the gallant Gen. Kearney, who lost his life at the Second Battle of Bull Run.[1]

Arriving at our position we Ordered Arms and awaited the signals as described in the General Orders. [See Appendix 4, Item 7.] Preciseley at noon the booming of Artillerey announced the arrival of the new Lieutenant General, Commander of all the Armies. A large concourse of General and Staff Officers accompanied him as he rode to his position opposite the centre of the line on the rising ground to the South of us. When he had arrived at his position, which was designated by a flag, and wheeled his horse towards us, the bugle sounded and the whole line of forty thousand men Presented Arms, then came to a Shoulder and stood motionless while General Grant, Gen. Hancock and all the retinue rode first to the right then down in front of each line.

The bands of each Brigade struck up "Hail To The Chief" as he approached the right of each. It took some time to ride down in front of each line and back again to the reviewing position. Again the bugle sounded and each Regiment closed ranks, wheeled into columns of divisions at half distance and took up the line of march. As my Battallion was already on the right and facing South when the ranks were closed all we had to do was to move off directley to the front. First and leading the entire corps was General Barlow (commanding the First Division) and Staff, then my little Battallion of three companies, led by our diminutive martial band, my Officers all mounted. We marched up to the marker, wheeled and passed in front of the Reviewing Officer.

It was the first time we had seen General Grant, who raised his hat in acknowledgement of our salute as we passed by. The men did themselves credit, each one feeling the responsibility of the situation. They marched with a precision and alignment which was *perfection*. Having passed the point of interest, we broke into a Double Quick and out of the way. Then followed the main body of Troops. The bands of each Brigade marched and played in front until they reached the Reviewing Officers where they wheeled to the left and around again and remained in that position, playing until the Brigade had passed, where they ceased playing and fell in rear of their Brigade. The same operation having to be gone through with each successive Brigade as they passed.

We had reached our camp before the last Regiment and Batterey had passed in front of the Reviewing Officer. That was a proud day for me and for my Battallion, as General Hancock was the first who had given us the position to which our arm of the service was entitled. It was also a memorable day, being the first that the Army of the Potomac had seen of the future hero of the war, General Grant, and to me the review was peculiarly impressive.

Not six months from that day the II Corps was again reviewed, this time by General Hancock alone. It was the morning after the battle at Reams Station, and, although the Corps had been reinforced by many thousands of the "flower" from the defences of Washington, less than *seven thousand men* appeared that morning for duty. The tears streamed down the Generals face as he exclaimed, "Where now is my old II Corps"? It was their last battle.[2]

The V and VI corps were reviewed in like manner by General Grant in the days that followed. When all this was done, the final steps preparatory to opening the campaign were completed. Meanwhile we were as active as ever in my camp.

Not a night passed but I was aroused from once to half a dozen times by the Orderlies from Head Quarters bearing orders from the General. It was his custom in those days to get up about six in the morning, take his breakfast quietley and move around his Head Quarters until about 11 O'clock. Then he would mount his horse and ride over to the different Head Quarters of his Division Generals, around the camps or out to the outposts, returning about 4 in the afternoon when he took his dinner and returned to his tent. The Staff would begin whispering to each other that the "trouble" was about to commence and prepare themselves accordingly. Soon the General would make his appearance, hands full of papers and rage about like a caged lion, calling first on one then annother of his staff. Soon mounted Orderlies would be flying in all directions carrying orders and dispatches from one end of the Corps to the other. This was generally kept up until the late hours of the night or early hours of the morning no matter what the weather. The more inclement the night, the more it seemed to please him and the staff and orderlies were sure to be kept on the go splashing through the mud and rain until daylight, when things would again assume their normal condition.

Once having learned this peculiar characteristic of the General, I concluded to save myself the trouble of undressing at night and so laid down on my blankets with everything on but boots, catching what sleep I could during the day.

One incident occured during our stay at Camp Peach Orchard which I must not forget to mention, the execution of a Soldier for desertion. It was a beautiful day. The execution took place on the high ground to the Southeast of our camp and about a mile distant. The First Division was mustered to witness the execution, which we could observe through our glasses. It was a sad spectacle and our men as well as myself were deeply affected by the scene. The plaintive strains from the band playing the "Dead March in Saul"[3] came floating to us on the springlike breeze. The solemn procession and the final drop could be distinctly seen by us. To die in the heat of battle on the field of honor is glorious, but such an ignominious death is too sad to reflect upon.

Nothing now seemed to be wanting but orders to move. The chaplain, Rev. Mr. Prichard, was sent back to Washington with the Colonel as none of the Battallion commanders seemed to desire his Society and because we had no room for him. All the extra and superfluous baggage was boxed up and sent to Washington also. After numerous inspections the Army was brought down to a fighting trim. General Orders No. 18 was issued, indicating that the decisive day was near at hand.

Head Quarters Second Army Corps.
April 25, 1864.

General Orders
No. 18

The following general rules will be strictly observed by all Commanders during the campaign:

1.—The practice of burning refuse material whenever the command breaks camp will not be tolerated.

2.—Columns on the march will habitually throw out advance and rear guards, skirmishers and flankers.

3.—Whenever a division goes into bivouac or camp, at least one brigade will be in line of battle, the remaining brigades being disposed in column or in line at the discretion of commanders.

4.—Divisions, Brigades, and Regiments must preserve their proper intervals on the march to avoid useless fatigue for the men. A halt of from five to seven minutes should be made every hour. The men should always be required to fill their canteens before marching.

5.—Troops, artillery and trains will be habitually closed as far as practicable toward the head of column at every halt, leaving the road clear. The simplest plan is for each regiment of a brigade to move up parallel and opposite to the leading regiment, which should not halt in the road. Every exertion will be made to lessen the depth of column.

6.—Whenever circumstances admit, the troops will march on the side of the road, giving the road to artillery and trains.

7.—Streams not bridged, and difficult ground, will be invariably passed by the troops in *close order*. The practice of allowing the men to pick their way at such times, is a most fruitful source of straggling.

8.—Commanders are enjoined to have frequent roll-calls on the march and one invariably on arriving in camp. The result of this last roll-call giving the number of unauthorized absentees per regiment, will be transmitted to these Head Quarters.

9.—No calls will be sounded nor will any bands be permitted to play except by special authority from Corps Headquarters. Existing orders with reference to the discharge of fire arms will be strictly enforced. On the march, pickets marching off will no longer be permitted to discharge their pieces.

10. Commanders of Divisions and Independent Brigades will send a Staff Officer to these Headquarters at the conclusion of each march or movement to receive orders and to notify the Major General Commanding of the location of their commands.

11. Commanders must have their Headquarters in rear of and near to their commands.

12. The Provost Marshal of the corps and Division Provost Marshals will keep an accurate record of all men arrested by them for straggling or other misdemeanors and at the close of the campaign or before, an abstract of this record will be published as a part of the military history of each regiment.

By order of Major General Hancock.
W. P. Wilson
A. A. A. G.

On the morning of May 3rd General Walker informed me confidentially that the Army would move that night and directed me to ride over to Maddens, 4 miles distant, in order that I might see General Brooks and become familiar with the ground. As my Battallion was to lead off in the movement and as the nights were dark, it was important that I should thoroughly know the ground. He also informed me that upon my arrival that night at Maddens, which was on the outposts of our line, General Brooks Brigade would act as an escort for my command and train to Elys Ford, 12 miles distant, where we were to lay a ponton bridge for the passage of the II Corps. So I rode over to Maddens, had an interview with General Brooks and returned to my camp.[4]

Then came the order which opened the campaign. My directions were to strike tents and move as soon as it was dark enough to be hidden from view of Clarks Mountain, situated a few miles to the South of us, on which the enemy had a lookout which could observe all our movements by daylight. A portion of my command was left to come on later while I proceeded with all that was necessary to build a bridge at Elys on the Rapidan.

Head Qurs 2nd Corps May 3, 1864

Order— Extract

The command will move tonight as follows:

The 4th Brigade 1st Division will move at 12 PM in the direction of Elys ford, followed by 10 boats of the bridge train with the necessary Equipage.

10 boats of the bridge train will move to Madden at 8 PM parking in rear of Brooks Brigade. The remainder of the bridge train will move to Elys ford as soon as the road is clear of troops. This part of the train will move from its present camp before daylight so as to be hid from the sight of Clarks Mountain.

By order of Major Genl Hancock
CHMorgan, Insp Genl, Chf of Staff

Major Brainerd
Com of Engr Det

The route led through Maddens and Richardsville. Promptly at half past 7 O'clock my tents were struck and the command marched. The shades of evening were falling fast but we moved rapidly to the road which led to Maddens. This road was corduroyed for a distance of about two miles and on either side of the corduroy, which was just wide enough for one team to move, was a ditch and then a swamp.

What was my constirnation on arriving at this road to find it already filled to its whole length with Artillerey and their supply wagons. They had moved too soon and were now blockading the road which belonged to me. As it was all important that my command should be *first* or the whole movement might result in failure, I immediateley rode to Gen. Hancocks Head Quarters and reported the state of the case to Col. Morgan, Chief of Staff, who at once reported to General Hancock.[5]

Then there was a storm. Col. Morgan returned and ordered me to *take the road* no matter what the consequences and should *anything* obstruct the way to "*tip it off*," as they had no business there. He sent a Staff Officer with me and my train was soon in the road. Such pulling and hauling as followed beggars description. Taking a company of men to the front, we made for the right of way. Everything, no matter what, Artillerey, baggage, or forage wagons, that obstructed us, was promptly dumped into the ditch on either side of the roadway.

My orders were to reach Maddens at midnight. All my Officers and men knew this and hence they worked with the determination of men in desperate circumstances. Occasionally an Officer of Artillerey would protest against our proceedings and threaten all sorts of vengeance, but the presence of the Staff Officer with a few vigorous remarks had a very quieting effect and we went on our course as usual. I think we must have turned over at least twenty wagons and not a few Artillerey pieces and caisons on our march across the causway.[6]

At last we reached the open road and were glad enough to move on our way unobstructed. The Artillerey Officers learned a lesson by that nights experience which was not soon forgotten. After this we crossed a road moving at right angles where I met General Barlow at the head of the First Division and passed the intersection of the roads just in time to prevent a confusion with his Troops.

My command was put to a double quick and soon the lights of General Brooks Brigade appeared in view. Preciseley at 12 O'clock midnight, I stepped into General Brooks tent and soon his Brigade was on the march ahead of us and following my command came the First Division.

Very few events of my military experience have afforded me so much pleasure as my prompt arrival at Maddens on the time designated, for now I comprehended the importance of being on time. Had any mishap or disaster occured in consequence of my command being behind time, I alone would have been blamed. And in those days no excuse would have been accepted for a blunder which might retard or disarrange the movement.

We were soon on the march, Brooks Brigade first then my command. We trudged through the darkness in silence. Nothing could be seen but a few men directley in front and nothing heard but the rattling of their canteens. Our march was rapid with very few and brief haltings. Finally, as day light appeared, the sight of the long line of muskets filling the road away to the front and rear inspired us all with a feeling of confidence.

At 6 O'clock we reached Elys Ford and without stopping for refreshment, commenced laying a bridge across the Rapidan. The other Troops meanwhile cooked and drank their coffee. Spaulding with his Flying Trains was on the ground before me. He was escorted by a Brigade of Cavalry and had built a bridge of canvass boats and crossed the transportation of the cavalry while the men dashed into the stream and swam their horses across.

A thin line of the enemys Cavalry pickets occupied the opposite bank but, seeing the overwhelming force coming upon them, they fired a few rounds and ran. A few of our Cavalry were wounded but none killed. On arriving at the opposite bank our Cavalry commenced the advance and Spaulding speedily took up his bridge and followed. My arrival occured just as he was taking up his bridge.

At 7 O'clock my bridge was finished and the Troops commenced pouring across. All day long, in close order and at a rapid rate, the columns moved over. At about dark the whole Corps of 40,000 men had crossed into the Wilderness and were rapidly brought into line.[7]

Lee meanwhile, discovering the point of our attack, was hurrying forward his Army from Orange Court House to meet us and dispute the passage. So sudden had been our movement and so well were all the plans carried out, that a large majority of our Army had accomplished the crossing before his advance struck us. For while the II Corps was crossing at Elys Ford, the V and VI Corps were also crossing at other points.[8]

Soon after dark and after the Infantry had crossed came the Artillery and the transportation, ambulances, amunition, baggage, and forage wagons [and] six hundred six mule teams belonging to the II Corps. In one continuous stream they crossed all night and the next day until 2 p.m. Not an accident or delay occured, not a man or an animal injured, until the last man and the last wagon had passed over.

Early in the morning the rattling of musketry announced that our advance had met the advance of the enemy and that the battle had commenced, a battle that proved to be one of the most awful and desperate of any known to history. According to orders received in the morning to take up my bridge as soon as everything had crossed to the South bank, at 2:15 p.m. I gave the order and in 56 minuites by my watch the whole of the bridge equipage was on the wagons and moving into the Wilderness.

Arriving at an opening I halted and for the first time in sixty hours my men sat down to cook their coffee unmolested. The day was hot and dusty and the men well tired out from loss of rest and constant work for three days and two nights. My orders, received from Major Duane, Chief Engineer, were to follow the trains of the II Corps to Todds Tavern but Todds Tavern was not in our possesion and was destined not to be. While we were drinking our coffee, annother order came from Major Duane to park my train with the train of the II Corps at Chancellorsville. We were soon on our way to that place, the fight within a mile of us going on with desperate energy.

We were now fairly in the Wilderness, a desolate sandy place, overgrown with pine, chiefly second growth, with pleanty of underbrush. Our Army was all across the river and all the bridges were taken up behind us. Like Cortez when he burned his ships, there was no means of retreat and no alternative but to win or perish.[9]

Both Armies were fighting for the possession of the Brock Road which led through the Wilderness and into the open country beyond. Lee came down on the Orange Court House Plank Road and thus had the advantage of us. It was essential to our existance to have the Brock Road and Lee's endeaver was to drive us from that and back into the river. For that reason our Troops were hurried forward with all possible despatch and formed in front of [and parallel to] this road as fast as they arrived, thus extending our line to the left, our right resting on the river to prevent a flank movement upon us. The Wilderness was well supplied with bridle paths but only these two regular roads traversed through it directley.

Our march to the Chancellorsville house was close and tedious. We could not move rapidly as the road was so blocked up with moving Troops and trains. At times the fighting was going on frightfully near us, then it would recede again. We were now once more on the old battle field of Chancellorsville. The lines of breast works thrown up by the combattants on both sides were still visible, some of them being in very good order. We passed over several of these lines on our way. All night long we marched and halted and marched again. It was but a short distance, less than four miles, but the road was so blockaded and the woods to each side so impassible that progress was almost impossible.

When at last we arrived at Chancellorsville, where was found an open space in the forest, it was quite dark. The trains and Artillery of the whole Army were either there or concentrating at that point.

Wagons, mules, horses, and Artillerey were packed so closeley together in the thick darkness that one not accostomed to Army desipline would have thought it confusion worse than confounded. As soon as a halt was ordered and we had found the place where it was supposed we were to pass the night, we threw ourselves upon the ground, hoping to get a little, much needed rest. It was now about 11 O'clock at night.

I had just dropped to the ground and pulled my blanket around me when Spaulding appeared. He had found me in the darkness and came with orders for our whole command, all of the Battallions including his own (as we were all bivouaced there) to move *at once* and report to General Griffin commanding a Division of the V Corps by *daylight* the next morning. To men who had not rested for three consecutive nights this was not pleasing news but we were all on our feet with alacrity and cheerfulness and on our way again. Our trains were left at Chancellorsville under guard and the 50th Regiment N.Y. Engineers was now going into the line to act temporarily as Infantry.

The desperation of our situation was now patent to all, for it was an unusual thing to place the Engineers, on whom so much depended, in such a position. It was evident that on the morrow every available man in the Army would be needed and so it was. For that eventful Friday, the 6th day of May, was the one on which the *back bone* of the Rebellion was broken.

In the cool grey of the morning the firing was again renewed. Soon after sunrise we arrived at the foot of a rising ground, over and beyond which was our line face to face with the enemy. We halted in the valley for a few moments rest and coffee then our whole Regiment fell into line and *loaded*. The stray bullets of the enemy struck the ground in front and whizzed over us.[10]

Then followed a remarkable scene the like of which I had often read about but never before witnessed. Brave men with not a drop of cowardly blood in their veins, men who had witnessed many a scene of blood and carnage and knew not what it was to disobey an order or fail in the slightest duty in presence of an enemy, dropped out of the ranks by twos, by threes, by the score, and stepped to the rear to attend to the demands of nature. No one laughed, no one blamed them, for all felt that peculiar feeling experienced only in the presence of *death*. It was a striking illustration of that strange peculiarity of mans nature which submits under *apprehension* of danger, when the actual reality, met face to face, serves only to stimulate and embolden him.

It was all over in a few moments however and our Regiment marched boldly up the hill and over into our position in line of battle, our right resting on the road [Orange Turnpike], our left joined to the Troops of the V Corps. We soon had a breastwork of logs in front of us and were as anxious as any for an assault upon our position. In our immediate front was General Crawfords Brigade of Griffins Division of the V Corps. We were brought in to form a second line at this important and exposed point as it was of vital importance to hold the road.[11]

With the rising of the sun the battle increased in earnestness and fury, first to the right of us, then to the left of us, then on our front. They charged and charged again only to be met and repulsed. Every foot of our position extending for more than six miles was tried as by fire and found impregnable. Sometimes they would charge all along the line, then withdrawing, make a feint with a brigade or two at one place in order to deceive us while a whole division or perhaps two would mass and pounce upon us in annother. Sometimes a temporary advantage would be gained by them when our Troops would be massed, charge and drive them back again to be in turn repulsed and driven back by the enemy.

All day long we could hear that peculiar "*Rebel Yell*" which they gave when they charged, sometimes on our right or front and sometimes dying away far to the left of us. The sun came up hot and the woods, which were now dry, were fired in a thousand places. The smoke from the clouds of powder and the denser clouds caused by the burning woods became stifling, suffocating, blinding. Two hundred thousand men, inspired with the desperation of demons, were fighting in a wilderness of fire.

Every moment some souls were leaving that atmosphere of hell, their bodies to be consumed by the devouring elements. Hundreds of wounded on both sides, unable to crawl away from the swiftly approaching flames, could only lay and moan and roast and die. To add to the miseries of the battle, no water could be found as no springs or running brooks were in the Wilderness. Many died from thirst, many from excitement and sun-stroke. The Ambulances were all filled and hundreds of wounded and dying were placed in forage and ammunition wagons. The Surgeons could do no more than give a hasty first dressing to the seriously wounded, while those who were able were allowed to walk or crawl to the rear.

I have read many descriptions of the Battle of the Wilderness, but all fail to give any conception of the awful scene. I said there was no water in the Wilderness. There was one little stream, almost dry, called Piney Branch Creek, but it was of no use except to the Troops who were in position on each side of it, though muddy water was carried from it to considerable distances.

During the day, while there was a comparative lull in front of us, I visited the Brigade of Crawford. The 146th N.York men in this brigade had suffered severely. Col. Jenkins was either killed or taken prisoner the night previous, they knew not which. At any rate he was not heard of afterwards for many months, when the evidence of his men established the fact of his death but his body was never recovered or recognized owing to the nature of the ground.[12]

All the long day the fight raged without any seeming advantage to either side. Towards sunset there was an ominous lull on our front. Our men were allowed to stack their Arms and cook some coffee. I was sitting on the ground sipping a cup full, my horse tied to a tree near by, when suddenly a most terrific yell startled us. It was the old Rebel Yell. A sharp fusilade of musketry followed and we all jumped for our Arms. At the same instant, General Griffin came tearing up to where I stood, his horse in a foam and he in such a state of excitement as to appear beside himself. He shouted so that all the men could hear, "Major!! fall in! fall in!! move to the right! double quick!! The whole of the VI Corps has broken and the Rebels are comeing right over us!"

The excited manner of the General, the yelling of the Rebels and the sharp rattling of musketry started every man to his feet and a general stampede for the stack of Arms. In the hurry to get their Arms the stacks were knocked over and many of the pieces went off. Bulletts whistled around and among us in a liveley manner, coffee pots were tipped over and dire confusion reigned for a few moments. All on account of the excitment of General Griffin.

I shouted the order to "Fall In" and rushed for my horse, who had partaken of the general excitement and was rearing and plunging in a frantic effort to get away. The bridle strap was drawn so tight that I could not untie it, so taking out my knife I cut the strap but before I could approach him to mount he had darted back into the woods. There was no time to loose and I ran to the head of the line, leaving Heaton to get the horse the best he could.

We started off to the right in a "double quick," crossing the road.

It was the work of but a few moments for our Regiment to occupy the line of breastworks and in a wonderfully short space of time they had entireley recovered from the effects of the panic. We closed up the ranks and the men attended to orders as coolly as if on parade. A painful sight met our eyes.

Sure enough, there was the VI Corps or a Division of it coming pell mell back towards our line, a broken, disorganized mass of men. On they came, out of the woods like so many bees, across the open space in front and over our breastworks. It was not many minuites before the whole body of the demoralized Division had passed over us and their Officers were making frantic efforts to rally them. Some of the Officers seized the flags, appealed to the men, called on them by name to rally for the good name of their Regiment. Some cursed and others pleaded with them. I saw some Officers beseeching their men with tears in their eyes for "Gods Sake" not to abandon their colors. Finally seeing a strong line of fresh, undaunted Troops in front of them, they began once more to collect in squads, then in companies. It was not long before comparative order and desipline was again restored.

The grey backed enemy, who had caused all this consternation by their sudden onslaught and terrific "Yell," made their appearance at the edge of the clearing where we gave them one good round of musketry and the force of their onslaught was broken. Observing our strong line prepared to meet them, they did not care to come farther and, as it was now getting quite dark, they disappeared again under the black shadows of the woods and night soon folded its mantle over the two contending hosts.[13]

Chapter Twenty

The Wilderness, Fredricksburgh and Spottsylvania May 1864

The last chapter closed with a description of the rout of the old VI Corps, or one Division of it, on that ever memorable Friday, the 6th of May, 1864. Had the assailants who make that last charge known how near they were to victory or how little remained between them and General Grants Head Quarters, they might not have halted, contented with their partial success. For as it happened, our Regiment was all that remained between them and the Commanding General, who, with General Meade, was sitting on the ground leaning against a tree less than half a mile to our rear.

The Commanding Generals of our Army, surrounded by their Staff and all of Head Quarters transportation, were sitting by this tree in consultation when an Officer rode up to them in great consternation and in an excited manner informed them that a Division of the VI Corps had been broken and were now falling back in a complete rout, that the enemy were rapidly following them and in a few moments more would occupy that very ground. General Grant quietley removed his cigar from his mouth and replied that he did not believe a word of it. The two Generals then went on with their consultation as though nothing had happened. This memorable reply by General Grant was taken by the Army as an index of his true character.[1]

All night we remained behind our entrenchments under arms. We built no fires, for that would betray our position, and we did not know but they might at any moment renew the assault, notwithstanding the darkness, made denser by clouds of thick smoke settling over us like a dense fog. During all that night the troops were kept in constant motion. Weak points were strengthened, ammunition brought to the front and such of the wounded as could be secured were carried to the rear. The night dragged slowly along. The day had been a dreadful one and all felt anxious for the morrow.

When daylight [May 7] at last appeared we expected and were prepared for the customary daylight assault but they came not. At about 8 O'clock we were relieved from our position by fresher Troops and ordered off to the left towards a little creek for the purpose of constructing a number of small bridges across it.[2]

We were passing the spot which we had held the day before when suddenly and without warning the old Rebel Yell again filled the air. Our men came pouring back upon us in clouds followed closeley by the enemy. This time it was not a rout but a preconcerted movement, for almost instantly the men had come back over our log breast works, the artillery, which had been concealed, opened with dreadful effect upon the on comeing Rebels. Cannister and schrapnell at short range soon made fearful havoc among them. It was short work, short but decisive. They saw the trap when it was too late and, before they could again place themselves under cover of the woods, the open space which a few moments before was filled with their advancing Troops flushed with victory, was now covered with their dead and dying men, a bleeding, crawling mass of men too horrible to look upon.[3]

That was the last onslaught they made upon us that day. The firing was kept up continually but neither side made any very extended demonstration while both were occupied in manouvering for position, burying the dead and caring for the wounded. During the day our Regiment built more than twenty little bridges across Wilderness Run so that our Troops could move from one end of the line to the other with ease and rapidity.

General Burnside made his appearance among the Troops looking bright and cheerful. The Army of the Potomac had not seen him since Fredricksburg, as he had been off on distant fields of duty. As he rode along the lines he was everywhere received with cheers and words of welcome by the soldiers. Notwithsanding his misforture, he was still beloved by the old soldiers of the Army of the Potomac, beloved for his manliness and generosity. The applause of the soldiers seemed to please him much, as well it might, for no other of the defeated Generals were ever received with such manifestations of pleasure and good will.[4]

While operations on our immediate front seemed to quiet down after the tremendous assault and disastrous repulse of the morning, the fight still continued throughout the day on the centre and off towards the left. All movements were made in anticipation of further attacks upon our front and as we now had considerable artillerey in position, an attack would have been welcomed. However, as I have said, the day passed without any important demonstrations from either side.

My command, together with the ballance of our Regiment and a portion of the 15th, were again in reserve. The lull in operations gave me an opportunity to look about a little, an opportunity which was improved by riding over the different portions of the line, care being taken not to go away too far from the main command in case we might be suddenly needed.

I was much interested and impressed on approaching a common looking farm house situated in an opening in the woods, to learn that Stonewall Jackson was burried near by. His grave was situated in the heart of the Wilderness on a knoll, unmarked by stone or board. It was hard to realize, as I stood beside that loneley grave, that the little mound of earth before me hid from view all that was mortal of the man whose great deeds had filled the world with wonder and amazement. Twas he who came so near annihilating our Army less than two years before within a few miles of this very spot and whose sudden and mysterious death alone prevented the consumation of the project. I lingered for a long time at the grave of that wonderful and eccentric man. Nor could I leave the spot without having experienced those peculiar feelings of awe and respect for the memory of the genius which, though that of an enemy, possessed the faculty which inspired his Soldiers with a religious enthusiasm, resulting in most wonderful victories and made his name a terror to ourselves.[5]

Soon after returning to my command, orders were received to return to Chancellorsville House where we had left our trains parked a little more than two days before. We were soon on the march but on arriving at the road we found it litterally filled with Troops, Artillerey and wagons. Being the only available road in our possesion, all the impedimenta of the Army was obliged to move upon it. Fresh troops moved in one direction while Regiments thinned, worn out and bleeding moved in annother. Artillerey was going and comeing. Ambulances were filled to overflowing with the wounded and dying, some groaning, some too far gone to utter more than a faint cry for water, many a one whose eyes were glazed in death and all bandaged, bleeding, mutilated—it was a pittiful sight.

The Ambulances could not contain them all. Scores of Army

wagons were brought into requisition for the purpose. As fast as the forage and amunition wagons were emptied of their supplies they were at once filled with the bleeding men. Darkness set in upon the seething, sweltering mass of humanity. Thick dust arose in clouds enveloping the whole as with a heavy mantle. Through this mass of dust and humanity we slowly trudged along. Bye and bye the moon arose, its rays cast a weird light upon the moving host, who appeared like so many spectral forms flitting across and through the grimmy cloud of dust and smoke.[6]

At about midnight we arrived at the Chancellorsville House. An opening in the woods surrounded the deserted mansion and into this opening we marched and *packed* ourselves. At the command "Halt. Rest." all dropped down in their places in the deep sand and were soon fast asleep. My saddle was my pillow, old Grey lay down at once like a man and was soon enjoying his rest. For my part I was too tired and my mind was too full of the events of the preceeding days to go off at once into slumber but gradually I lost conciousness, though for a long time I realized that men and horses were passing within a few feet of my head.

At 3 O'clock [May 8th] we were aroused and ordered to move on. In a minuite we were all on our feet but my *revolver* was gone. Some vandal had taken it from its sheath while I lay in an unconcious state. Our trains had been ordered away during our absence at the front and were now a few miles to the East of us on the road towards Fredricksburgh. The sun came up scorching hot and with it the fighting was resumed all along the line. We made our way slowly through the masses of impedimenta and at 3 O'clock in the afternoon came up with our trains, which we found irregularly parked in a little field near the river Ni, a small stream one of four named respectiveley: Mat, Ta, Po and Ni, which formed a junction a few miles beyond and made the stream known as the Matapony.[7]

This little spot was on the edge of the Wilderness and we were refreshed beyond all expression at the sight of green grass and running rivulet. It was a very Oasis in the desert. The animals as well as the men were nearly crazed with delight on beholding the fresh water. A house near by was occupied by an old man who boasted of having two sons in the Rebel Army. We took pity on his weakness and spared him for mercys sake. Here in this delightful spot we were allowed to remain all the ballance of the day and all the night while the fierce battle was raging all about us. How sweetley we slept that night, the first for many nights. No downy, feather bed ever felt half so restful as did our blankets on the dry ground that night.

The next morning [May 9th] we awoke stiff and sore but wonderfully refreshed by sleep. At 6 O'clock we were again in motion, with orders to move to Aldrichs House. The sky was of brass as it had been all the days before. The fight was renewed with the usual desperation that characterized this most wonderful of battles. Early this morning brave old General Sedgwick, Commander of the VI Corps was killed. The whole army loved him and deeply lamented his loss. *Wadsworth*, *Hays*, and now *Sedgwick* had offered up their lives thus far in the desperate conflict.[8]

Our losses were terrible and still the numbers of wounded increased and as yet there was no prospect of a termination of the battle, no means of getting our wounded away and all communication with the North cut off and no base as yet secured. It was a struggle for life or death, victory or annihilation.

Arriving at Aldrichs House, we found fresh masses of wounded. Here we remained that day [May 9] as there was no use for us as yet. We were allowed to stand under Arms as a reserve with the men allowed to stand for orders while the Officers did what we could to alleviate the distress of the wounded. I will not attempt a description of what I saw that day. It would be but a repetition of the story of agony and death. We gladly parted with whatever we had that would administer to their comfort and remained at this duty far into the night, when we lay down again, more heart sick than ever before of the dreadful scenes we had witnessed.

The battle was now working from the right towards the left. Each night the Corps on the right was withdrawn and moved to the rear and brought again to the front on the left. This was done for the purpose of flanking the enemy and necessitated a like counter movement on his part. Grant was always on the offensive, which was his only salvation for had he once ceased his offensive and assumed a defensive attitude his Army would have been lost.[9]

The weather came off excessiveley warm and our Soldiers, who were abundantly supplied with clothing, commenced an indescriminate casting away. For miles around and through the Wilderness, the ground was carpeted with overcoats, drawers, shirts and blankets until every man was reduced to summer fighting trim. All the inspections we had before the campaign commenced failed to rid our Soldiers of their superflous baggage and now they threw it away of their own accord. Even of their knapsacks they stripped themselves and not one in a thousand could now be seen with more than a rubber and woolen blanket folded into a roll, the ends tied together and slung over the shoulder. All they now wanted was their musket, cartridge box, haversack, canteen & blanket. Subsequentley the Richmond papers boasted that they had picked up enough overcoats, blankets, knapsacks and under clothing to supply their Army during winter.

Early the next morning [May 10th] we received orders to leave Aldrichs but the order was soon countermanded. The day was hot and the battle was resumed as usual. Towards noon we were notified to be in readiness to move at a moments notice and my men and Train were all in line, ready and anxious to be on the move. Our left had now reached Spottsylvania in the open country but the enemy were already there and occupied the high ground which commanded the whole open country. It was necessary to have possesion of this point in order to successfully flank the enemy and around this high and commanding ground the battle raged for many days. The Battle of Spottsylvania was the most desperate and bloody of all the succession of bloody battles of the campaign.

At 11 O'clock, May 10th, I received an order communicated to me verbally to proceed to Fredricksburgh at once and build a bridge across the Rappahannock at that place. Fredricksburgh was *eight miles* away. The sun was hot and the road *dusty*. In less time than I can write it the animals were put on a trot. The men threw their Arms and accoutrements onto the wagons and ran alongside. For 8 miles we scarceley stopped for breath and as we approached the hights back of the City overlooking the Town, River and country, so familiar to me two years before, I could not restrain my enthusiasm and involuntarily shouted with joy. The shout was taken up by the men and we passed rapidly through the line of works built by the enemy and from which they had successfully repulsed our assaults on the dreadful 11th, 12th & 13th of December, 1862.

We went down through the town and to the river without stopping, to the very spot where I was wounded in attempting to build a bridge from the other direction. Rapidly we unloaded our bridge material, the men springing to the work with a will. At 4 O'clock p.m. the bridge was finished and ambulances filled with wounded men commenced crossing to the North side of the stream, thence to Acquia Creek, and by steamer to Washington, where willing hands and loving hearts were anxiously awaiting to administer to their comfort and recovery. Thus in less than five hours my command had marched 8 miles and constructed a bridge 420 ft. long, a feat the like of which can not be found recorded in the annals of our war and I very much doubt if ever it was equaled by any Engineer Troops in any Army whatever.[10]

If it was a relief to us to see the wounded at last on the way to comfort and attention in the hospitals at the capitol how must they have felt? Poor fellows, some of whom had been drawn around in the ambulances and wagons for more than a *week* through the hot sun and dusty roads, their wounds bleeding and festering, their tongues parched and altogether suffering indescribable agony.

All that night and the following day the ambulances and wagons filled with wounded filed across the bridge and we were all heartily glad when the last of the melancholly train had passed over and out of sight on their way towards the Potomac. Many hundreds of those too seriously injured to stand the removal by wagons were provided for in the houses of the city, which were taken possesion of by our forces and transformed into hospitals.[11]

You must remember that we were now seperated from the main Army by some ten or twelve miles, they having swung around to the left and in front of Spottsylvania Court House where the fight was still in progress for the possession of that high and commanding position. Our isolated and exposed situation rendered us liable to be *raided* by the enemys cavalry and caused considerable apprehension for our safety at Head Quarters and orders were sent me by a staff officer to use every precaution against surprise. Accordingly, as soon as my bridge was finished and a sufficent Guard left for its proper attention, I distributed a strong Picket Guard on every road leading to my bivouac and the remainder of the command was kept under Arms during the night.

Just at dark I rode over into the city and found the people in a high state of excitement over rumors which appeared to come from authentic sources that a large body of Rebel Cavalry was advancing and would pounce upon the City during the night with the object of capturing our supply trains which had now begun to come into the Town. As the large number of wounded and their attendants were already there, my anxiety was naturally very great as I knew very well that the destruction of the Ponton bridge would be their first object. Consequentley there was no sleep for that night.

Every preparation was made for a vigorius defence, including a little earth fort which we threw up and which we determined to defend to the last or until reinforcements could be sent me. Anxiously awaiting the expected assault, the night wore slowly away and morning dawned but no enemy appeared. Subsequentley we learned that a large force did approach within a few miles of the city but, either through some misunderstanding of the Officers or learning of the preparations made to receive them and finding that the movement was deprived of its chief requisite, surprise, the enterprise was abandoned.[12]

The next day I established my camp and made things as comfortable as possible. Then we had a delightfully refreshing rain, the first we had had since the opening of the campaign. The grass came out beautifully green and was enjoyed by both men and animals. The transformation from the dingy, dusty, smokey, horrible Wilderness to the bright green fields & cool atmosphere, laden with the fragrance of spring flowers, with the rolling country around and behind us and the broad, clear, Rappahannock river on our front, filled us with delightful emotions that cannot be expressed.

The next day, May 12th, Miss Dix, with a large corps of nurses crossed our bridge into the Town for the purpose of nursing the wounded men. Miss Dix was a kind of Florence Nightengale in our Army, devoting her labor, time, and fortune to the care of our sick and wounded Soldiers.[13]

Having leisure, I went over into the City to visit the Hospitals, all the Churches and very many of the largest and best residencies that had been appropriated and were filled with the wounded of both Armies. Friend and foe alike received all the care and attention possible. Men dressed in blue and grey lay beside each other, no longer enemies now, but friends and brothers in sympathy and suffering.

As I passed among them with a pleasant word for such as could converse, I noticed that our men were as a rule cheerful and hopeful, while the most of the Enemy were gloomy and morose. Generally they accepted our attentions with as good grace as could be expected.

All this while the battle around Spottsylvania was raging bitterley. The roar of the battle was distinctley heard in Fredricksburgh and fresh relays of wounded were constantly arriving. On this day General

Hancocks Corps made their famous assault upon the Rebel lines and captured *eight thousand* prisoners. It was the first really decided advantage we had gained during all the long fighting.[14]

We now had showers nearly every day. Early on the morning of the 13th we saw a long, dark mass of men filing over the hills to the Southward and approaching the city. On and on they came, in a seemingly unending line. At first we knew not what to make of it but presentley the news flew about that the prisoners taken the day before by the II Corps were now approaching under a strong escort of armed men. Bye and bye the head of the long column passed through the Town and commenced crossing over the bridge. It took several hours for them all to cross over.

As they passed near us I could distinctly observe every man. Among the prisoners were three General Officers who were allowed to ride in an ambulance, the remainder walked.[15] Some regiments were well dressed, some indifferentley well and some in rags and tatters, hatless and shoeless with knees and elbows exposed to view, over all a dirty, grey, dingy color. The majority presented a very unkempt condition. It was a grand sight to see so many of the enemy in our hands who but the day before had been fighting us. Finally they had all crossed and passed away to the North to await exchange.[16]

Then came an order from General Head Quarters for my whole command to report to Spaulding at Salem Church, this order was first sent to General Benham who, with his Head Quarters, was quietly resticating at Belle Plaine on the Potomac.

Meanwhile, the covering chess of the bridge was so nearly worn through by the incessant passing and repassing of the numerous supply trains carrying up fresh supplies of food, forage and amunition, as to become dangerous. As I had no more chess and must of necessity cover the bridge with *something*, I proceeded to demolish an old depot building near by, the material from which served as an ample covering.

I broke camp at 4 p.m. and marched with nearly all my command (except a small guard) to Salem Church, arriving there at dark. I found Spaulding with his detachment there and slept in an Army wagon with him. We all took a good soaking during the night. The next day, the 15th, a report was circulated that a large force of the enemy was moving on our flank with the object of cutting off our supply trains by getting between the main body of our Army and Fredricksburgh. Troops were sent to counteract the movement and all the trains were ordered back into the City. Back they went "helter-skelter," ourselves being among the number, arriving at my old camp ground that we had left the day before during the afternoon.[17]

The next day [May 16] a detatchment of the 15th Regiment came up from Belle Plaine to relieve my command of the charge of the bridge and I was ordered to once more rejoin my command, the old II Corps. Accordingly turning over the bridge to the Officer of the 15th, I moved my camp across the river and halted on the plain beyond the Town. Here I was met by an Officer who ordered me to remain where I was until further notice as General Hancocks Corps was in motion and his future position was uncertain.

Our Battallions, all except Spauldings, were now altogether at this place. Spaulding himself was with me temporarily. Owing to the fatigues and extra exertions of the men since the opening of the campaign, quite a number were on the sick list and for the first time since leaving Harrisons Landing the camp diarrea seized me also. For several days I was as weak as a kitten.

Reinforcements were continually arriving, consisting chiefly of the Heavy Artillerey Regiments which had for so long a time been posted in and around the defences of Washington. They were all fine looking men, well disiplined and neatly dressed, their bright Arms and accoutrements and clean uniforms contrasted severeley with the appearance of our men who had been through the dust and smoke of the Wilderness. These fresh Troops were desperateley anxious to get to the front, as they had never seen much fighting but, poor fellows, they soon saw enough of it and some of their Regiments were fearfully decimated soon after.[18]

Soon after the arrival of the 15th, annother bridge was constructed so that we now had two across the Rappahannock and the time was occupied in rearranging our entire transportation for all the trains.

On the 18th I rode up to General Hancocks Head Quarters, about 10 miles, and reported to Colonel Walker, Assistant Adjutant General. As I rode up to Head Quarters General Hancock had just ordered an assault by his entire Corps upon the key to the enemys position on the hights of Spottsylvania. This gave me annother opportunity of witnessing an assault without being immediateley exposed, although the shell from their batteries came frequentley unpleasantly near my head.

The Troops moved forward in fine style on a double quick. On they went over the rising ground up towards the long line of yellow fresh earth on the summit. Our batteries in their rear played vigorously over their heads in order to keep the enemy down. On and up went our brave men until, within a hundred yards of the line of earth works, there came rolling out and down the hill a thick white cloud of smoke which was followed by the familliar rattle of immence volleys of musketry and soon the whole hill side was completeley enveloped in the grey whitish cloud of smoke. The cheers of the men from both sides and the incessant rattle of the musketry rendered the scene most intenseley exciting. I could but hold my breath and pray for their success.

For a few minuites all was observed in mysterey. Moments seemed like hours. Then the cheering ceased and dark masses of our men were seen through the openings in the uprising smoke returning as they went but with awfully suggestive gaps in their ranks. The assault had failed. Soon the smoke cleared away and disclosed the ground for a long distance thickly strewn with our dead and dying men. It was an awfully grand spectacle, one often repeated around that ground which has been justly styled "Bloody Spottsylvania," whereon our Army lost more men in killed and wounded than on any other field during the campaign before our arrival at Petersburgh.[19]

Turning from this scene I returned to the hospitals which were thickly strewn a short distance to the rear and where our poor men were receiving their first dressing. Here were sights which might well make an angel weep to behold. Scores of men came limping and crawling in by themselves, scores of others were brought in on litters and laid down while the surgeons with sleeves rolled up and knife in hand and cloths bespattered with blood were *operating* upon wars victims who were stretched out upon the numerous tables.

Moaning in agony, the man is laid upon a table, one administers chloroform and soon the moans ease. The limb is made bare. With quick and accurate cuts the knife has performed its duty. Then the dull rasping of the saw for a few seconds only and the arm or leg is laid down on one side among others and the patient is removed to make room for annother. Skillful hands have caught up the arteries and veins and sewed up the gash but the poor victim must drag out many months of convalescense and hobble through life a maimed and ruined man.

For several hours I witnessed these scenes, doing what little I could to alleviate the suffering. Occasionally a friend or an acquaintance was brought in to be operated upon and, passing around among the dead, I recognised more than one, now cold in death, whose name, hand and cheerful voice I had often met in the communion of friendship. With a heart more like lead than of flesh and blood, I turned away from the scene at last and rode sorrowfully back to my camp, darkness having spread its thick mantle over the earth, covering all its sins, its sorrows and anguish long before my arrival. Glad indeed was I to find my Johnny awaiting me with his ever present cheerfulness and a cup of warm coffee for my refreshment. Glad was I to greet my comrades still living, for it did seem that day as though all my acquaintances were destined to be killed.

The next day I rode to Belle Plain & back, a ride of 26 miles and the day following, the 20th, an order conveyed through Spaulding was received directing me, as I was senior Officer in command, to have all the Battallions rejoin their several Corps as soon as possible.

CHAPTER TWENTY-ONE

Bowling Greene, the Pamunkey, North Anna and Cold Harbor May & June 1864

On receipt of the order to rejoin the II Corps, I immediateley commenced dismantling and loading the upper bridge. This duty occupied the entire night but, as the moon was up, we had no difficulty. We left the City at about 3 a.m. on the morning of the 21st and by 7 O'clock we had joined the II Corps, which was then on the move, as was all the Army, executing annother flank movement. Our progress was not very rapid, as it was necessary to keep well closed up in order to be ready for an attack should one be threatened during the movement. At dark we bivouacked six miles north of Bowling Greene. The skirmishers of the two Armies carryed on a desultory firing all the while.

In conversation with Col. Walker that evening, he informed me that he had not taken off his clothing for 17 days and nights and he was not an exception to any of Gen. Hancocks Staff. It was here that we received reports of the defeat of Generals Butler and Siegel on distant fields.[1]

During my absence at Fredricksburgh, Captain McGrath had been left with a small detatchment of my men at Head Quarters and he succeeded in getting things in a most beautiful muddle at the Quartermasters and could get no rations for his men, who were all glad enough to have me return and, as they said, "straighten things."

The next morning bright and early we were again on the march. The firing was also resumed along the line to the right and rear of us. We passed through Bowling Greene, a Town of considerable size before noon. General Hancocks Head Quarters had meanwhile gone a long distance in advance.

In company with Col. Batchelder, the Quartermaster of the Corps,

I rode on ahead of the command and entered the Town, which was really a beautiful place.[2] We stopped at a house where we saw signs of life and asked the good Lady if she would accomodate us with a dinner. She consented and went about getting it. Meanwhile her daughter, a buxom young Lady of perhaps twenty five summers, entered the room and for half an hour entertained us with choice Southern expressions and ideas.

According to her notion, the *Yanks* were little better than barbarians and cannibals. She abused us in round terms and to her hearts content while we sat quietley smoking our pipes out on the piazza. When her Mother announced dinner ready, she sat at the table with us and before we had completed our meal she became not only reconciled to our presence but actually made herself quite agreeable. We paid for our dinners and left the house on good terms with both mother and daughter.

We had not gone far from the Town when an Officer met us on his way back to the villiage with orders from General Hancock forbidding any Officers or men from entering the houses of the inhabitants under penalty of severe punishment. Col. Batchelder and myself had made a fortunate escape and enjoyed our little adventure wonderfully.

We overtook Head Quarters in the afternoon just as General Hancock had succeeded in gaining the position that he coveted and which, he said in great glee, he could hold against the whole Southern Confederacy and for which he had made his forced march. We were now about 40 miles from and on the direct road to Richmond. General Hancock had beaten General Lee in the race for the desirable position which we now held and had already strongly fortified. In less than two hours after their arrival our men had thrown up substantial and formidable earth works of the loose sand with their tin cups and plates alone, without the aid of shovels. Both sides came to attach much importance to the formerley much dispised earth work or rifle pits.

General Lees advance felt of our position and finding us too strongly posted to hazard an attack and comprehending the object of our movement being to get on his flank and rear wiseley determined upon and preceeded quickly to put into execution his next best plan. Making a show of resistance on our front, he rapidly transferred the main body of his Army across and secureley entrenched himself on the south side of the North Anna River.

The force left in our front did not long deceive our Generals. Lee had at last found his equal in strategy. Long before daylight the Troops of General Hancocks Corps were on their rapid march to the North Anna. My command left our bivouac at 7 O'clock on the morning of the 23rd, this time *following* the Artillerey. We were 12 miles from the North Anna. Early in the afternoon the head of our columns reached the river and found the enemy in possession of two quite extensive earth works on the north bank of the stream. They had fired and destroyed the Rail-Road bridge across the river and were entrenching themselves on the other side.

General Birneys Division was ordered to assault and take the works on the north side of the river. That General speedily disposed of his Division for the fray and gave the command to carry the works. The order was obeyed with alacrity and was completeley successful. Several hundred prisoners were taken with but slight loss on our side. General Hancock said it was the handsomest assault he had ever seen. When I arrived upon the ground it was all over and our men in possesion of the works. The Rebels had a batterey in a bend of the river on the other side which enfiladed the position and made it very uncomfortable to show ones head outside the works.

We bivouaced in a field near by and were soon fast asleep, tired and weary from our 12 mile march and excitement. Early next morning we were up and doing and we constructed two bridges across the stream, strange to say, without opposition from the enemy, except from the Batterey above mentioned which threw shells at us all the day. Captain McDonald built one bridge above my two, all of which were below the Rail Road bridge which continued burning.[3]

The enemys line was nearly a mile from the river at the point opposite the bridges and our forces crossed and confronted them in their line of breastworks. Why General Lee did not take up his position on the bank of the stream and dispute the passage is a mystery which I am unable to explain.[4]

Having completed my bridges, I was sent for by General Hancock and by him introduced to General Burnside. The latter recognized me as the Engineer Officer who did him some service at Fredrecksburgh in 1862 and remarked that in his opinion the position of the Engineer Troops on the morning of Dec 11th was the most desperate of anything recorded in history. He asked General Hancock if he would allow me with my command to come up to his position, which was several miles to the right, and build a crib bridge for him. General Hancock consented to let me go that night providing he would let us return in the morning. A Staff Officer of Burnsides was to accompany me to show the position. Accordingly I sent for two companies of my Battallion to take their coffee and be in readiness to move on my return or to follow up the bank of the stream as soon as ready, provided I had not returned.

I then set out with the Staff Officer. Just then a heavy thunder shower came up and wet us all through to the skin. The spot where General Burnside wanted a bridge was in a deep gorge of the stream, with rocky banks on each side rising almost perpendicularly to a hight of about 100 ft. To build a ponton bridge here was impracticable. The current of the stream at this place was very rapid with rocks pleantifully distributed, showing their heads and points above the surface.

A portion of General Burnsides Troops occupied the South bank but were in momentary fear of being attacked and driven back unless larger masses of men could be brought over. The passage by wading was of slow progress as the rapid current was breast deep and but few men could keep their feet to get across. Very many in attempting it were carried down the stream and with difficulty landed on either shore while some were drowned in the attempt. The shower of the afternoon had considerably raised the stream and darkness was comeing on.

After the hard days labor we had accomplished, the prospect of a nights work in water up to the armpits was not pleasant to contemplate, but as soon as the men arrived they went at it with a will. Trees were chopped down, cut to the proper length and notched, some cutting, some rolling them to the water, while others in the water placed them in position and others found stone and filled the cribs to sink and hold them in their places. The cribs were 6 ft. wide by 10 ft. long placed lengthwise with the stream and 14 ft. apart. On and across these cribs were placed heavy logs and they in turn covered crosswise with smaller ones thus forming a corduroy road, high and dry above the water.

All night long we worked thus, with the rain pouring and thunder roaring, the gorge black with darkness. Occasional flashes of lightning revealed the men hard at work in the rolling torrent, the noise of which drowned the voices of the Officers and made altogether one of the most wierd-like scenes that can be imagined.

When morning dawned and the clouds rolled away we had a substantial structure completed and General Burnside was enabled to cross his Corps to the other side, dry shod and without difficulty. In the morning I reported to General Burnside that his bridge was finished and he seemed so well pleased and grateful that he returned his thanks with many expressions of kindness, inviting me to breakfast with him. I sat down under his tent fly, which was all he had for covering and enjoyed a cup of coffee and, as I was wet, a draft from his flask also. Then, taking my leave of him, returned with my men to our bridges.

Returning to our camp we dried ourselves in the sun and built annother bridge above our two former ones for the use of the II Corps. The fight was in progress all along the lines during the day without any marked advantage to either side. At noon I returned to camp and lay down for a few hours sleep. Then annother fierce thunder storm arose, accompanied with a high wind which blew down all our tents, mine included, and gave us annother thorough soaking. This was truly delightful business.

The next day, May 26th, finding that nothing could be gained by remaining here, General Grant determined to withdraw the Army from the south bank of the North Anna and make annother flank movement. This intelligence was communicated to me early in the morning by Col. Morgan as I rode up to Head Quarters, according to custom, to receive orders.[5]

Now this movement was one of great delicacy, requiring the utmost care and caution. Could the enemy once know of our intention and discover of the least weakness in his front, he would of course advance in force and drive the remainder of the Army into the river. Hence it was necessary to keep up a strong show during the day and quietley, silentley, withdraw during the night.

In order to be sure of saving our bridges, General Hancock desired me to build a floating, wooden bridge sufficient to cross the last picket line. Then, after the main body of the Troops with the Artillerey and ammunition wagons had crossed, take up the pontons and load them, leaving the floating bridge of refuse material until the Pickets had crossed, when this could be cut away and allowed to float down the stream. It was expected that our Pickets would be followed to the very bank of the river by the advancing lines of the enemy.

As I was very much fatigued and knew of the duties which would devolve upon me during the comeing night, I decided to leave the construction of this *extra* bridge to *Captain McGrath*, particularly as he was anxious to show me what he could improvise in that line. So, giving some general instructions, I left for my tent for a couple of hours rest.

On my arrival there I found several acquaintances, Officers from different Regiments, who had stopped to see me for a few minuites while their men were crossing the bridges and to buy a few rations for themselves. Now, as it happened, I was well supplied with rations having an entire wagon drawn by six little white mules for my personal accommodation. This gave me an opportunity to carry a good supply of provisions and always a barrel of whiskey for the use of my men when they underwent unusual exposure in the water, as they were often required to do. My friends from other Regiments had had scarceley anything to eat for several days, and of course I sent them away happy, with their haversacks and canteens well filled.

After an absence of two hours I returned to see how Captain McGrath was progressing and I was just in time to prevent a calamity. The Captain had secured two pieces of square timber of length sufficient to reach across the stream from shore to shore. Placing these along the bank about four feet apart, he had covered these timbers with boards, nailed down. As I arrived on the bank he was having one end of this raft held to the shore by several men, while several more were shoving off the other end in order that the force of the current might carry that end to the opposite shore. Here he thought it would remain and form a passage way for foot soldiers. He did not comprehend that the force of the current would inevitably bend the structure, thus shortening the raft. It would get loose and, floating down with the swift current, carry away our ponton bridges, only one hundred feet distant, to sure destruction.

It was too late for me to stop it and for a moment I thought the bridges would sureley be destroyed. Fortunateley there happened to be a piece of rope lying on the bank. Seizing one end of this, I ran out on the raft and secured it then ran back and took a turn with the other end around a tree. By this time the farther end of the raft was swinging across the stream at a fearful rate. The rope straightened as the other end struck the shore. Then the force of the current, which was very great, bent down the middle of the raft, completeley submerging it and drawing the rope out as taunt as a violin string. Smaller and smaller it grew until I expected every moment to see it snap. That was an anxious miniute. To say that I felt like flying does not express my feelings. Had that rope broken then, our bridges would have been swept away and the II Corps left on the opposite bank of the stream with no means of escape. What a terrible, awful calamity that might have been, involving results that might have determined a different result to the whole campaign, perhaps of the whole war. And all through the stupidity of this Officer. But the rope *held* and for a few moments every man who beheld the scene held his breath.[6]

Firmly securing one end of the structure, I ordered the other cast loose and the miserable thing, which came so near involving the loss of the II Corps, swung around harmlessly to the shore. Of course the project was abandoned and I inwardly resolved never again to leave a duty to that subordinate that I could perform myself. I reported to General Hancock that the project was not feasible for want of time and he accordingly made a desposition of his Artillerey to cover the taking up of the bridges and issued the order for the withdrawal of his Corps.

Just after dark the Corps commenced falling back. First came the Artillerey, which was posted on the high ground in rear of us and shotted, ready to give the enemy a warm reception should they attempt to molest us while taking up the bridges. Then came the ammunition and supply wagons and then the Infantry. By a little after midnight all of the II Corps except the strong Picket line had quietley and safeley withdrawn.

Everything was cleared beyond the line of Artillerey accept [*sic*] the wagons to load the bridge material and my command. Then I ordered all the superflous material loaded and every thing about the bridge that could be spared, taken up. Everything was *loosened*, so that the bridge could be taken up in a very short time as soon as the last of our men had come over.

Then came the dilicate duties of the Corps Officer of the Day. Fortunateley the night was dark. The Troops, according to orders, had left their camp fires burning, throwing on an extra quantity of wood just before leaving. Cautiously the Officer of the Day rode to each separate picket post and, whispering to the men, ordered them to run to the river for their lives and run they did, without further invitation, for until then they did not know that the main body had been withdrawn.

Back they came and over the bridge (only one being now left) almost breathless, in squads of two, three or four at a time, until, just as day light appeared, the Corps Officer of the Day came, alone, the last man of the II Corps to cross the bridge.

Then we sprang to our work as if our lives depended upon it. In *twenty seven minuites* by my watch, the last plank and last boat was loaded on the wagons and at preciseley 4 O'clock on the morning of the 27th we passed up the hill and through the batteries of Artillerey which stood, unlimbered and shotted, frowning towards the Rebels, who came up rapidly in persuit of our flying pickets but, seeing the array of solid iron waiting for them, concluded to persue no further and silentley withdrew.

Now annother flank movement was in progress. Whither we were going, no one outside of the commanding Generals Head Quarters knew. Nor did we care much, for now we felt that we had a leader in whom all felt the utmost confidence in his ability and the certainty of his leading us to ultimate victory, though his presence excited no enthusiasm. An hour or two after our withdrawal from the North Anna found us all together again on the road to Chesterfield. That is, all the Battallions of our Regiment.

The weather was enervating in the extreme but we marched on rapidly, passing through Chesterfield, where the entire corps trains and Artillerey were parked in a field under the immediate personal directions of General Hancock, who wanted to show his subordinates how to do it. It was wonderful to see what a small compass he packed them into. The General exhibited his familiarity with all the minor details which are so necessary in the education of a great General. I doubt the war produced a General so thoroughly competent to manage a Corps as successfully as General Hancock. He was and is my "beau ideal" of a perfect Soldier.[7]

At 2 O'clock p.m. we were on our way again. The Troops and trains pouring out of the field and stretching away on the road for several miles. At 2 O'clock in the morning we reached the intersection of the Hanover Court House Road and again bivouaced. Here I enjoyed three hours sleep in my Army wagon. As our supplies were nearly exhausted, we were all put on half rations, which means that we had half as much to eat in order to make our rations hold out twice as long. Our base of supplies, which had been at Fredricksburgh until we left Spottsylvania, was now being transferred to White House on the York river. During the interim we were again without any base and hence the short rations.[8]

At 5 1/2 a.m. the head of the column moved, but our command did not get out on the road until nearly 7, except one battallion which moved in advance of the corps to build a bridge across the Pamunkey river. The Troops moved in three parallell columns, one in and one on each side of the road. It was a beautiful sight to see as the Army moved on in three great streams, like three blue rivers sweeping on in their irresistable course towards their destination.

The Pamunkey is a deep, muddy stream about two hundred feet wide at the place where we crossed. Here we built two bridges for the use of the II and V Corps, the other Corps crossing on annother bridge further up. On the South side of the stream was a large plateau of ground on which nearly all the Army rested. Here we went into bivouac and remained the entire day. A general advance was ordered at noon but, for some reason unknown to me, the order was countermanded. The opportunity to rest and sleep was fully improved and we awoke on the morning of the 30th much refreshed.

We were early under Arms and moving slowly to the front while fighting extended along the whole line. Our position was advanced about three miles and we were but little more than 13 miles from Richmond. Every time our Troops took up a new position, a strong line of breast works was at once thrown up. This time a large portion of our front extended through woods so that the breast works were constructed chiefly of logs. We were now near our old ground which we had occupied when under McClellans command, being not far from Mechanicsville. One portion of our line, the right, nearly reached that place, while the left was being advanced to the south. As [this advance] now protected White House, our transports landed there and a new base was established. We were glad enough to draw four days rations again.

The enemy disputed our progress inch by inch but we gradually pressed them back towards the Chickhominy river which lay between them and Richmond. The next day, by dint of hard fighting, our line was advanced annother mile. The positions of the Corps from right to left was now as follows: II, IX, V & VI. The Artillerey kept up a heavy cannonade all day. One of our trains was sent back to the Mattapony to build a bridge for reinforcements which were comeing forward to us. During the night the entire position of the Corps was changed. The II Corps moved from the right to the left of the line and so ended the memorable month of May.

On the 1st of June, much apprehension was felt for the safetey of our right, which was reported in a very exposed position, and it was feared a general attack would be made upon it. But the day passed with the usual Artillerey firing and nothing more. Just at night we pulled out on the road and marched to Old Church, where we bivouaced for the night.

June 2nd we marched again but not far. We passed General Grant & Meade, with all the General Officers, who were holding a consultation by the road side, having taken the benches from a little church near by and placed them under the trees while the Troops marched on through the dust on either side of them. We halted near Cold Harbor.[9]

The enemy strongly intrenched himself at Gaines Mill and along the high ground to the north of the Chickahominy on the very ground we had formerley occupied under McClellan. It was evident from all appearances that he would here make a most desperate stand. To be defeated here would subject his capital to the terrors of bombardment. On the other hand, could we succeed by a general assault in breaking his lines, the annihilation of his army was certain, as he would be driven back into the Chickahominy, whence escape was impossible. His entrenchments were formidable, almost impregnable. The hazard was great but General Grant concluded to take the chance. The Troops were disposed accordingly and orders issued for a *general assault* all along the line at daylight on the morrow.

General Hancock sent for me to move my little battallion up to his Head Quarters during the evening. I found him together with the staff just in rear of Cold Harbor Tavern. A plateau of open ground stretched out in front of this Tavern about a quarter of a mile to a piece of woods. In this woods the line of the enemy was visible. A slight depression from the plateau formed a partial protection from their guns, which swept the plain, and into this were huddled Head Quarters with my train and battallion. Here we lay until morning.

The enemy kept up a continual Artillerey firing, the shells striking the plateau in front and richichottry over us. It was an anxious night, for we all felt the importance of the next mornings proceeding. The XVIII Corps, which had joined us but a day or two before, were posted on the extreme right of the line, which extended some miles to the westward. The signal, two guns from the XVIII Corps at daylight, was to be the command for the whole line to advance.[10]

Slowly the darkness disappeared and the grey of the morning advanced. A lull, the stillness which always precedes the hurricane, seemed to fall like a mantle over the country, which was also partially obscured by a thin haze or fog, through which could be seen our long line of men silentley awaiting the signal to advance. For some time, a few minuites perhaps, but which seemed like hours, the stillness prevailed.

Then came the signal, two guns fired in quick succession, which came booming from away off to the right, followed by a wild cheer and the instant response of a hundred cannon and thousands upon thousands of muskets. Annother moment and our part of the line was in motion. They were soon out of sight. The Artillerey fired rapidly and incessantly. The air above us was filled with whisteling bullets and schreiking shells bursting over our heads, the fragments flying about us in all directions.

Noticing a slight commotion among the Officers with General Hancock, I soon learned that one of the Staff who was standing near the General had his leg dreadfully shattered by a fragment of a shell. Soon four men came over the crest of a rising ground carrying upon a litter the bloody, mangled body of the gallant General Tyler.[11] Then a solid shot came bounding into the midst of us throwing up the dust a few yards in front of me and almost instantly striking and taking off the leggs of two of the mules in my train, which was parked in a solid body in the little valley. Then one of my men of Co. B was struck with a bullet but not mortally wounded. Almost at the same moment annother Staff Officer, Major Garnett, was wounded. Then came annother litter carrying Colonel Kellog with whom I had been acquainted.[12]

The roar of the battle drowned our voices. My Johnny looked out from behind a tree to see how I was faring just as a shell struck the tree in front of his head scattering bark into his eyes and mouth. Almost at the same moment an Artillerey caison filled with powder standing within a hundred yards of us was struck with a shell and exploded with the noise of a dozen cannon. Looking around in that direction we saw a column of white smoke ascending towards the clouds, expanding as it ascended, until it disappeared far above, while scattered around on the ground near by lay the blackened and charred remains of *six* Artillereymen. Poor fellows, gone in a moment of time to their last reckoning. Annother of my men was struck and severeley wounded and a stream of bleeding men commenced to pour back and down into the valley.

General Barlow had succeeded in gaining a position which enfiladed the whole line of the enemy, capturing *17* pieces of Artillerey and turning them upon their former owners. His Division had been terribly decimated in the operation however, and he sent back frantic appeals to General Hancock for reinforcements but General Hancock had none to send him. The whole Army was in line and there was no reserve. General Barlow was speedily driven back and away from the guns he had captured by the reserve which Gen. Lee had posted in a convenient position.[13]

In less time than it has taken me to write this, our Troops were repulsed from one end of the line to the other. The assault was a failure. The most bloody, most desperate assault of the entire war, without any exception. The assault at Cold Harbor stands out separateley and alone in the history of the war. Our losses in *ten minuites* exceeded in killed and wounded, *ten thousand men*. It was an awful sacrifice without any result. The losses of the enemy did not approach any where near this proportion.[14]

The firing was kept up on our retreating Troops until they, in desperation, with no other tools than their cups and tin plates, had thrown up an earthwork in their front for their protection. Earth works which in less than 24 hours had grown to formidable proportions and had been made almost impregnable. About 9 O'clock the firing began to lull and soon after ceased almost altogether.

For the first time since he assumed command, confidence in General Grant was shaken. There was no demoralization visible but a settled commotion which showed itself plainly in the faces of every Officer and every man in the Army, that ten thousand men had been sacrificed for nothing. The horrors of that bloody assault have made such an impression on my mind that I can never forget it. I will not attempt to describe the terrors of that day, but willingly leave them to your immagination and, however vivid that may be, you can scarceley exaggerate them.

It was reported on good authority that General Grant had decided to renew the assault on the following morning and in preciseley the same manner all along the line, but that he was dissuaded from the attempt by General Hancock and Meade [along] with the other Corps Commanders and that he very *reluctantly* consented to reconsider the order. The command was not given, but of one thing I am certain (and I form the opinion from what a hundred Officers prominent in command asserted) that *had* the order been given, not an

Officer, not a Soldier in the Army of the Potomac would have moved. The whole Army would have stood still, not in mutiny but in blank amazement.[15]

CHAPTER TWENTY-TWO

Cold Harbor, the James River and Arrival at Petersburgh June 1864

The general assault on the Confederate line at Cold Harbor had resulted disastrously to us but the fighting did not cease. For ten days we remained in nearly the same position, both contending Armies adding strength to the positions which they held. At some points the lines approached within a few yards of each other, while at other places the lines separated to a considerable distance.

No where along the entire length of earthworks was it safe for any man to show even so much as a hand above the parapets. The air was filled with whistling bullets yet not a man could be seen, nothing but a dull line of fresh yellowish earth from which during the day asscended a succession of little white clouds of smoke and at night, bright flashes of fire.

For a number of days General Grant seemed to be in doubt whether to commence a siege of the City by means of regular approaches or to again execute a flank movement. All indications pointed to a regular siege. During the afternoon after the assault my command was put into the trenches and in company with the infantry worked all night throwing up intrenchments. Just that night the rebels made a furious charge upon one position of our line but were repulsed. The firing was kept up in the most spiteful manner by both sides until long after dark. At daylight the next morning the same vindictive firing was renewed. At one place near the right of the II Corps where the lines nearly approached each other the bitterest portion of the fighting seemed to rage.

The dead who had been killed during the charge of the day previous still lay, stark and stiff, between the lines. Among the number thus exposed lay the gallant and noble Colonel Porter of the 8th New

York Heavy Artillery. The surviving Officers of his Regiment offered a liberal reward to any man or body of men who would venture to go out and secure his remains. Actuated more by motives of love for their former commander than the mere love of gain, one after annother of his men volunteered to perform that duty. They went out only to be cut down. The attempt was found to be too hazardous during the day, but at night it was hoped it might be performed. So, during the darkness of the night many went out for that purpose but they never returned and when day light of the following morning appeared, a line of dead men completeley encircled the dead body of their Colonel. At last, by common consent, the effort was abandoned.[1]

The next day, June 5th, my command was ordered back from the trenches and instructions given us to commence the making of Gabions. This certainly had the appearance of a siege. We were glad enough to get out of the pits and make gabions, a duty in which our men were well skilled. The gabions that we made during the day were used at night. For several days we were kept at this duty.

We had now no idea but that we were to approach Richmond from this point. Our works had assumed enormous proportions. A short distance to the left of the spot I have described and separate from the main line, a portion of a Brigade had thrown up an earthwork with which, on account of its isolated position, we could have no communication. The ground leading to this little citadel was open and completeley exposed to a sweeping fire from both directions by the enemey.

The little garrison held out manfully but it was feared they were short of rations and ammunition. We could not afford to abandon it as the position, if once in possession of the enemy, would give them a decided advantage over that portion of our line. Colonel (afterwards General) McKenzie said to me that he would like to visit the little outpost and would like to have me accompany him so that we might learn of their necessities and devise some means of connecting the outwork with the main line. Besides which, said he, if one of us gets struck the other can pick him up and bring him in, whereas, if one went alone and was hit, he would have to lay there.[2]

So we went up to the spot nearest the earthwork. The prospect was appalling but did not shake McKenzie, though I must confess that I heartily repented the undertaking. "Good Bye Brainerd!" shouted McKenzie and away he went like a deer across the open field. I saw little clouds of dust spit up in front and on both sides of him as the bullets of the sharpshooters struck the ground and flew around. I expected every moment to see him fall, but on he sped. The distance was not great, not more than two hundred and fifty feet, but *such* a *gauntlet*. Dead men lay here and there on the ground but over them and on he went. With a bound his form raised a moment on the parapet and disappeared, but not before I was on the way.

In my younger days I was counted a good runner and, in my hi-bob-a-ree days, he was a swift one who could overtake me on the way to the goal.[3] But never until that day was I put to my *utmost* speed. I know not how long it took me to get across that open country. It *seemed* like many minuites. What I do know is that my strides astonished even myself. I know too that my hair stood straight up, at least I judged so by the feeling of the *goose-flesh* on my scalp. Then the wicked whistling of the wind about me forced home to my conciousness the fact that hundreds of unseen bullets, meant for *me*, were searching for their victim. As I remember it now, it seems that I went across that space in about five bounds, but as that would give something over fifty feet to the bound, I think perhaps the estimate a little exaggerated. However, the last bound brought me clear of the parapet and into McKenzies arms.

After a few miniutes for breathing we had a chance to look around us. The earth comprising this little fort was thrown up high and thick. Through the parapet a few feet from the top, holes were pierced for the muskets and at each loop hole stood a man with his musket in place watching with eager earnestness for the least show of an enemy from the line opposite. As we looked through the loop holes to the right we could see the two, long irregular lines of earthworks stretching away in the distance, sometimes, approaching within a few yards of each other, sometimes receeding to a greater distance. Behind these two lines of earth stood the two contending armies. Clouds of smoke were arising and the noise of musketry was incessant but not a living man was to be seen.

Between these two yellow lines of earth a sad spectacle presented itself. Groups of bodies lay scattered along to the far distance. Silent and motionless they lay in all conceivable shapes and positions. Not many yards away lay the lamented Porter. I knew him by his uniform and stout person, bloated and disfigured though he was, his arms uplifted, his leggs stretched out, his blackened face with white foam oozing from his mouth, turned upward toward the blazing, schorching, June sun. Around him in different stages of decomposition, lay the bodies of the brave men who had died in the vain effort to recover his dead body. It was indeed a sickening, harrowing spectacle. The stench from these bodies now filled the air with a sickening odor and I was glad enough to turn away from the sight.

After getting the information we desired, we started again to run the gauntlet of the return. This time McKenzie generously told me to go first, as the second would probably have the hardest time of it, the attention of the sharp shooters being drawn by the first one. I suggested that we both go together and away we went. If getting over was hard, I think it was worse getting back. The enemys Sharpshooters seemed to comprehend that the two Officers who went out to the fort would soon return and therefore concentrated their fire upon this point. I had dreaded to go out but dreaded much worse to return.

We shook hands, said good bye and started. My description of the first journey will answer for the second, except that a greater number of men fired upon us. To tell the truth, I never expected to get back alive but we went through *unscathed* though the bullets fairly whistled around us and [we] were received by our friends inside the main line with hearty cheers.

The next day, June 6th, General Grant was obliged, in the interests of humanity, to ask for a truce of two hours in which to bury the dead and for two hours silence reighned until the burrying parties had performed their sad duties.[4]

When the fight was resumed as usual, a new species of warfare commenced. The lines were so close together and the earthworks of such thickness that ordinary shell and solid shot penetrated but a short distance into the earth and were buried without doing any material harm. In order to make it as uncomfortable as possible for the Rebels, a large number of small mortars were brought up and placed in our lines. These mortars were so small that four men could handle and move them from place to place with ease. A small charge of powder would lift a shell over into their works where it exploded carrying death and destruction around.

At first these mortars proved very annoying and destructive to the enemy. In many places the shells did not need to be thrown more than a hundred feet. We could destinctly hear the Rebels yell when these little messengers exploded among them. Soon, however, they constructed bomb proofs for themselves and were partially protected. Taking the cue from us, they also brought up some mortars from Richmond and the war was for several days carried on with bombs and mortars.

It was a beautiful sight to watch these bombs as they ascended and descended in regular showers during the nights that followed. Back and forward they went and came, their bright semicircles of flame streaming through the dark sky. It was a Fourth of July celebration on a large scale and kept up for many days and nights.[5]

The usual sundown charge was made daily by the enemey. The one on the evening of 8th was especially desperate. The sun during the days was very hot and enervating. The cool of the evening or morning was the most favorable time for active operations.[6]

One evening, just as I returned to my bivouac in the little valley, I was called upon by Lt. Col. W.H. Baird of the 126th New York accompanied by his Adjutant, Lt. M.V. Stanton. Then came Capt. Wm. A. Berry of the 2nd N. York Artillerey. It was just before sundown and the Rebels were shelling our position at a fearful rate. My quarters were in the little valley I have before mentioned.

I was laying on the ground just below the brow of the little elevation while the shells and solid shot from the enemys batteries struck the ground on the open field in front and bounded over. Baird & Stanton came running in and laid down on the ground beside me, followed by Berry. I was well acquainted with these Officers, but had not seen them in a long time. Baird I had seen but once since the Pennisula Campaign and Berry not since the day of our arrival at the North Anna. They all came for the same purpose, to see me and get something to eat and there we lay until Johnny, stooping low to avoid the flying shells, had boiled us come coffee. To stand upright was dangerous.

The cannonading was kept up vigorously until 9 O'clock that evening. Meanwhile, as we were waiting for the coffee to warm, Baird called my attention to a new shell that he claimed as his invention. The object being to make a shell more destructive than any then in use by casting it thinner in places, so that when the explosion within occured, a much greater number of pieces would fly around. To illustrate his idea, he made a sketch and gave it to me.[7]

We lay there for a long time with the shells screeching over us discussing the merits of Bairds new implement of destruction. Capt. Berry and Lt. Stanton joining in the discussion. After enjoying Johnnys coffee, we lit our pipes and went over together the incidents of the campaign, relating the adventures which each had met, the dangers encountered and hair breadth escapes we had made since last we had seen each other. In this manner we enjoyed ourselves until long after dark. At last they took their departure after I had filled their haversacks with provisions and each ones canteen from the contents of my blueheaded barrel. We then shook hands and separated, each one to his separate field of duty. Little did I then think that this was destined to be our last meeting.

Less than two weeks from that night, as I was riding back from Petersburgh to City Point one hot afternoon, noticing a hospital on one side of the road, I dismounted to see if any of my acquainces were among the wounded. I found several of my old friends among the crowd but all doing well. On the opposite side of the road was a new grave yard with many fresh graves. I strolled over carelessley, thinking that there *might* be some among the number with whom I had been acquainted.

I wandered around for some minuites reading the names of the dead, the simple inscriptions written in pencil upon pieces of cracker boxes and stuck into the ground at the head of the graves. I was about to turn away when my eye fell upon three fresh graves with the names of *Stanton, Baird & Berry* written upon the three pieces of board at their head. Could it be possible that my three friends, who had left me but a short time before in such excellent health and spirits were now lying *here*? I rubbed my eyes and tried to imagine that this was but a dream, but the terrible reality was *there*. My three friends lay side by side in their shallow graves, all killed and burried within two days of each other. Baird on the 16th, Berry on the 18th and Stanton on the 19th, and this was the first intimation that I had had of their death, though I had heard that Baird was wounded.[8]

With a heart full of grief I mounted and resumed my journey. I must confess that a feeling of superstition took possession of me, a belief that *all* of my friends were to be sacrificed in the struggle and for many weeks I could not rid myself of this feeling.

The morning after the visit of my friends, my command was ordered back from our exposed position. Corps Head Quarters had previously moved and our new camp was a pleasant opening in the woods where we set up our tents once more and for two days enjoyed comparative quiet.

My Quartermaster, Lieutenant Granger (and an excellent one he was) went down to White House on the York river for rations. On his return, it was reported to me that Quartermaster Sgt. Wheat was carrying on a separate speculation of his own. Wheat had managed to buy a huge cheese at White House and by smuggling it up in one of the wagons (all of which was against positive orders), he was now retailing it out to the Soldiers at the rate of 80 cents per pound. As the cheese cost him but 30 cts per pound he was making quite a nice little speculation. I ordered his arrest and the confiscation of his cheese, not quite a quarter of which had yet been sold. The offender, with his cheese, was brought up to my quarters and then he was required to cut it up and divide it among the companies free of charge. It turned out to be a loosing speculation for Sergeant Wheat and he never repeated the experiment.

A strong line of earth works was now run off for some distance inside of and at right angles to our main line. This indicated annother flank movement but in which direction we knew not. During the night of Saturday, June 11th, my orders came to move at daylight on Sunday morning in the direction of Mrs. Higgins house and to cut a road through the woods to facilitate the movement of the II Corps. The direction of the new flank movement was now apparent. Our base of supplies at White House was abandoned. None other could be established except it be at some point on the James River.

At daylight on Sunday morning my command moved, following the route which I had previously reconnoitered under instructions from General Hancock, and, arriving with axes, my men struck into the woods, felling the trees as we advanced. At night we had made a wide road for more than four miles good for the passage of Artillerey and heavy trains. We then moved on to Dispatch Station, where I was overtaken by Col. Morgan, Chief of Staff, who informed me that the General was much pleased with the road we had made. The enemy knew nothing of the movement so well had our Army been disiplined in the delicate manoeveres of a flank movement.

The picket fires were kept burning brightly during the night while the Troops were falling back and moving in solid masses to the left. A sufficient picket guard was left to keep up the firing and the camp fires but, when daylight appeared, not a solitary Yank was to be seen in all the long line of earthworks.

Subsequentley we learned that General Lee was much puzzled as to our movements. He did not know where to expect us next, whether on his right or left flanks. The new line of works was manned in order to cover our passage of the Chickahominy but the enemy did not follow. Before sunrise on the following morning we were again on the march. The main body of the Army had now overtaken us.

Arriving at the Chickahominy River we found a long bridge already constructed under a slight fire of the enemys outposts. One man was wounded of Maj. Fords Battallion which had constructed the bridge assisted by Spaulding and Beers command. This was the longest bridge that had yet been made during this campaign.[9]

From this point we made a forced march, not halting even for rest or refreshment. Just at dusk we marched into a splendid clover field where I supposed we were to bivouac for the night but no halt was ordered. On we went at a quick pace until 10 O'clock at night, when we arrived at a point below Harrisons Landing on the James River, a place called Wilsons Landing or Wilsons Wharf. Not a mouth full had we to eat all that day. A cup of coffee sufficed us and we dropped on the ground with no covering but natures canopy and were soon fast asleep.

The next morning, June 14th, we were roused up at daylight to find that we had been sleeping in another very beautiful clover field. In a short time everything was in motion. Measures were taken at once to throw a Ponton bridge across the river a few miles below us, opposite Fort Powhattan, that being the narrowest spot to be found for several miles. Spaulding, with all the other detatchments of our Regiment but mine, with all the Pontons and bridge material of the Brigade, was despatched to that point, where a bridge was constructed which crossed all the Army except the II Corps to the South side of the James River.[10]

Our Gun boats commanded the river above and below us and the enemy made no demonstrations on our rear until all were safeley over the stream. The James river at this place was a formidable stream to cross. It was about one quarter of a mile in width and had the right bank been held by the enemy a crossing would have been next to impossible. It is evident that General Lee never anticipated our crossing to the South side of the James, for he had nothing there to prevent us. The City of Petersburgh was fortified in the most thorough manner, the character of the works thrown up for its defence being second only to those around Richmond itself.

General Hancocks Corps was ordered to cross the river *above* Fort

Powhattan, not waiting for any bridge to be laid but to cross on the transports that were then lying in the River and hurry on to reinforce the Troops, or a portion of them then under General Smith, and join in an assault upon Petersburgh, the idea being to surprise the small garrison then in possession and capture them before Lee could hurry to their rescue from Richmond.

So it was now a race for Petersburgh. The day was occupied in getting the remnants of the Corps together and awaiting the arrival of the transports. During the afternoon General Birneys Division was transported across, but the Landing was a slow process. A *wharf* would expedite matters and an old one nearly destroyed by fire was found. Could this be repaired, our Troops could be landed with much greater rapidity.

Soon after 7 O'clock in the evening I was ordered to report to General Birney. Verbal instructions given me by Col. Morgan were that General Birney would furnish all the men I might need for the purpose. Soon after 9 O'clock, my entire command, except the transportation and a small guard, were across the river. A Tug took us across in two trips. The night was cloudy and dark, the country strange.

My first point was to find General Birneys Head Quarters. This involved a walk of several miles. When at last he was found, he informed me that I might have all the men I needed to build the approach. I represented that it would require a large amount of lumber for the purpose and then asked him to send a thousand men to tear down a few barnes that I had passed on my way and to have them bring the material to the wharf. The men were ordered at once on that duty under their own Officers while I returned with my men to the river bank.

Soon the material born upon the shoulders of the men began to arrive. I soon found that it would be a work requiring more time than was anticipated to construct a wharf from such material as I was getting. Half a dozen Pontons would do better service than all the barnes in the State. However, we went to work with a will but were making slow progress until about 3 O'clock in the morning, when off on the water I heard a voice calling my name. The Captain of a transport had half a dozen Pontons in tow with orders to leave them with me wherever I might be found. I looked upon this in the light of a Providential Interferance. By daylight, with the Pontons and the material sent by General Birney, a passable and quite permanent landing was constructed.[11]

Soon after, General Hancock came over in a transport and seemed much pleased with the undertaking. Then the Troops began to pour over by Regiments, Brigades and Divisions. The transports came up to the wharf and soon the Troops were off while the boats went back for annother load.

My transportation and rations were all on the other side and by the time I saw my army wagon again 24 hours had passed since a morsel of food had passed my lips. In addition to my empty stomach, I had taken a fearful cold during the night and could hardly speak when morning came, so taken altogether I suppose I was not in a very amiable mood that day.

The day of the 15th was occupied in transporting the Troops and trains. After dark my train was taken across the river. At 2 O'clock that night all except my army wagon had crossed the James. This being my Head Quarters I left it on the North bank until General Hancock should move his.

Just as I was leaving for the North side of the river, an Officer asked me if I would deliver an errand for him at Head Quarters. Of course, I assented. He then asked me to say to Col. Morgan that the rations for which they had telegraphed General Butler, *he thought* were now on the way as he saw a steamer comeing down the river evidentley from Bermuda Hundred. The Officer referred to was Col. Batchelder, Chief Quartermaster of the Corps.

As I crossed the river and walked up the rising ground towards my Quarters, I passed General Hancocks Head Quarters, saw Col. Morgan and delivered the message as it had been given to me. On looking up the river we saw the steamer comeing down. The Signal Officer was instructed to signal the advance of the Corps on the oposite side of the river to "Halt," and this was done by signal.

In less than *ten minuites* the steamer came down opposite us but instead of stopping she kept right on her course. It was now evident that this was not the steamer that was expected with rations and the Corps was at once signalled to advance, the delay occasioned by the whole operation did not occupy *15* minutes.

What was my surprise when a few days afterward we received the New York papers to see in the Times a long article from the pen of Mr. Swinton on "How We Lost Petersburgh," attributing the entire blame to mistaken information conveyed to Head Quarters by an Engineer Officer.

THE GREAT CAMPAIGN.

MILITARY REVIEW OF THE SITUATION BEFORE PETERSBURG.

THE STRATEGY OF A WEEK.

MOVEMENT FROM THE CHICKAHOMINY TO THE JAMES.

THE ACTIONS OF WEDNESDAY, THURSDAY, FRIDAY AND SATURDAY.

RICHMOND AND THE PROSPECTS OF THE CAMPAIGN.

From Our Special Correspondent.
Army of the Potomac, Near Petersburgh,
Tuesday, June 21, 1864.

[Swinton's narrative is picked up in the middle of the story]

II.

HOW WE FAILED TO TAKE PETERSBURGH.

When, early on Wednesday morning, the Second Corps had effected the passage of the James River at Windmill Point, Gen. Hancock was met by a dispatch from the Commanding General, directing him that, if provisions had arrived, he should ration his men before proceeding toward Petersburgh. Just at this time he received information, seemingly reliable, and conveyed, I believe, by an engineer officer who had just come up from the pontoon bridge, to the effect that the transports had arrived with supplies. This fact Gen. Hancock communicated to the Commanding-General, stating that he would soon begin issuing rations, and would then move forward as directed.

The information proved to be a mistake—a lamentable mistake—the transports had *not* arrived. Yet this simple error caused a delay in the movement of Hancock's corps of *five hours and a half*. His column joined Gen. Smith's troops at 1 o'clock on the morning of Thursday. But had he been up earlier by the difference noted of five and a half hours, he would have reached the front at 7:30 on the evening of

> Wednesday, *which was precisely the moment at which* Gen. *Smith made his attack!*
>
> [*New York Times*, June 23, 1864]

Of course, every one at Head Quarters knew who the Engineer Officer referred to was and so the joke, for such it was considered then, was on me. But while this might be considered in the light of a joke among us who knew, I felt that it would be quite natural for many at home to enquire *who* this Engineer Officer might be. Would my name become public in connexion with so serious an affair, it would not be a very pleasant thing to comtemplate.

I therefore determined to ask General Hancock for a Court of Inquiry and accordingly I visited General Hancock in his Quarters and requested him to place me under arrest.

"What for, Major?" said he. When I explained, he said, "Never mind Major, I will fix him. This is not the first time this correspondent has misrepresented affairs in this Army, but it will be his last. He must leave this Army. Dont let it trouble you, Major." And so I left the General feeling much better.[12]

Not many days after, Mr. Swinton was politeley informed in an order from Army Head Quarters that his presence would no longer be tolerated in the Army of the Potomac on the ground of general untruthfullness and forwarding to his paper mischievous reports. So, Mr. Swinton left the Army. He was an excellent correspondent and has written the best history of the Army of the Potomac yet published, but he was at times a little indescreet and this led to his final dismissal from the Army.[13]

Early on the morning of June 15th, Head Quarters moved across the river and rapidly towards the City of Petersburg. I was ordered to follow at 11 O'clock. That day we left the James River, making a *forced march* towards the City. We halted not for food or drink but rushed on our way almost at a run. Such a march we never made before. Just at sun down, having marched full 20 miles in the hot sun on dusty roads we came in sight of the spires of Petersburgh and drew up in line near General Hancock as the Corps commenced its first assault upon the Rebels line of works, part of which were carried and held by us.

Epilogue

It must have been quite a thrill for Major Brainerd to see the church spires of Petersburg. It was an exhilarating time for him that began with the crossing of the Rapidan and ended with the crossing of the James. The Army of the Potomac had been locked in deadly combat with the Army of Northern Virginia for six weeks. The campaign was not the success that Lt. Gen. Ulysses Grant had hoped for, but reached an objective that the men could be proud of.

There would be more fighting in the coming ten months as the Union army placed Petersburg and Richmond under a siege that was the beginning of the end for the Confederacy. While we do not have Major Brainerd's words to guide us, we can glean some facts from the records.

Once the campaign of siege began, the engineers helped to design and build the works investing Petersburg. Between June 19 and July 29 the ponton trains went into camp near City Point, the supply base of the army and the site of Lieutenant General Grant's headquarters. There, the engineers repaired their equipment and placed it securely under guard. Detachments of engineers made ten thousand gabions and twelve hundred fascines for the siege works and constructed forts, covered ways, roads, and bridges. Between June 22 and 30, Major Brainerd commanded a battalion that went into the rifle pits in front of the Jones house and served the II Corps as infantry.[1]

On the Fourth of July, the 1st Battalion, 50th New York Volunteer Engineers, Major Brainerd's unit, hosted a gala dinner. Those attending the event praised it as a "fine affair that did honor to its donors."[2]

On July 10, headquarters consolidated all detached engineer battalions under the command of Lt. Col. Ira Spaulding, 50th New York

Volunteer Engineers, and the regiment was once again together. The regiment mustered with all their combat equipages early on the morning of July 30 and was on alert during the mine explosion that resulted in the Battle of the Crater. After that assault on the Confederate works surrounding Petersburg failed, the engineers returned to their camps.[3]

In September 1864, Major Brainerd was in charge of several companies of the 50th working between Fort Sedgwick and the Norfolk railway. These men corduroyed roads in the covered ways. Rain hampered this work and caused the engineers to labor continually to keep the roads open to wagon traffic.[4]

On September 30, 1864, Major Brainerd commanded 280 men who served as infantry in and around Fort Bross, a part of the second Union line near the Norfolk and Petersburg Railroad. This was during the Battles of Chaffin's Farm near the Richmond defenses and Poplar Springs Church near the western end of Lee's lines around Petersburg. Two days later the regiment formed up along the Jerusalem Plank Road. From there, they went to the western end of the Union lines to construct new forts, extending the siege lines as a result of the battles that had gained new ground. Major Brainerd commanded a force of engineers at Fort Dushane, and work proceeded to build up the forts, embrasures, covered ways, redoubts, and gun emplacements in that section of the line.[5]

Major Brainerd received a brevet lieutenant colonelcy signed by President Abraham Lincoln for gallant and meritorious services during the campaign before Petersburg to date from August 1864. In November 1864, he took sixty days' leave of absence and received pay for four months. Brevet Lt. Col. Wesley Brainerd traveled to Chicago to be at home when his wife gave birth to their first son, Irving Gage Brainerd, on December 4, 1864.[6]

The 15th New York Volunteer Engineers, the other engineer regiment in the Volunteer Engineer Brigade, lost most of its men when their two years' enlistments expired in the middle of 1863. The regiment was understrength for some time until November 1864, when it underwent a complete reorganization. Seven new companies joined the regiment's rolls. Some of these new companies had enlisted for the 50th and some for infantry regiments.

As soon as these new men brought the 15th up to strength, Major and brevet Lt. Col. Wesley Brainerd received a promotion to colonel, dated from October 15, 1864. He then took command of the regiment in January 1865.[7]

As 1865 began, Union troops surrounding Petersburg and Richmond were well aware that they were going to make the final push to destroy the Confederacy as soon as weather permitted. On January 7, 1865, three companies of the 15th New York Engineers transferred to General Terry's command and shipped out for the Fort Fisher Campaign in North Carolina. These troops were under the command of Lieutenant O'Keefe.[8]

On February 5, 1865, a detachment of the 15th New York Engineers arrived with a thirty-ponton-boat train from Washington. That same day, the entire engineer brigade, consisting of the 15th New York Engineers, the 61st Massachusetts, 18th New Hampshire, 1st Maine Sharpshooters, and Michigan Sharpshooters reported to the front to Major General Parke, commanding the IX Corps. This was during the Battle of Hatcher's Run, another bid by the Union command to extend the army's lines farther to the West.[9]

During this movement, Colonel Brainerd commanded the engineer brigade, as General Benham was in Washington. From February 5 to February 11, Colonel Brainerd kept his men in the trenches covering for the troops off to the west of the siege lines, who were fighting to extend the lines. At the end of this time, the brigade returned to the City Point defenses.[10]

Between February 11 and April 1, 1865, the Engineer Brigade drilled, performed camp duty, and worked on the fortifications at City Point. All of the regiments except the 15th New York Engineers transferred out to line brigades where they joined other units. General Benham held the City Point defenses with the 15th alone. During this time the regiment sent detachments to Deep Bottom and Bailey's Creek on various assignments.

On March 25, part of the command turned out to man the City Point defenses during the Confederate attack on Fort Stedman. On March 29, the regiment went into the works as the campaign that would end the war began with skirmishing and engagements at several sites.[11]

On April 2, the final campaign began, and Colonel Brainerd and the engineers marched out to the support of Major General Parke, IX Corps. The infantry regiments stormed into the captured Confederate works and saved them from recapture. Colonel Brainerd was in reserve with 950 engineers and 750 dismounted cavalry. Later that night they supported General Wilcox. The Union army entered Petersburg the next day, and the engineers repaired the burned bridges connecting the town to Richmond. Colonel Brainerd and his command repaired the principal bridge and the railroad bridge for the passage of artillery and infantry. They also threw a ponton bridge across the river to speed the movement of troops for the pursuit of General Lee and the Army of Northern Virginia.[12]

The 15th New York Engineers then returned to their camp at City Point. The 50th took part in the pursuit of Lee as far as Farmville. The brigade reunited on April 13 at Army headquarters at Burkesville, and for the first time in many months the Volunteer Engineer Brigade was together.[13]

All fighting in the Virginia area ceased after General Lee's surrender of the Army of Northern Virginia on April 9, 1865. The engineers continued to build some bridges and repair roads in the area of operations. Finally, there was little to do but pack up and move all the trains and equipment back to their Washington base. Colonel Brainerd took command of the troops marching through Richmond on May 5, 1865, and they were in Alexandria on May 12. On May 25 the engineers marched proudly down Pennsylvania Avenue for the grand review, with Colonel Brainerd at the head of the 15th New York Engineers. They went into camp at Fort Berry, and finally, after almost four years of war, Col. Wesley Brainerd mustered out of the army on June 14 and returned home.[14]

After the war, Wesley Brainerd moved to Evanston, Illinois, and there, a third child, Belle, was born on August 22, 1866. Wesley went into the lumber business as principal in the firm Soper, Brainerd and Co. They did business on the corner of Polk and Beach Streets with the capacity of one hundred thousand board feet a day at the mill. This business thrived from 1865 to 1876.[15]

During the Great Chicago Fire in October 1871, Wesley took charge of a large group of citizens digging ditches to prevent the fire from engulfing Evanston. Soon after, a volunteer fire company formed, and Wesley became fire marshall, the first one in Evanston. Another first in Evanston were the board sidewalks which Wesley managed to have built in concert with his father-in-law, Eli Gage, with Soper, Brainerd lumber, no doubt.[16]

In 1873 Wesley became interested in the iron-smelting process and managed the Brighton Smelting Works. The Chicago and Colorado Mining and Milling Company was incorporated in 1876 with the help of Lyman Gage and other financial backers, and Wesley became president and manager. That same year, Wesley moved to the new state of Colorado to develop the company's mining property at Camp Talcott in Ward district, Boulder County.[17]

It was at Camp Talcott that Wesley came into his own. His experience with mechanical devices, his natural and innate ability to solve problems, and his inventive and inquisitive nature made him one of the foremost innovative mining engineers in Colorado. He held patents on many early mining inventions, and the mining complex that he managed became the most advanced and modern in the state by the turn of the century. All of the lifts in the various mines ran by electricity generated from a central powerhouse. The power source was water piped in from a lake two miles away via a wood flume and steel pipe run that wound through the mountains. The 740-foot vertical drop created a pressure of 320 psi. It was the first such operation in existence. Telephone lines and switches connected the entire 800-acre complex. Wesley's own milling operation in the mountains produced the lumber for the flumes. The Colorado and Northwestern Railroad had a stop at Brainerd Station on mine property.

Wesley was also a prospector, discovering over 60 claims himself, all of them profitable. The Talcott complex also operated as a stock ranch, and Wesley carried on this interest by developing a similar full-blood cattle operation on a ranch in Nebraska.[18]

In the 1880s the editors of the Century Company invited Wesley to contribute to a new series on the war being published in *Century Magazine*. Wesley's contribution was a short article, "The Pontoniers at Fredericksburg," in which he defended the roll the engineers played during that campaign and explicated his role in the Skinker's Neck reconnaissance. The series was later published in book form under the title *Battles and Leaders of the Civil War*.[19]

Wesley was an active Mason and a charter member of the Colorado Commandery of the Loyal Legion, serving as its commander in 1894–95. Although probably a War Democrat during his war years, Wesley became a prominent Republican in the immediate postwar years. Later in life he declared himself an Independent and was not affiliated with any political party. He was elected to the Boulder, Colorado, Board of Trade and the American National Institute of Engineers.[20]

Irving Gage Brainerd, for whom the memoir was written, became a clerk in the First National Bank in Chicago. He later managed the stock ranch in Colorado and was assistant superintendent of the mining operation there. He never married and died in Pueblo, Colorado, on December 9, 1904. He was buried in the Gage family plot at Rosehill Cemetery, Chicago.[21]

Belle Brainerd married Emil Phillipson on December 13, 1888, but they divorced after producing three children: Brainerd Fisher Phillipson (b. 1890), Marie Augusta Phillipson (b. 1896), and Belle Lefler Phillipson (b. 1900). Belle Brainerd died in Omaha, Nebraska, on June 22, 1905. Her son was raised by his father in New York City. Wesley and Maria adopted Belle's two daughters, and, after Wesley and Maria died, the children became the wards of Lyman Gage, Belle Brainerd's uncle.[22]

Lyman Gage, Maria's brother, also worked at the First National Bank of Chicago, first as cashier (1868–1882), then as vice-president (1882–1891), and finally as president (1891–1897). During this time he served as president of the American Banker's Association. Lyman became a national figure when he served on the board of the 1893 Chicago World's Fair, becoming the leader in raising money for the event and overseeing the twenty-three-million-dollar budget. He then served as secretary of the treasury under Presidents McKinley and Roosevelt (1897–1902). In 1902 he became president of the United States Trust Company. He retired in 1906 to southern California, where he died in 1927 at age ninety.[23]

Amelia Maria Gage Brainerd traveled to Colorado with Wesley and lived in the camps and at the mines during the summer. They took rooms in Denver during the harsh winters. They retired to southern California in 1906, and there Maria died on October 28, 1908. at Point Loma. Wesley accompanied her body to Chicago, where he put her to rest at Rosehill.[24]

A writer described Wesley at the turn of the century as "a man of fine physique, in whose countenance kindness, amiability and benevolence glow. To all public enterprises of a helpful nature he is liberal and enterprising. He is exceedingly hospitable and happy is the guest who comes beneath his roof."[25]

Col. Wesley Brainerd survived his wife and children and died at age seventy-seven at Point Loma on August 19, 1910. He too is buried in the family plot at Rosehill.

On April 9, 1995, during the 130th Anniversary/Appomattox Surrender Re-enactment at Rosehill Cemetery, William Hencken, Wesley Brainerd's great-great-grandson, and seven members of the 1st Michigan Engineers and Mechanics led by Corporal Jerry Feinstein paid tribute to Col. Brainerd at his gravesite with a reading from the memoir and a twenty-one-gun salute. So, eighty-five years after his burial, Col. Wesley Brainerd finally got his military funeral. May he rest in peace.

Col. Wesley Brainerd's Family Genealogy

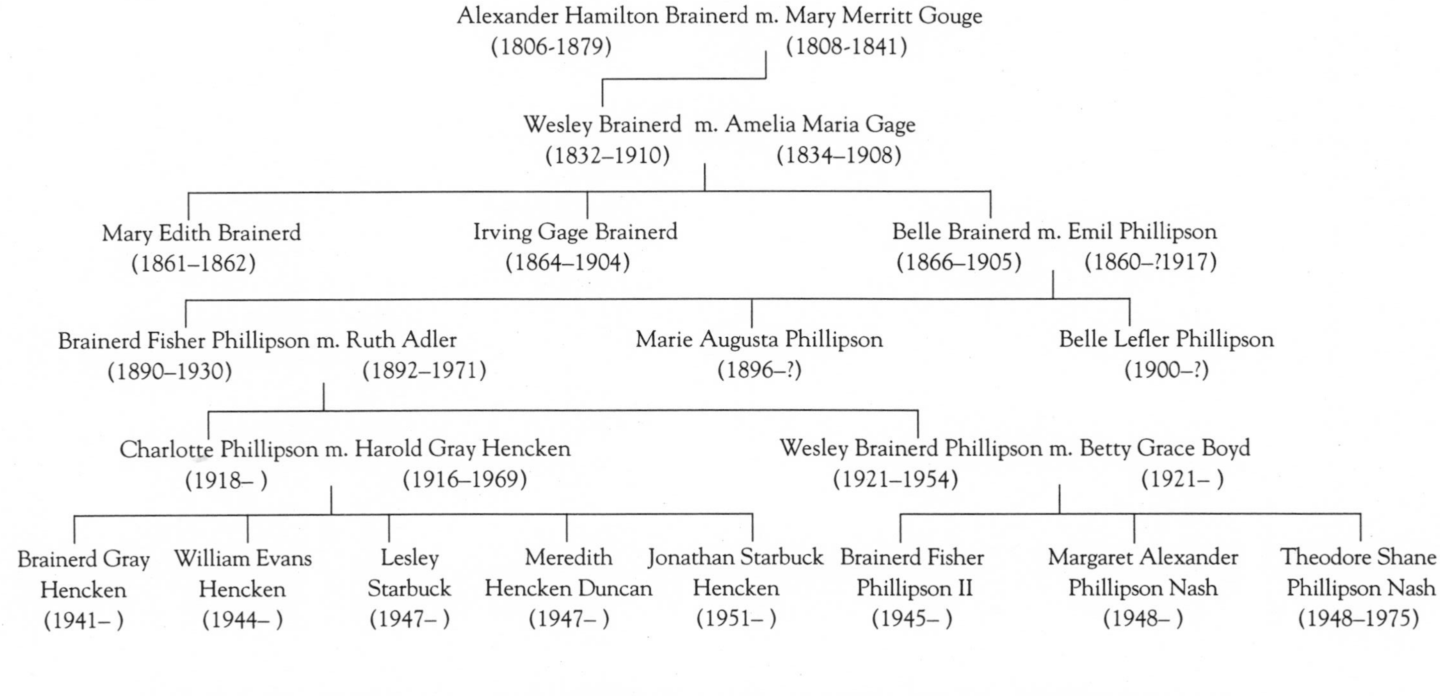

Among the current generation of the Colonel's descendants, those born in the late 1960s to the mid-1970s, there is a namesake. He is Wesley Brainerd Phillipson II (1972–), son of B. F. Phillipson II (1945–). Wesley has one sister and nine cousins.

Appendix 1

Service Records

The following information was compiled from: Phisterer, *New York in the War of the Rebellion*; USMA; ARAGNY; Boatner, *The Civil War Dictionary*; or Pension Records, National Archives and Records Administration unless otherwise noted. Not all veterans filed for pensions so some of the entries are expanded with postwar information and some are not.

ALEXANDER, George. Joined the 97th New York Infantry on October 2, 1861, was wounded at Spotsylvania on May 12, 1864, and discharged for disability from these wounds on December 14, 1864, as a captain.

ALEXANDER, Barton Stone. Graduated from the United States Military Academy seventh in a class of fifty-six in 1842. Alexander was assigned to the U.S. Corps of Engineers and contructed forts and lighthouses before the war. He served as an engineer with distinction during the war, was the first commanding officer of the Volunteer Engineer Brigade, and received four brevets to brigadier general. The 1842 class at West Point produced twenty-three generals during the war.

BAIRD, William H. Mustered in as major, 126th Regiment New York Volunteer Infantry, after previous service in the 38th New York Volunteer Infantry, on August 9, 1862, surrendered and paroled September 15, 1862, at Harper's Ferry, Va. (Antietam Campaign), dismissed, November 8, 1862; disability removed, June 26, 1863; lieutenant colonel, November 5, 1863; killed in action, June 16, 1864, at Petersburg, Va.

BEERS, Edmund O. At age thirty-two enrolled at Elmira, N.Y., on August 1, 1861, for three years. Captain from September 16, 1861; major from February 27, 1863; mustered out October 26, 1864, City Point, Va.. Breveted lieutenant colonel, U.S. Volunteers, from March 13, 1865. Beers died April 17, 1919, of nephritis in Washington, D.C., and was buried in Elmira, N.Y.

BERRY, William A. At age twenty-two enrolled at Staten Island, N.Y., for three years in the 2nd Regiment New York Heavy Artillery. First lieutenant, Company A, August 22, 1861; captain, February 14, 1862; killed in action, June 18, 1864, near Petersburg, Va.

BEVERIDGE, John L. Served in the 8th Illinois Cavalry as major, September 18, 1861, to November 2, 1863; colonel, 17th Illinois Cavalry, January 28, 1864, to February 7, 1866. John L. Beveridge was the son-in-law of Philo Judson, chaplain, 8th Illinois Cavalry. Beveridge went on to become sheriff of Cook County, Illinois, U.S. congressman from Chicago, and lt. governor and governor of Illinois. J. N.Reese, *Report of the Adjutant General of the State of Illinois*, vol. 8 (n.p., 1901). Abner Hard, *History of the Eighth Cavalry Regiment, Illinois Volunteers* (n.p., 1868).

COWAN, George W. At age twenty-three enrolled at Sylvania, Pa., for three years as private, Company C, on August 30, 1861; sergeant from December 27, 1862; re-enlisted January 1, 1864; mustered as 1st lieutenant, Company B, February 18, 1865. Mustered out June 13, 1865.

DOXY, Stephen G. At age thirty-seven enrolled at Suspension Bridge, N.Y., for three years as private, Company E, on December 17, 1861; discharged, September 23, 1862. Enlisted September 21, 1862, at Washington, D.C., for three years as private, Company C, 15th New York Engineers; dishonorably discharged for desertion, March 17, 1864, at Washington, D.C.

EMBICK, Frederick E. At age twenty-six enrolled at Williamsport, Pa., for three years. Major, 50th New York, from August 23, 1861; lieutenant colonel, 106th New York Infantry, from September 30, 1862; colonel, 106th New York from August 4, 1863; dismissed September 23, 1863; disability removed, December 14, 1863; recommissioned as a colonel, not mustered, December 19, 1863; honorable discharge, from September 21, 1863, at Foxville, Va. The 106th was a fighting regiment, losing 303 men during the last two years of the war.

A Frederick Embich is listed as entering West Point in 1855 but not graduating with the class of 1860.

FOLLEY, George W. At age twenty-four enrolled at Rome, N.Y., August 15, 1861, for three years. First lieutenant, Company C, from September 27, 1861; captain from November 27, 1862, succeeding Brainerd. Resigned May 18, 1863.

FOLWELL, William W. At age twenty-nine enrolled at Washington, D.C. for three years. First lieutenant, Company G, from February 13, 1862; Captain, Company I, from December 12, 1862; Major, February 1, 1865. Mustered out with regiment, June 13, 1865, at Fort Berry, VA.

FORD, George W. At age thirty enrolled at New York City on April 17, 1861, for one month as first sergeant, Company F, 7th Militia; mustered out June 3, 1861. Enrolled at Elmira, N.Y., on August 22, 1861, for three years in the 50th. Captain from September 18, 1861; major from June 3, 1863; resigned March 25, 1865. Breveted lieutenant colonel from August 1, 1864, and colonel, U.S. Volunteers. Ford was with the 7th New York Militia when it marched to Washington to secure the capital from attack in April 1861.

FULLER, Wallace S. At age nineteen enrolled at Rome, N.Y., August 26, 1861, for three years. Private, Co. C, August 31, 1861. Reenlisted January 15, 1864. Mustered out with company June 13, 1865, at Fort Barry, Va.

GIFFORD, H. J. Gifford served in the Gansevoort Light Guards in Rome, N.Y., as 2nd sergeant. At age twenty-five he enrolled in the 13th New York Volunteer Infantry on April 25, 1861 at Rochester, N.Y., for two years as drummer of Company A. He was promoted to 2nd lieutenant May 27, 1861. Discharged August 29, 1861, for promotion to 1st lieutenant, 33rd New York Infantry, Company D. Promoted captain, February 5, 1862. Attached 49th New York Infantry, May 22, 1863. Served in 49th Infantry, Companies G, A, and E. Mustered out July 3, 1865, brevet major.

GILBERT, Porteus C. At age twenty-two enrolled at Geneva, N.Y., on July 26, 1861, for three years. Captain from September 17, 1861; resigned July 6, 1862. After his service in the 50th, Gilbert served as a contract surgeon for the army on four separate occasions for from three to nine months each. His widow, Harriet Sayles Potter Gilbert (married August 30, 1860), stated in a pension deposition that Porteus Gilbert lived an intemperate and dissolute life for many years, was unable to practice medicine because of that, and was a salesman for a drug company in Oneonta, N.Y. They were later separated because of his alcoholism and for many years before his death did not communicate. Gilbert was in and out of treatment centers, insane asylums, and indigent care facilities for many years, even serving another stint in the army at Fort Clark, Texas, in the late 1870s. Late in life, however, Gilbert reformed, joined the Episcopal Church, and opened a clinic in Saratoga, N.Y., for the treatment of morphine and alcohol addiction. He claimed a pension for liver and spleen enlargement and died of cardiac syncope on June 11, 1898, at Saratoga, N.Y.

HEATON, John W. At age nineteen enrolled at Rome, N.Y., on August 30, 1862, for three years. Private, Co. C, promoted artificer, no date, mustered out with company June 13, 1865. Heaton served as WB's orderly.

HEWITT, Charles N. At age twenty-six enrolled at Geneva, N.Y., on July 26, 1861, for three years. Assistant surgeon from August 16, 1861; surgeon from June 28, 1862; mustered out with regiment, June 13, 1865. Married April 22, 1869, to Helen R. Died July 7, 1910, in Red Wing, Minnesota. Charles Hewitt's pension application filed March 19, 1909, was attested to and signed by William Watts Folwell, a former captain in the 50th.

HOLLINS, George Nichols. Born Baltimore, Md., September 20, 1799, died Baltimore, Md., January 18, 1878. Entered the U. S. Navy at fourteen and served in the War of 1812. Promoted captain in 1855 and commanded USS *Susquehanna* at Naples at the start of the Civil War. Resigned to go south and was appointed commander, CSN, June 20, 1861, and captured the steamer *St. Nicholas* nine days later. Commanded at New Orleans when that city was threatened by Farragut, but was relieved of duty and brought to Richmond to serve out the war there in the Confederate Navy Department. Faust, *Historical Times*, 365.

HOYT, Henry O. At age twenty-four, enrolled at Philadelphia, Pa., August 20, 1861, for three years. Second lieutenant, Company C, from September 17, 1861; quartermaster from June 11, 1861; relieved October 16, 1863; dismissed October 29, 1863.

JAMES, Edward C. At age twenty enrolled at Ogdensburgh, New York, August 14, 1861, for three years. Adjutant, 50th New York, from August 23, 1861; major, 60th New York, from May 1, 1862; lieutenant colonel, 106th New York, August 29, 1862; colonel, 106th New York, from September 30, 1862; discharged for disability, August 4,

1863. This is an extraordinary record. James went from being a twenty-year-old adjutant to colonel in thirteen months. Note that James was succeeded by Frederick E. Embick as colonel of the 106th.

LANG, John H. T. At age thirty enrolled at Buffalo, N.Y., January 5, 1863, for three years as private, Company C; corporal, May 1, 1863; sergeant, March 16, 1864; 1st lieutenant, December 18, 1864; mustered out, June 13, 1865.

LUDLAM, James D. Served in the 8th Illinois Cavalry as adjutant, September 18, 1861, to July 28, 1862; 1st lieutenant., Company F, July 28, 1862, to August 4, 1862; captain, Company F, August 4, 1862, to March 1, 1864; major, March 1, 1864, to January 5, 1865. Reese, *Report of the Adjutant General*.

MURRAY, John B. At age thirty-seven enrolled in Seneca Falls, N.Y., on August 1, 1861, for three years. Captain, 50th N.Y., from September 30, 1861; major, 148th New York Infantry, from September 17, 1862; lieutenant colonel from October 26, 1863; colonel from October 16, 1864; mustered out with regiment, June 22, 1865, at Richmond, Va. Breveted brigadier general, U.S. Volunteers, from March 13, 1865.

The 148th was late to get into combat, serving on garrison duty until November 1863. From then to the end of the war the 148th lost 267 men, mostly at Cold Harbor and the Siege of Petersburg.

Murray died October 8, 1884, of apoplexy caused by malarial poisoning contracted while with the 50th on the Peninsula. He was very active in the GAR and was hailed in an obituary as being responsible for initiating the laying of wreaths on veterans' graves, which led to the celebration of Memorial Day. He was buried in Seneca Falls, N.Y.

NORTON, Charles B. At age thirty-seven enrolled at New York City on May 28, 1861, for three years. Acting 1st lieutenant and quartermaster, 39th New York Infantry from June 6, 1861, not commissioned; quartermaster, 50th New York, from August 15, 1861; resigned June 11, 1862; assistant quartermaster and captain, U.S. Volunteers (staff officer, Quartermaster Corps), June 11, 1862; quartermaster, acting lieutenant colonel August 20, 1862; relieved January 6, 1863; resigned January 26, 1863. Breveted colonel and brigadier general by the State of New York for meritorious services in the field during the war, to date from March 13, 1865. Norton died January 29, 1891, in Chicago.

O'GRADY, Bolton W. At age twenty-seven enrolled at Syracuse,

N.Y., on August 21, 1861, for three years. Captain from September 18, 1861; resigned April 14, 1863.

PATTEN, John E. R. At age thirty-five enrolled at Elmira, N.Y., on August 14, 1861, for three years. Captain from August 26, 1861; resigned July 18, 1862. Married Sarah Maria Noble on November 28, 1851. Died at Hornellsville, N.Y., June 3, 1904, of apoplexy.

PERKINS, Augustus S. At age twenty-three enrolled at Washington, D.C., on December 18, 1861, for three years. First lieutenant from December 18, 1861; captain from July 18, 1862; killed, December 11, 1862, at Fredericksburg.

PERKINS, Henry W. At age twenty-seven enrolled July 26, 1861, at Elmira, N.Y., for three years. First lieutenant from September 13, 1861; discharged December 23, 1862, to accept commission as captain/AAG, USV.

PERSONIUS, Walker V. At age twenty-five enrolled at Millport, N.Y. (or Pa., there is a Millport in both states), on August 22, 1861, for three years. Captain from September 16, 1861; mustered out September 20, 1864, at Elmira, N.Y. After the war Personius was a merchant in Motts Corners, Elmira, and Brookton, N.Y. He was granted a veteran's pension but denied an invalid pension (migraine headaches) and died September 24, 1914. There is nothing in the record that states that Personius was ever married.

PETTES, William H. Graduated from the United States Military Academy twenty-third in a class of forty-five on July 1, 1832. He resigned his commission in 1836 after service in the Florida Indian War. At age forty-eight, enrolled at Buffalo, N.Y., on September 10, 1861, for three years. Lieutenant colonel from September 18, 1861; colonel from June 3, 1863, succeeding Stuart; mustered out with regiment July 5, 1865. Pettes died in Maryland on February 29, 1880.

POTTER, Hazard A. At age forty-eight enrolled at Geneva, N.Y., on July 26, 1861, for three years. Surgeon from August 16, 1861; resigned June 7, 1862.

PRITCHETT, Edward C. At age forty-eight enrolled at Geneva, N.Y., on August 24, 1861, for three years. Chaplain from September 10, 1861; mustered out September 20, 1864, Elmira, N.Y. Time sequence would suggest that Pritchett served out his term of enlistment and did not resign.

SEAMANS, Byron R. At age twenty-four enlisted August 20, 1861, at Rome; mustered in as corporal, Company C, September 16,

1861, to serve three years; died of disease (typhoid), October 22, 1861, at Halls Hill, Va.

SIMPSON, Wallace B. At age twenty-one enrolled at Vienna, N.Y., on September 4, 1861, as private, Company C, for three years; sergeant, December 16, 1861; re-enlisted January 1, 1864; mustered out June 13, 1865, at Fort Barry, Va.

SKILLEN, Charles H. Enrolled in Rome as captain on May 2, 1861, and by May 17, 1861, was the 14th's lieutenant colonel. He was killed in action on June 27, 1862, at Gaines Mill during the Seven Days Battles. The 14th New York Volunteer Infantry Regiment was mustered in on May 17, 1861, in Albany, N.Y., for two years. Company G was recruited in Rome.

SMALLEY, William O. At age twenty-seven enrolled at Geneva, N.Y., on July 30, 1861, for three years. Captain from August 16, 1861; resigned March 7, 1863. Married Mary Clymena Tuttle June 22, 1859. Died of consumption September 15, 1868, at Honeoye Falls, N.Y. Survived by wife Mary and daughter Lottie, born May 13, 1860.

SPAULDING, Ira. At age forty-three enrolled at Niagara Falls, N.Y., on August 5, 1861, for three years. Captain from August 29, 1861; major from October 13, 1862; lieutenant colonel from June 3, 1863; mustered out with regiment, June 14, 1865, at Fort Berry, Va. Breveted brigadier general, U.S. Volunteers, from April 9, 1865.

STANTON, Martin V. At age twenty-six enrolled in the 126th Regiment New York Volunteer Infantry at Waterloo, N.Y., for three years. First Sergeant, Company G, August 22, 1862; surrendered and paroled at Harper's Ferry on September 15, 1862 (Antietam Campaign); 2nd lieutenant, January 6, 1863; 1st lieutenant, March 4, 1863; wounded in action, February 6, 1864, at Morton's Ford, Va., and May 6, 1864, at the Wilderness, Va., and June 16, 1864, before Petersburg, Va.; died of wounds June 18, 1864, near Petersburg, Va.

STRYKER, John, Jr. At age twenty-one enrolled at Rome on May 2, 1861, for two years, 14th N.Y. Infantry. Second lieutenant from February 6, 1862; 1st lieutenant from February 6, 1862, captain from July 20, 1862; mustered out May 24, 1863, at Utica, N.Y.

STRONG, John B. At age twenty-three enrolled at Orcutt Creek, Pa., on September 9, 1861, for three years as private, Company A; Company C November 1, 1861; corporal, March 16, 1864; sergeant, January 1, 1865; mustered out June 13, 1865, as John B. Strang.

STUART, Charles B. At age forty-seven enrolled at Washington,

D.C., July 26, 1861, for three years. Colonel, 50th New York Volunteer Infantry (later Engineers) from August 15, 1861. Resigned, June 3, 1863.

TYLER, John T. At age twenty-two enrolled at Rome, N.Y., September 5, 1861, as private, Company C, September 7, 1861, to serve three years; died of disease, November 7, 1861, at Washington, D.C.

VERNAN, John S. At age forty-two enrolled at Washington, D.C., as private, Company E for three years; corporal, March 18, 1862; sergeant, March 24, 1862; died of disease, March 27, 1864, at Engineer Brigade Hospital, Washington, D.C.

WARREN, Freeman. At age twenty-eight enrolled at Elmira on September 3, 1861, as corporal, Company C for three years; discharged for disability, January 29, 1863, at Emery Hospital, Washington, D.C.

WOODBURY, Daniel P. Graduated from the United States Military Academy sixth in the class of 1836. Woodbury served in the artillery and the engineers, received three brevets to major general, and died in Key West, August 1864.

YATES, Henry. At age twenty-nine enrolled at Albany on February 18, 1862, for three years. Second lieutenant from February 18, 1862; died aboard steamer *Daniel Webster* in New York Harbor, May 23, 1862.

Appendix 2

Newspaper Articles

There were thirty-nine separate newspaper articles tipped into the text of the memoir and four lying loose. Of these, eight have been incorporated into the manuscript, nine are here in the appendix, two are included in endnotes, and the rest have been deleted for reasons of redundancy or irrelevance. Those included here add to a better understanding of WB's narrative and the times in which he lived.

Item 1

"The Sugar Party" was tipped into the memoir at the end of chapter 1. It is a fine example of the entertainments enjoyed by a small town society before the Civil War. There are two articles in the *Roman Citizen* reporting "Sugar Parties" much like the one described here; March 30, 1859, and April 13, 1859. The latter article is almost a rewrite of the event in this article, but much shorter. This article might be from the rival paper in town, the *Sentinel,* which could have covered the party in more detail. This article was not identified by newspaper or date.

THE SUGAR PARTY

During Wednesday night, the gale which had unintermittingly raged all week, subsided, and Thursday came in with a mild breath and a mottled sky, thus giving that necessary concomitant to all outdoor gatherings—a fine day.

THE STARTING

At 9-30 A.M., a happy company of about two hundred invited

guests, comfortably circumstanced in four cars, under charge of "Judge" Haselton, started for the Sugar Bush of Col. Savery, as previously announced. Included in that number were the Rome Brass Band and the Gansevoort Light Guard of this place, and not a few invited guests from Utica and elsewhere. All were full of joy and anticipating the sweetest possible time. Our pleasant village had hardly faded from our view before we arrived at a "getting off place," which seemed to be a short distance from anywhere, which has since been christened "Sugar Crossing." At this point the thoughtful Colonel had provided an assortment of conveyances for all of the company, and it was a luxury to see the bloom and beauty of Rome as they filed along the road in vehicles containing from one to twenty. The Gansevoort Light Guard here formed, presented arms to the cortege, and then marched up to the "bush," a distance of less than half a mile, escorted by the Rome Brass Band.

AT THE BUSH

Arrived at the Sugar Grove, which we have previously described, a sort of "squatter sovereignty" occurred in the twinkling of a eye. Cauldrons and kettles with the materials for a fire at different points throughout the grove had been provided, and around each of these a joyous little colony sprung up, making the giant old maples bend their heads inquiringly down to discover the cause of so much jollity. There was sap in abundance, and the defeat that this saccharine "institution" experienced will never be forgotten. During the forenoon the sun shone out cheeringly and that grove presented a strangely picturesque scene. The tall sweetners with their branches interwoven above us, the merry party below, the numerous campfires with their smoke floating lazily away before the gentle breeze, the happy group around, each making the very air ache with the fun, life and excitement they could not contain, presented a scene which must be experienced to be realized: it cannot be described. Add to this the enlivening strains of the martial band of the Gansevoort Light Guard and the maple sugar notes of the Rome Brass Band, and the sweetness of the occasion may be partially imagined. But when the sugar "began to come,"—when the bands played it, ladies smiled it, men talked it, soldiers drilled it, and everybody ate, looked and acted it, the intensity of sweetness became overpowering, and the sight of a barrel of vinegar, by way of a contrast, would have been indeed a relief.

THE DINNER

But the generous host and hostess of the festive occasion could not allow us to live sugar entirely. They had provided a "dining hall" with a table 100 feet in length, at which the party sat down, now increased to about 400, from Taberg, Camden, Adams and vicinities.

The table was provided with a tempting array of the substantial edibles which were properly discussed, and then succeeded with the luxuries of life. After two tables had been satisfied, each seating about 200, an extremely interesting affair occurred which we shall call a

SURPRISE PRESENTATION

Unknown to our generous host and hostess, whose munificent hospitalities their friends have repeatedly enjoyed, the company had, by a previous concert of action, provided an elegant silver pitcher, valued at $100, with the following inscription: "Presented to Col. R.G.Savery and Lady, by their friends, at his Sugar Bush, April 7th, 1859." Dr. A. Bickford, of Camden, was nominated as President, and Mr. E.H.Shelley in a neat, appropriate and feeling speech presented this handsome token of the esteem of the friends of our host and hostess. It was a perfect "surprise." The Colonel with much feeling accepted this valuable and deserved gift. He made a few heartfelt remarks in acknowledgement and promised a more formal response through the papers ere long. The thing was happily conceived and beautifully carried out.

During this time and afterwards Mr. L.W.Moulton, whose taking ways are well known, succeeded in getting some fine daguerreotype views of the grounds, the crowd, the Band, and the "Guard." By to-morrow, no doubt, they will be on exhibition at his rooms at Elm Row.

THE TARGET SHOOT

While enjoying sweets of life the Sons of Mars did not forget the sterner duties. A tent was prepared for the "Guard," near which they formed, drilled about the grove, and then occupied a portion of the afternoon in a Target Shoot. A target was set up 300 yards off, at which a large quantity of wild shooting was done, and through which some few balls were passed; one of the number hitting the centre.

During the afternoon the sky began to betoken rain, but fortunately none of any amount fell to seriously mar the festivities.

HOMEWARD BOUND

At about 5 P.M. the train started on its return, and arrived in Rome just previous to a slight shower. After their return, the Gansevoort Light Guard, which is becoming a well drilled, fine appearing company, performed some fine evolutions in our main streets: going through with "the street firings" in a soldierly style.

Thus concluded a huge merry-making, for which, in behalf of all present, we tender our warmest thanks to Col. Savery and Lady, with the wish that they may long to enjoy repetition of this, their 2d Second Annual Sugar Party.

THAT SUPPER

Col. Savery gave a supper at the Willett House, Tuesday evening, to the Rome Brass Band, the Staff and their ladies, and some other invited guests. At 11 P.M., the guests sat down to a bountiful repast, provided by "mine" excellent host of the Willett, (Wm. J. Riggs,) and discussed the eatables and good things for nearly an hour.

At the close of the eating, Col. Savery with wine glass in hand, arose and prepared as a toast,

"George Washington, the father of his country."

Drank in silence.

D. E. Wager was called upon to respond. The individual referred to arose after some hesitation and exhibition of bashfulness, and excused himself from speaking, as the call was entirely unexpected, the lateness of the hour, and that after the repast just finished, if he had had a pretty speech to recite when he came into the room, it was now owing to press of other matter (as the printer would say) crowded out. Besides it was not his province, nor in his line to respond to such a toast—that the response should be made by one of the military, or one of the Band; he "the speaker" was not a fighting man nor a musician although he had the reputation of being a great "blower." Our reporter has an indistinct or rather muddy recollection of what else was said, but as near as he can now remember, the speaker gave a toast eulogistic of, and complimenting the Rome Brass Band, and called upon the leader, Mr. Shelley, to respond. Mr. S. arose, and he too, like the preceding speaker, was to FULL for utterance. He said he was no talker, yet he sometimes could give persons fits, (Mr. S. is a tailor); he should not respond, but would give as a toast

"Col. Savery—he has come the rigs over us tonight."

This called up Col. S., who had been for the previous hour sandwiched between two ladies, and alternately dividing his attention between them and the viands on the table; in fact he seemed to our reporter, to be as undecided which way to turn his attention, as was the quadruped that starved to death between two stacks of hay. The Col. was averse to arising, (as we should have been had we been in his place) and he declined to respond, but called out Dr. Cobb, the Surgeon in the Staff. The vest of the Dr. like the waist coats of the rest of us, was too tight for comfortable breathing, or long talking; he declined to respond, but gave us a toast,

"Martha Washington, the mother of our country,"

And called upon our reporter to respond, who declined, but stated that as the father and mother of our country had been toasted, he would give,

"Mary Washington, the mother of George, the grandmother of her country!"

H. C. Case, was called upon to respond, but his military clothes sat so close to him, and gave him such fits that he could not speak either; his cramped situation, therefore, would naturally induce a train of tho't of those whose garments were more expanded—as a man on a sultry day, naturally thinks of ice water and the polar regions—so those in "tights" are ever panting for more room in their clothes; at least so we judged by the toast which he gave, which was as follows:

"The ladies of the 46th Regiment."

No one seemed to have gallantry enough to respond to this toast, (to the gentlemen's shame be it spoken.) This hesitation to respond, on the part of the military guests present, maybe, jealousy, owing to the fact that they considered the company toasted by Mr. Case, as a rival company of infantry!

We presume, however, that the above was not the reason, for it is fair to presume that the military present neither knew or cared whether the company "went afoot or horseback." A toast from another guest to the Gansevoort Light Guard, called out Capt. Skillin, who thanked the audience kindly for the kind expression of their feelings, and gracefully bowed his acknowledgements, in a manner, that only a military man can imitate.

A number of other sentiments were offered, but as our reporter took no notes, and now merely quotes from recollection, which the next day was none of the best, we cannot give.

About 50 or 75 persons sat down to the repast,which was hugely enjoyed, and from which they arose with hearts far lighter than their bodies. It was a joyous time, and will long be remembered.

ITEM 2

This article from the *Roman Citizen* accompanied WB's narrative of his attempts to raise a company for Colonel Stuart's regiment. Charles B. Stuart was not a graduate of West Point but had somehow managed to obtain the title of 'General.'

FRIDAY, AUGUST 16, 1861.

A NEW COMPANY FROM ROME

Capt. Wesley Brainerd of this village has received a letter from Gen. C.B. Stuart of Rochester, inviting him to recruit a company here for the new Engineer Corps, which has been organizing for some weeks past, and is expected to be mustered into service on the 25th inst, at Elmira. The Regiment is to be armed and equipped as a rifle regiment, and included among its duties will be occasional skirmishing, scouting, &c. But its main labors, and those peculiar to its organization, will be the re building of railroads and bridges destroyed by the enemy, the running of trains, or the performance of any other kind of mechanical and engineering work that may be necessary.

As the principal duties in this corps will be of a professional character, a position in it is more desirable than one of equal rank in an ordinary infantry regiment.

The style of men required are able-bodied and experienced military, civil and mechanical engineers, mechanics, boatmen, lumbermen, etc. There are a considerable number of young men in this village and the vicinity, of the classes above indicated who should be glad to join the Engineer Corps of Gen. Stuart.

Gen. Stuart is an able and efficient officer, a graduate of West Point, and has held a responsible position as a Government Engineer. He will make a capital commanding officer. The uniform of the regiment is now ready, and consists of first quality army cloth, trimmed with green. Two companies will be armed with rifles and sword bayonets and the other eight with rifled muskets, which are preferred for all purposes except scouting. The regiment is to be employed mainly in the engineer service, and when so employed the officers will receive ten per cent additional pay, and the men fifty per cent a day extra.

Patriot Citizens of Rome! Will you not respond to this call of your country, in this her time of need. So far our village has done but little. Let our young men buckle on their armor, and go forth to the field of battle to strike sturdy blows in defence of the Union, and for the punishment of the traitors, who now seek to subvert our liberties. Let the Romans now manifest their patriotism by promptly supplying a Company for Gen. Stuart's Engineer Regiment.

All our citizens know Wesley Brainerd. He was for years 1st Lieutenant, and afterwards Captain of the Gansevoort Light Guard in this village, and is a prompt, capable, and efficient officer. He is anxious to strike a blow for his country in this her time of peril, and has consented to give up his business, and to separate himself from his young and amiable wife to risk his life upon the battle field in defence of his country and the rights of Freemen. The muster rolls, and other necessary papers will be on hand on Monday next, at the armory of the Gansevoort Light Guard, where all who desire to join are invited to appear. As there is only ten days from this date to perfect the organization of the company, there must be no delay.

ITEM 3

This article appears in chapter 2 and elaborates on WB's narrative of his final departure from Rome for Elmira with the last of his recruits.

ROME AUG. 30TH 1861.

REVOLVER PRESENTATION

Capt. Wesley Brainerd, with a second squad of men, left this village for Elmira on Friday of last week, to be attached to Col. Stuart's Regiment of Engineers, and from there will go to the seat of war, and strike earnest blows for his imperiled country. He was escorted to the R. R. Depot by the Gansevoort Light Guard, Lieuts. Rowe and Northup, commanding, the whole headed by the Rome Band.

In the Armory just prior to starting, an elegant Revolver with 200 rounds of patent cartridges for the same, was presented Capt. Brainerd, as a slight token of the friendly regards of the many friends he leaves behind. The presentation was made by E. H. Shelley, in a few brief but

well timed remarks. The following is a copy of the letter which accompanied the "small arms" to which it refers:
To Capt. W. Brainerd,

Dear Sir: The undersigned a few of your friends, having with very great satisfaction, observed your assiduous efforts to raise a Company of Volunteer soldiers, and having learned of your success, and of your probable speedy departure from home to join our noble army, for the defense and maintenance of the Union, desire to place in your hands some slight memento of our regard for you as a neighbor and a soldier. The piece of "small arms" herewith delivered, though inconsiderable in value, we trust will be accepted by you as a pledge of the deep interest we take in the present struggle, and in all who are engaged in it.

You, and the brave men who have joined you, go from us with the earnest prayer on our part, that all your efforts, as soldiers, may be effectual in aiding the success of the government in subduing forever the spirit of rebellion.

Very truly your friends and fellow citizens:

B.J. Beach, L.E. Elmer, A. Sandford, E.H. Shelley, Daniel Cady, A.Ethridge, J.D.& S. Moyer, H.M. Lawton, Geo. Batchelor, W.W. Wardwell, H.D. Spencer, Foot Falley & Co., C.M. Dennison, Abbot & Redway, J.V. Cobb, R. Sandford, R. Walker, J.D. Ely, M.H. Fiske, J.B. Elwood, S.P. Lewis, J. Carley, A.H. Pope, G.I. VanAlen.

Capt. B. made a suitable response. He said this token of friendship on the part of his neighbors was most grateful to his feelings, and hoped that he should be able to discharge his duties in the field in such a manner as to do honor to his native town. Three cheers with a 'tiger' were then given to Capt. Brainerd and then the same number to the brave fellows who go with him.

An immense crowd accompanied the volunteers to the depot, and the tear started unbidden in many a manly eye, as the gallant boys were bidden "God speed," in their patriotic undertaking. May Heaven shield them from the bullets of the enemy in the day of battle, and return them safely to their homes and friends.

ITEM 4

This article is from the *Roman Citizen*, Friday, September 20, 1861, and relates a trip that Mr. Sanford, the editor, and A. H. Brainerd, WB's father, made to the camps of the 50th New York at Elmira, New

York. The perspective of the editor as to the state of the regiment is necessarily upbeat for the benefit of the many relatives of the soldiers who would read the article.

TWO DAYS AT CAMP LESLEY

For the purpose of getting a better knowledge of the every day duties of a military camp, and also to visit the Rome members of Capt. Brainerd's Company, with A.H. Brainerd, Esq., we left this village on the 11:30 train on Friday last, for Elmira, where Col. Stuart's Regiment is encamped. The weather was all that could be desired. We went by rail to Geneva, thence by steamboat up Seneca Lake some forty miles, and thence to Elmira via. Elmira and Canandaigua R. R. arriving at 10 p.m. We bestowed ourself with mine host Blossom, of the "Brainard House," one of the finest Hotels of the place.

On Saturday morning we paid an early visit to Camp Lesley, so named by Colonel Stuart, in honor of Mr. Lesley, chief clerk of the War Department. It is situated one mile from the village, on the farm on which the State Fair was held a year or two since. The meadow contains some forty acres of land, and is beautifully located for a military camp. For ten or fifteen rods from the road fence there is a gentle descent, then a gradual ascent, on which is a noble old orchard. On this ascent the encampment is pitched. It consists of some two hundred and eight army tents of all descriptions, laid out in streets and avenues, in perfect order and in strict accordance with military discipline—presenting a charming view to the eye of the spectator. Some of the tents are furnished with a board flooring, but by far the greater portion of them had nothing but a little clean hay upon which the soldiers lay, as many of them expressed it, to "enure themselves at once to the sober realities of a soldier's life."

Perfect neatness and order prevails in all parts of the Camp, the sanitary regulations for which, under the direction of the Regimental Surgeon, are rigidly enforced. Not an offensive odor could be observed within the bounds of the Camp—all the Tents are daily inspected by an officer, and nothing that can conduce to the comfort, health and happiness of the men is neglected or omitted. The Regiment is now nearly filled up, and the number in Camp is about nine hundred.

Col. C. B. Stuart, who has command, is well known to the citizens of the State. He is a hearty, robust man of about sixty years. He has at different periods held the office of State Engineer, had command of the Brooklyn Navy Yard, and has had a large experience in the construction of many of the most important Rail Roads in this State and the Canadas. He is a man of untiring industry, and a rigid disciplinarian. The amount of business he will dispose of in a given time, is astonishing. He has the confidence and respect of both officers and privates, and is emphatically "the right man in the right place." He richly deserved the high praises his Regiment has received from distinguished military visitors, who pronounce it in point of numbers, discipline, and the moral and physical character of the men, one of the best they have ever seen.

The Regimental Staff and all the Commissioned officers, are men of high social position, and several of them are graduates of West Point and other celebrated Military Schools.

The Regimental Surgeon is Dr. H.C. Potter, of Geneva, assisted by Dr. C.N. Hewett, of the same place. Dr. Potter has the reputation of being one of the most skillful Surgeons in the State, and under a blunt exterior is possessed of a heart that overflows with the "milk of human kindness." He is ably seconded by Dr. Hewett, and under their joint care the men rest assured that they will receive every possible care and attention.

Dr. Potter is deeply interested in the cause. One of the Captains of the Regiment, P.C. Gilbert, is his son-in-law, and his young and beautiful Bride of only a few months, spends much of her time in the Camp, and will accompany the Regiment to Washington, so as to be near her husband in case sickness or accident should befall him. Capt. Gilbert was with his wife on a bridal tour in Germany when the war broke out, but hastened home at once, raised the company he now commands, and is ready to peril all for the good of his Country. His wife is styled "Daughter of the Regiment," and is held in the greatest respect by all the men, over whom her influence is most potential for good.

In the Chaplain, we were most happy to recognize our old friend Rev. E.C. Pritchett, many years since a resident of Rome, a man of genial temperament, and admirably adapted to the responsible position he is called to occupy.

Thus far, the Regiment has been fed at the barracks located about sixty rods south of their camp. On Saturday last, Lieut. Folley had charge of this department. He had the entire supervision for the day, sees that the soldiers are properly fed, and reports the entire numbers of rations consumed during the day. We took dinner with the soldiers.

The bill of fare consisted of boiled pork, potatoes, capital bread, rice, &c., with the proper fixings. The quantities of food comsumed are enormous. On Saturday morning the chief cook informed us that he cooked fifteen hundred pounds of meat alone, and all other things in proportion.

The health of the Regiment is excellent, but five of the members being in the hospital, and all of these in a convalescent state.

The duties of the Sabbath are much the same as other days, with the exception of drilling. At about 9 o'clock a.m., Sunday, Gen. Van Valkenburgh, of Albany, visited the camp, and after a short conference with the commandant, it was whispered that marching orders had arrived, and considerable excitment insued. The surmise was correct. In a short time the Regiment was mustered by companies, the men notified that they were ordered to the seat of war, and were then asked if they were ready to obey the call. The question was responded to by an earnest, united and enthusiastic shout. They were all ready to go.

At 3 p.m., under the care of one of the surgeons, the men marched to the Chemung river, which ran on one side of the camp, and the entire Regiment, with the exception of those on guard, were required to go into the river and make thorough ablution. The boys enjoyed it hugely, and as cleanliness is indispensible to health, the bath has a most salutary effect on the men.

At 4 1/2 o'clock, the whole Regiment was paraded for religious service. The scene was imposing, and at times great solemnity prevailed. The men formed two sides of a square, the spectators forming the other sides. The singing was by a Choir, formed of the Regiment, and "Old Hundred" and the "Doxolagy" was rendered with an unction we have never heard excelled. The Sermon by the Chaplain, was earnest and to the point, and the soldiers duty to his Country and his God, was impressively enforced. During the prayer, we noticed that the wife of Col. Stuart, Capt. Gilbert, and many others of both sexes devoutly kneeled, thus setting a good example to the soldiers.

In conclusion, we are happy to state that the Company from this village, commanded by Capt. Wesley Brainerd, is one of the best in the Regiment. It numbers 101, and is composed in the main of stalwart, hearty men, who are anxious and ready to strike the traitors and help to crush the rebellion.

As a matter of interest to many in this County who have friends in the Regiment, we give below the names of the Regimental Officers and Commandants of Companies, and the muster roll of Capt. Brainerd's Company in full:

REGIMENTAL OFFICERS.

Colonel—CHARLES B. STUART.
Lieutenant Colonel—William H. Pettes.
Major—F.E. Embick.
Quartermaster—Charles B. Norton.
Adjutant—E.C. James.
Surgeon—Dr. H.A. Potter.
Assistant do.—C.N. Hewett.

Captains.
Geo. W. Ford, P.C. Gilbert, B.W. O'Grady, P.E.R. Patten, Wesley Brainerd, E.O. Beers, Ira Spaulding, J.B. Murry, W.V. Personius, Wm O. Smalley.

Daughter of the Regiment—Mrs. Captain Gilbert, Daughter of Dr. Potter.
Chaplain—Edward C. Pritchett.

Roll of Capt. Brainerd's Company:
Captain—WESLEY BRAINERD.
1st Lieutenant—George W. Folley
2d do.—Henry O. Hoyt.

Sergeants.
Orderly Sergeant—John J. Carroll; 2d do.—Simeon H. Brown; 3d do.—Geo. N. Burt; 4th do.—James Griswold; 5th do.—Nicholas Drewey.

Corporals.
Freeman Warren, Reuben Griswold, Escourt C. Wells, Daniel Swartfinger, Charles N. Eddy, Peter McKenna, Arthur B. Avery, Joseph Cook.
Wagoner—Edward Delos Thornton.
Drummer—John B. Squires.
Fifer—Isaac T. Seamans.

Privates.
Alexander Allen.
James Odell.
Charles Brainard.
Ira Jefferson Campbell.
Thomas Colopy.
Barthemon Burk.
Chester Covell.
Geo. DeLorain Smith.
John Cross.
Benjamin A. Snow.
William Edy.
Robert E. Thayer.
Wallace S. Tuller.
Charles H. Waterman.
Jacob Haff.
George Young.
John George L. Henry.
George Younge.
John Lynts, Jr.
Floyd Marshall.
Thomas Meek.
Freeman Ellis.
Byron R. Seamans.
Charles Benedict.
David Reese.
Wallace R. Simpson.
James Hillman.
Owen Crandell.
John Baldwin.
Joseph Henry Younge.
John O. Golden.
David Fitz Gerald.
Edward A. Lyman.
Irwin C. Tackenor.
Franklin Shepard.
Richard B. Hard.
James M. Brookins.
Charles W. Hicks.

Thomas McDonnal.
Albert W. Ellis.
Hiram E. Butler
Emanual Marshall.
John T. Tyler.
James Harris.
Wm. Blakesly.
Charles Purdy, Jr.
Henry W. Lyman.
Judson Odell.
Mather Platler.
George W. Goodspeed.
Abram Harrison.
Noan S. Rumsey.
Peter Belcher.
Addison Stone Ashley.
Richard H. Gardner.
Floyd Ashley.
Phillip Worth.
Wm. H. Hulslander.
James Pendergrast.
James A. Boyce.
John B. Strong.
Avery Dawley.
Sylvanus S. Bixby.
Francis A. Wood.
George N. Cown.
Kimball S. Wood.
Orsen B. Welch.
Chester Myers.
Chester P. Rolph.
Willis H. Cole.
Charles S. Price.
Franklyn Graham.
Charles Mackinson.
Samuel Doney.
Charles L. Whitman.
George V. Canfield.
John N. Harvey.

George Kye.
Samuel Welch.
Filander B. Dunlap.
Walter McKinney.
Wm. A. Heath.
N.B. Hughes.
Wm. Harer.

The visit was in the highest degree satisfactory to us, and we have no doubt the Regiment will do honor to itself and the Country when it shall be brought face to face with the enemy.

The Regiment was to have left for New York on Wednesday last.

ITEM 5

This article was tipped into the memoir at the end of chapter 4 and has a complete discussion of the equipments needed and the procedure followed for building a ponton bridge. It is from the *Roman Citizen* and dated January 17, 1862.

ARMY CORRESPONDENCE.

WASHINGTON, JAN. 4TH, 1862

A Sandford, Esq. Dear Sir: I assume that the doings of the 50th are sufficiently interesting to its many friends in and about Rome, to merit an occasional column or so in the old Citizen, so welcome and so rarely seen here. As the regiment has remained in Washington, apparently inactive, for two months or more, one might naturally inquire to what purpose? The question admits a ready answer, for, like the whole army of the Potomac, we have been getting ready to do something, and the "On to Richmond" people may console themselves that when this army moves again it will move in earnest. But how, where, or when, that movement shall be made, all here seem willing to trust our leaders. During the time we have been here, our regiment has been practiced in the various duties of engineers in active service, such as making fascines and gabions, constructing roads, building trestle and pontoon bridge, etc. We have also thrown up a six gun battery near the camp, on the east branch of the Potomac, which is considered by military men a fine specimen. Our men, however, are not excessively fond of digging, they prefer employment in some work requiring more skill than in handling the pick and shovel. The Pontoon drill pleases them, for it is both exciting and interesting. A pontoon bridge, as you know, is built upon boats. These boats, or batteaux, are 31 feet long by 5 feet wide, and 3 1/2 feet deep, built scow fashion. Each batteaux weighs 1,250 pounds, and has a buoyant power of about 15,000 pounds; 34 of these are attached to one train. In forming the bridge, these batteaux are placed 20 feet apart, and every alternate one anchored up and down stream. Across the batteaux are laid five string pieces, called balks; these are five inches square, and are lashed to the batteaux; the chess, or flooring, is next laid over the balks. These chess are 1 1/4 inches thick, 1 foot wide, and 15 feet long; they are not nailed to the balks, but held in their position by side rails laid over them, one at each side, and lashed to the balks underneath, leaving a carriage way ten feet wide. This forms a temporary and portable bridge, capable of sustaining the heaviest artillery or baggage trains. Troops, in passing over, take the route step, as the cadence step would produce a swaying motion, highly prejudicial to an upright position. We have (or will have when the train is completed,) one thousand feet of this bridge, the whole of which can be thrown out in one hour's time, taken in, packed on the wagons and started off in another hour. A full engineer company consists of 150 men. By combining two companies we have a surplus to load and unload the wagons. Before commencing the bridge, the company is divided into squads, each squad in charge of a non-commissioned officer. Each squad has its own particular duty to perform. The boatmen, cablemen, balkmen, lashers, chessmen, side rail men, and mawlers. They are drawn up in line on the shore, and at a given signal, take their positions. The first boat is placed in its proper position; the cablemen board her and pay out the cable; the balkmen march forward with the balks, place one end on the boat; the lashers lash them, and the boat is shoved off the length of the balks, when the other ends are fastened; on come the chessmen with the flooring; the boatmen pass back and take another boat; the second boat is boarded in like manner by the cablemen and lashers, the balks arrive, the boat shoved off, the flooring laid, the side rails fastened, the mawlers smooth any irregularities. More boats, more balks, more chess, and more side rails, and so the bridge seems to push itself out like a huge telescope, and before we are hardly aware of it is across the stream. You must know that there is considerable science displayed in the construction of this bridge, as simple as it may appear when reduced to military order and precision. There is another mode of constructing the bridge which is sometimes necessary, that is, to put it together along one side of the stream, float it to its proper position, fasten one end, and allow the other to swing across with the current. To transport this train, with all its boats, lumber, and trappings which are necessary, requires 85 teams of 6 horses each, or about 500 horses. The teamsters form a separate company under charge of an officer, who enjoys the title of "Head Teamster;" and this is a Pontoon train.

Our regiment will be divided into divisions of two companies each. Each division will have charge of a train, and be assigned to separate corps of the army. The first train will go out in charge of Co. E, Capt. Spaulding, and Co. C, which is our company.—When we are going remains as yet a profound secret. Madam rumor has consigned us nearly every place from Halifax to San Francisco, and from Fort Monroe to San Juan Island, but we shall probably know when we recieve marching orders, and not much before. We have lately recieved a new uniform of dark blue, and our new rifles are expected daily. They are a Prussian arm, and said to be a very superior weapon. We have also new Sibley tents. These tents are about 15 feet diameter, and 20 feet high, cone shaped, a la sugar loaf. In each of these tents 20 men establish their parlor, dining room, bed room, library, armory, kitchen, wood shed, and nursery. Here they lay, with feet to the centre, (the stove is in the center,) and heads radiating towards the circumference. Here's where you get your cool, fresh air, and plenty of it, and here's where you are sure to have a good conscience, and a ravenous appetite. Think of that, ye dyspeptics, and enter the army at once.

It is with feelings of gratitude I am able to inform the friends of the company of its perfect good health. One man has died since my last to you, his name was John T. Tyler, of McConnellsville. On the 1st of November last, we had twenty-five sick and in the hospital; to day there is but one. The weather is cool, but pleasant; we have had no snow, and very little weather that we would consider wintry. The men are all in fine spirits, and anxious to be called into active service.

And here I take great pleasure in acknowledging the receipt of a number of articles, contributed for the comfort and welfare of my company, by their friends in Rome and vicinity. To the ladies of the Presbyterian church, the Union Volunteer Relief Association, and to many others, the company offer their sincere and hearty thanks.

With great respect, I remain, your friend,
Wesley Brainerd.

ITEM 6

This article was tipped into the memoir in chapter 10 and contains an account of the crossing of the Rappahannock at Fredericksburg that adds to and elaborates WB's narrative. It is signed with the initials W.S. which could stand for William Swinton.

CROSSING THE RAPPAHANNOCK

Interesting Incidents and Full Particulars
Correspondence of the New York Times:
Fredericksburg, Va.,
Thursday Night, December 14.

I localize this letter Fredericksburg, but it is assuredly "living" Fredericksburg "no more." A city soulless, rent by rack of war, and shooting up in flames athwart night's sky, is the pretty little antique spot by the Rappahannock; erewhile the peculiar scene of dignified ease and retirement.

The advance of the Right Grand Division of the Army of the Potomac rests here tonight, after a series of operations which are certainly among the most extraordinary of the war. To those who retired to rest, uninformed of what night was destined to bring forth, the spectacle this morning must have seemed strange enough to be the improvisation of the magician's art. One hundred and fifty pieces of cannon covered the circular sweep of the heights of Fredericksburg; 150,000 men in battle array had sprung from the earth, and lay, ready for the advance, behind those heights. But to the initiated, who spent the night in vigils and knew what work crowded its busy hours, it was all intelligible enough. All night artillery came, and came with its ceaseless, heavy rumble, and as each battery arrived from the rear it was posted in the place selected for it by the Chief of Artillery. All night the perpetual tramp of men moving to the front filled the air. Pontoon trains unwound their long, snake-like forms, and were drawn, each boat by its team, down to the river's brink.

It has been determined, in council of war, held on Wednesday, that, instead of extending our lines of operations along the river from Falmouth to Port Conway, the entire army should be crossed at or near Fredericksburg.

Five pontoon bridges were to be thrown across the river—the first at the Lacy House, which lies directly opposite the end of the main street of Fredericksburg, half a mile below Falmouth; the second and third within a few hundred yards from the first. The remaining two were to be thrown over a mile and a half or two miles further down the stream, and on these the grand division of Gen. Franklin, the left, would cross, while Sumner's and Hooker's grand divisions, right and centre, would use the three upper ones.

It was about 3 o'clock this morning when the boats were unshipped from the teams at the river's brink. Swiftly and silently the Engineer corps proceeded to their work. A dense fog filled the valleys and water margin, through which the bridge-builders appeared as spectral forms. The recital of the special correspondent with the left will inform you of the details of the construction of the two lower, Franklin's, bridges. Work there was performed with perfect success; the engineers being allowed to complete the first without any interruption whatever, while the construction of the other was but slightly interrupted by the fire of the rebel sharpshooters.

We were not so fortunate with the upper bridges. The artificers had but got fairly to work, when at five o'clock the firing of two guns from one of the enemy's batteries announced that we were discovered. They were signal guns. Rapid volleys of musketry, discharged at our bridge builders, immediately followed. This was promptly responded to on our side, by the opening of several batteries. The fog, however, still hung densely over the river. It was still quite dark, and the practice of the artillerists was necessarily very much at random. The Engineer Corps suffered severely from the fire of the sharpshooters concealed in the town. The little band was being murderously thinned, and presently the work on the bridges slackened, and then ceased.

Meanwhile the firing from our batteries, posted about a mile from the river, was kept up vigorously. The effect was singular enough, and it was difficult to believe that the whole affair was not a phantasamagoria [*sic*]. It was still quite dark, the horizon around being lit up only by the flash of projectiles, which reappeared in explosive flame on the other side of the river.

Daylight came, but with it came not clearness of vision for onlookers. The mist and smoke not only did not lighten, but grew more opaque and heavy, hugging the ground closely. Our gunners, however, still continued to launch their missiles at a venture. The rebel batteries hardly returned our fire, and this chariness of their ammunition they preserved all day, not a dozen rounds being fired during the whole forenoon.

Toward 8 o'clock a large party of general officers, among them Gen. Burnside, the corps commanders, and many others of high rank, had congregated in front of and on the balcony of Gen. Sumner's headquarters, Phillips' House, situated about a mile directly back of the Lacy House. The performance could be heard, but not seen, the stage was obstinately hidden from view, and all were impatient that the curtain should rise. Aids and couriers came and went with messages to and from the batteries and bridges.

At 9 1/2 o'clock official notification was received that the two bridges on the extreme left were completed, and Gen. Franklin sent to Gen. Burnside to know if he should cross his force at once. The reply was that he should wait until the upper bridges also were completed.

Meantime, with the latter, but little progress was made. During the next couple of hours half a dozen attempts were made to complete the bridges, but each time the party was repulsed with severe loss. On the occasion of one essay, Capt. Brainard of the 50th New York Volunteer Engineers, went out on the bridge with eleven men. Five immediately fell by the balls of the rebel sharpshooters. Capt. Perkins led another party and was shot through the neck, and the 66th and 57th New York regiments, which were supporting the 50th and 15th New York Volunteer Engineers, Gen. Woodbury's brigade, suffered severely. It was a hopeless task, and we made little or no progress. The rebel sharpshooters, posted in the cellars of the houses of the front street, not fifty yards from the river, behind stone walls and in rifle-pits, were able to pick off with damnable accuracy any party of engineers venturing on the half-completed bridges.

The case was perfectly clear. Nothing can be done till they are dislodged from their lurking places. There is but one way of doing this effectually, shell the town. At 10 o'clock Gen. Burnside gives the order, "Concentrate the fire of all your guns on the city, and batter it down!" You may believe they were not loth to obey. The artillery of the right, eight batteries, Col. Tyler, left centre, seven batteries; Capt. DeRussy, left, nine batteries. In a few moments these thirty-five batteries, forming a total of one hundred and seventy-nine guns, ranging from ten-pounder Parrotts to four and half inch siege guns, posted along the convex side of the arc of the circle formed by the bend of the river and land opposite Fredericksburg, opened on the doomed city. The effect was, of course, terrific, and, regarded merely as a phenomenon, was among the most awfully grand conceivable. Perhaps what will give you the liveliest idea of its effect is a succession, absolutely without intermission of the very loudest thunder peals. It lasted thus for upward of an hour, fifty rounds being fired from each gun, and I know not how many hundred tons of iron were thrown into the town.

The congregated Generals were transfixed. Mingled satisfaction and awe was upon every face. But what was tantalizing, was that though a great deal could be heard, nothing could be seen, the city being still enveloped in fog and mist. Only a denser pillar of smoke, defining itself on the back ground of the fog, indicated where the town had been fired by our shells. Another and another column shows itself, and we presently saw that at least a dozen houses must be on fire.

Toward noon the curtain rolled up, and we saw that it was indeed so. Fredericksburg was in conflagration. Tremendous though this firing had been, and terrific though its effect obviously was on the town, it had not accomplished the object intended. It was found by our gunners almost impossible to obtain a sufficient depression of their pieces to shell the front part of the city, and the rebel sharpshooters were still comparatively safe behind the thick stone walls of the houses.

During the thick of the bombardment a fresh attempt had been made to complete the bridge. It failed, and evidently nothing could be done till a party could be thrown over to clean out the rebels and cover the bridge head. For this mission General Burnside called for volunteers, and Colonel Hall, of Fort Sumter fame, immediately responded that he had a brigade that would do the business. Accordingly, the 7th Michigan and 19th Massachusetts, two small regiments, numbering in all about four hundred men, were selected for the purpose.

The plan was that they should take the pontoon boats of the first bridge, of which, there were ten lying on the bank of the river, waiting to be added to the half-finished bridge, cross over in them and, landing, drive out the rebels.

Nothing could be more admirable or more gallant than the execution of this daring feat. Rushing down the steep banks of the river, the party found temporary shelter behind the pontoon boats lying scattered on the bank, and behind the piles of planking destined for the covering of the bridge, behind rocks, etc. In this situation they acted some fifteen or twenty minutes as sharpshooters, they and the rebels observing each other. In the meantime new and vigorous artillery firing was commenced on our part, and just as soon as this was fairly developed, the 7th Michigan rose from their crouching places, rushed for the pontoon boats, and pushing them into the water, rapidly filled them with twenty-five or thirty each.

The first boat pushes off. Now, if ever, is the rebels' opportunity. Crack! crack! crack! from fifty lurking-places go rebel rifles at the gallant fellows, who, stooping low in the boat, seek to avoid the fire. The murderous work was well done. Lustliy [*sic*], however, pull the oarsmen, and presently, having passed the middle of the stream, the boat and its gallant freight comes under cover of the opposite bluffs.

Another and another boat follows. Now is their opportunity. Nothing could be more amusing, in its way, than the result. Instantly they see a new turn of affairs. The rebels pop up by the hundreds, like so many rats, from every cellar, rifle-pit, and stone wall, and scamper off up the streets of the town. With all their fleetness, however, many of them were much too slow. With incredible rapidity, the Michigan and Massachusetts boys sweep up the hill, making a rush for the lurking-places occupied by the rebels, and gaining them, each man capturing his two or three prisoners. The pontoon boats on their return trip took over more than a hundred of these fellows.

You can imagine with what intense interest the crossing of the first boat-load of our men was watched by the numerous spectators on the shore, and with what enthusiastic shouts their landing on the opposite side was greeted. It was an authentic piece of human heroism, which moves men as nothing else can. The problem was solved. This flash of bravery had done what scores of batteries and tons of metal had failed to accomplish. The country will not forget that little band.

The party once across, and the rebels cleaned out, it took the engineers but a brief period to complete the bridge. They laid hold with a will, plunging waist-deep into the water, and working as men work who are under inspiration. In less than half an hour the bridge was completed and the head of the column of the right grand division, consisting of Gen. Howard's command, was moving upon it over the Rappahannock. A feeble attempt from the rebel batteries was made to shell the troops in crossing, but it failed completely.

Your correspondent found an opportunity to cross the river along with the party who first went over, in a boat, having been curious to take a closer view of the city which we have for nearly a month been observing over the river, not three hundred yards wide, without the power of visitation. As the rebels were in very considerable force on the heights back of the city, one could not extend his perambulations beyond the street fronting on the river. Every one of the houses which I here entered, a dozen or more, is torn to pieces by shot and shell, and the fire still hotly rages in a dozen parts of the city. A few citizens, a score or two, perhaps, male and female, presently made their appearance, emerging out of the cellars, whither they had taken refuge during the bombardment. Three women, white, whom we found in a cellar, told us that they, with a majority of the inhabitants, had moved out of Fredericksburg a fortnight or so previously, but that growing reassured by our long delay, they, with a good many others, had come back the evening before. The former inhabitants they report as now living in various parts of the environs, some in negro huts, and others in tents made with bedclothes, &c.

During the afternoon of the bombardment we observed a couple of white handkerchiefs waved out of the windows of a house in the city. This was taken by some for a flag of truce, and the Chief of Artillery was on the point of causing the shelling to cease. Gen. Burnside, however, decided that it was probably merely only the wonted rebel ruse, and ordered operations to be continued. We found out that the demonstrations were made by two of the women referred to, with the desire that we should send over a boat and convey them away from Fredericksburg.

Among other prominent objects during the bombardment was a large British flag, flying over the house of the English Consul. This personage, however, was not found in his house, when we entered the city, and the flag was taken possession of and brought away.

A number of rebel dead were found in various parts of the city, some exhibiting frightful mutilations from shells, and I took as a trophy a rifle, still loaded, out of the grasp of a hand belonging to a headless trunk.

The infantry in the city appear to have been Mississippians, South Carolinians and Floridians; those of them that we took prisoners were wretchedly clad, and mostly without blankets or overcoats, but they generally looked stout and healthy, and certainly in far better condition than they could have been were there any truth in the report of some deserters the other day, to the effect that for three weeks they had nothing to eat but the persimmons they were able to pick up.

Although we are not yet fully informed of the present positions of the enemy, there seems to be good ground to claim that General Burnside has succeeded in outgeneraling and outwitting them. His decoys to make them believe that we were about to cross our main force at Port Conway seem to have succeeded admirably. I suppose there is no harm now in my mentioning that among the ruses he employed was sending down, day before yesterday, to Port Conway three hundred wagons, and bringing them back by a different road, for the sole purpose of making the rebels believe that we were about to cross the river at that point. To the same end, workmen were busily employed in laying causeways for supposed pontoon bridges there, while the gunboats were held as bugaboos at the same place. Completely deceived by these feints, the main rebel force, including Jackson's command, seems to have been two or three days ago transferred twenty or twenty-five miles down the river. It must be remembered, however, that without the utmost celerity on our part they can readily retrieve this blunder by a forced march or two. Signal guns, at 5 o'clock this morning gave them the cue to what was going on, and doubtless they have not been idle during the intervening hours. Tomorrow will disclose what unseen moves have been made on the chessboard. W.S.

Item 7

This letter was written by L. E. Seymour of Company C to his brother soon after the Battle of Fredericksburg. It was reproduced in the Rome *Sentinel* in 1906 shortly after Seymour's death. It complements WB's account of the attempts made to build the ponton bridge across the Rappahannock from the viewpoint of an enlisted man. The article was not tipped into the memoir but was loose in the clamshell case containing the original.

Camp near Fredericksburg, Dec. 13, 1862.

Dear Brother: Thinking you will be anxious to hear from me I thought I would take this, the first opportunity, to write you. I am all safe and sound and not injured in the least. We left our camp Wednesday afternoon about 4 o'clock with our train and traveled almost to the river, where we lay until about 12 o'clock at night. Then we went down to the river right before the city. The rebs' pickets were on the other side; we could see them plainly, as the river is not very wide. We commenced to lay our bridges. We had batteries on a hill just behind us and we had a regiment of infantry on each side of us close to the

river. They lay down on the bank. We had laid our bridge almost across when a volley of musketry came pouring into us; it was awful. *Captain Brainerd* ordered us off. As soon as possible we ran ashore and hid. I and several others went up the river, creeping, and we soon got clear of the bullets. We managed to get up the hill and behind the batteries, and then we went off to the right, where stood a large brick house where Burnside was. Soon all our boys were there except those that were killed and wounded. Wilcox was the first man killed in our company, and Blakeslee was in our company. Our company and H were together laying the bridge, and A and J were about a rod above laying their [*sic*]. K and F were half a mile below us. Another bridge was laid by the regulars. We got together again and went down to try it over. The captain was wounded in the arm, the Captain of Company J was killed and two or three of our men wounded. We left it the second time and about an hour afterword went at it again. Watson was then wounded in the hand. We then left until afternoon, when old Burnside got everything ready and sent us down to work. As soon as we got there the rebs commenced again. They were in cellars and houses. Then our batteries played on them so heavily that it silenced them. Our batteries kept playing on them all the time steadily. Then a regiment of infantry came pouring down the hill and jumped in to the boats we had in the water. They shoved across and charged on the rebs, which made them skedaddle, and down came their flag. Soon our bridge was completed and we left the ground that night. Our troops kept crossing and the city is ours. I will give you a list of the killed and wounded as far as I can. Killed—Wilcox and Blakesley. Wounded—Captain Brainerd in the arm, Watson in the wrist, Pete McKinney in the shoulder, Dunlap in the shoulder, Bascomb slightly in the leg. That is all, I believe, in our company. In Company H there were two killed and six or seven wounded. In all the companies at work there were some hit. They are now shelling all the time and have kept it up ever since the bridge was laid. The cannons roar all the time. We are now back to our camp, about four miles from the enemy, and we can hear distinctly every cannon. I got a ball through my overcoat tail. It must have passed between my legs. Tell all inquiring friends that I am safe and sound. Write soon and send those boots as soon as possible.
Your affectionate brother,
L. E. Seymour.

ITEM 8

This is the complete text of Lt. Col. Barton Stone Alexander's article in defense of the engineers following the Fredericksburg controversy. It was tipped into the memoir in the middle of Chapter Eleven and appeared originally in the New York *Herald*.

Our Pontoons in Virginia.
Washington, D.C., 1863.

In your paper is an article on the subject of pontoon bridges. In the opening sentence of that article you state "that the terrible disasters at Fredericksburg and other places, which resulted in such slaughter and disgrace to the Union, were caused by the abominable abortions called pontoons."

You also go on to state that "in this particular all the world is ahead of us," that Great Britain, Germany and Russia have adopted a much better system, which came from an American, and which consists of light corrugated iron wagon bodies of such dimension and form as to serve as a wagon for the transportation of supplies—a boat with which the troops can cross rivers—a raft to carry heavy ordinance, but, most important of all, as a bridge.

You then ask, "Why is it that this or some better system was not, immediately after the late disasters, substituted for the miserable wooden scows, called pontoons, projected by imbeciles, under the Secretary of War, and which have carried such desolation to every family?"

If the people of this country could be made to believe that our recent disasters were caused by our system of pontoon bridges, that this system was projected "by imbeciles under the Secretary of War," and that this wicked abortion must continue to be made "to serve private interests, regardless of consequences," there can be no doubt but that they would at once demand a reform in this branch of our service; that the "imbeciles" should at once be disgraced, and that all "private interests" growing out of the present system should be sacrificed. Such, indeed, would appear to be the object of your article.

Now, Mr. Editor, I think you have been misinformed on this subject, and as I am responsible more than any other person for the system of pontoon bridges which we have in our service, and as the whole army knows that I am responsible, I must beg your indulgence while I undertake to correct the errors into which you may have inadvertently been led.

The questions which the public care to have answered are these:

1. Are we behind the age in our system of pontoon bridges? I maintain that we are not.

2. Is our system the best that is known to the nations of Europe? I maintain that it is.

3. Was due inquiry made as to what was the best system of pontoon bridges now in use in England, France, Russia, Austria, Prussia and the German States before our present system of pontoon bridges was adopted? I maintain that it was.

4. Can this system be improved? I answer, all analogy teaches us that it can be, and that it will be hereafter.

5. Was the mechanical ingenuity of our people invited to improve, or suggest improvements, before this system of bridges was finally determined upon? I maintain that it was.

6. Is the War Department, or the persons, whether "imbeciles" or not, who are acting under it, now ready to adopt an improved system of military bridges if one can be found? I maintain that it is both ready and willing to do so.

A simple statement of the facts in the case will be the easiest way to show the grounds upon which I base my answers to these questions.

In the month of September, 1861, General McClellan sent for me and told me that he wanted me to get up ten or twelve bridge trains for our armies. I had not asked for this duty; I had simply to obey orders. I knew something of the subject before undertaking it, and I had the assistance of Captain Duane, of the Corps of Engineers, who had given it as much attention, both practically and theoretically, as any other man in the country. The General told me that he did not wish to "tie my hands," that he thought it best not to give me positive orders in writing, usually done in military matters, as to what I should do in the performance of this duty. He advised me to look well into the system of bridges now in use in Europe, and adopt such, or such combinations of several, as I thought best, with any improvements that might be suggested, keeping in view all the time the character of the rivers to be crossed, particularly in Virginia; the fact that the country was wooded, and therefore afforded materials for more substantial bridges than could be made by pontoons; that there would be great mechanical skill in the army, and that it would be supplied with all necessary tools.

Such, in general terms, were my instructions, and I set about executing them. After examining the different systems of military bridges then in use in Europe, I came to the conclusion that the French was the best. I accordingly adopted the French pontoon as the basis of our system. For this conclusion I respectfully protest against being called an "imbecile under the Secretary of War."

My own researches had led me to believe that the French pontoon was the best boat then used in making military bridges. The experiments that had been made with it at West Point had led Captain Duane to the same conclusion; and, besides this, General McClellan was of the same opinion. Now, General McClellan at that time probably had more knowledge of pontoon bridges thna [*sic*] any other man in the United States. He had for many years been on duty, as an officer, with our only company of engineer soldiers at West Point. He there thoroughly studied the subject of pontoon bridges, both theoretically and practically. He afterwards travelled all over Europe, during the Crimean war, and examined the different kinds of military bridges then in use, giving the subject his particular attention. With all his knowledge about pontoon bridges, he coincided in opinion with Captain Duane and myself, and, if imbecility must be charged to the employes [*sic*] of the War Department, he must share it with us.

It should be stated, however, that I did not confine myself exclusively to the French pontoons. The opportunity to test the merits of other bridges was not suffered to be lost. We had an India rubber train; we drilled with it and with the French train, and there was no difference of opinion as to the superiority of the latter. I also had made an "advanced guard train" of canvass, after the manner of Russians. This was used on the Chickahominy, and is even lighter than the corrugated iron wagon bodies. In addition I also had a trestle train made, after the manner of the Austrians. But, most important of all, I did not forget the corrugated iron wagon bodies, which had for many years been used in our service in crossing the Plains. I had one hundred of these wagons. We made boats and rafts of them, but never succeeded in making a bridge. We experimented with them, however, until it was the concurrent opinion of all who witnessed the experiments, except perhaps the manufacturer, that they were not suited to form the supports of a military bridge for a large army. For crossing small streams with a small force, as had been the case in sending detachments of troops across the continent, when there was plenty of time, they had been used as a makeshift for a boat, and sometimes with success; but, as they are now made, it was found that we could not make a practicable bridge with them in any reasonable time. Doubtless they may be improved here after; but so may the art of printing. The enemy were knocking at the gates, and we wanted pontoons, just as we want the morning papers; we must have both, without waiting for the future improvements. I will add, in conclusion, that our pontoon trains have been eminently useful and successful. General Banks first used them at Harper's Ferry, in crossing the Potomac, before General McClellan opened the peninsular campaign. General McClellan was there and saw the bridges, but suggested no improvements. They were of great use during the siege of Yorktown in landing troops at Poquosin river, and in building several long bridges over Wormley's creek. At West Point General Franklin landed the whole of his division and all his artillery in a few hours with the aid of these pontoons. We again used them when we were advancing on Richmond in crossing the Chickahominy river; and in the retreat from Harrison's Landing nearly the whole of General McClellan's army, with all its *impedimenta*, passed over a bridge, at the mouth of this stream, some two thousand feet long, a feat which deserves a name in history. Again, after the battle Antietam, nearly the whole of our army passed the Potomac on these bridges. No accident has ever happened. Our troops have confidence in them; even the horses have learned that they are secure. When the engineers build a pontoon bridge no question as to its security is ever raised. The following is an extract of a letter from Brigadier General J.G. Barnard, the chief Engineer of the Army of the Potomac:

Camp Lincoln, Near Fair Oaks, Va., June 18, 1862.
Gen. J.G. Totten, Chief Engineer United States Army, Washington, D.C.:

> Our short experience has shown that the French batteau [*sic*] is of universal applicability for all purposes where there is water, and that it makes a perfectly reliable bridge. The Birago trestle bridge is dangerous, unreliable and only fit for advanced guards. We brought no India rubber pontoons with us. I should consider it preferable to the trestle bridge, but not reliable for the uses of a large army.

Such were the opinions of General McClellan's Chief Engineer, after our experience at Yorktown, at West Point and in crossing the Chickahominy.

But these pontoons are heavy; this is granted. They weigh one thousand three hundred pounds; not one ton, as stated in your article. They did not get to Fredericksburg when General Burnside wanted them; this is also granted. Had they been as light as anything except a telegraph message they could hardly have been there at that time; unless they had been forwarded in season. We wanted them at Yorktown, and we had them there. We wanted them at West Point, and could not have landed without them, and we had them there. We again wanted them both in crossing and recrossing the Chickahominy, and we had them at both places. They were afterwards wanted on the Upper Potomac, and they were there; and merely because, for some reason or other, they were not found at the Rappahannock when most wanted, it is no more fair to condemn the boats because they are heavy than it would be to condemn our cannon when the roads in Virginia will not permit them to be moved.
B.S. ALEXANDER, Lt.Col.

ITEM 9

This fine article was tipped into the memoir at the beginning of chapter 16. It is an observation of the Army of the Potomac after the Battle of Gettysburg and is an insightful portrait of a great army. It appeared in the Baltimore *American* on Tuesday, July 21, 1863, and was unsigned.

THE ARMY OF THE POTOMAC

It is very seldom in the history of the world, of one lifetime, at least, that the curious observer is privileged to look upon a military organization gotten up on so grand a scale, and so unmistakably composed of veteran soldiers, as the Army of the Potomac as it at present appears. If there ever was any real military elegance about it, as it is exhibited on holiday occasions in the cases of smart volunteer corps in the great cities, when such peace-warriors are bright with lace and gay with dancing plumes, that look has disappeared long ago; and in place of it is now seen the rustiest, dustiest, sun-embrowned, travel and battle stained set of soldiers the world has ever known. How can any one do full justice to these apparently interminable columns of cavalry and artillery especially, as they wind along the streets of the usually quiet village on their way to new conflicts and new triumphs?

All of nearly one color, apparently, as they enter amidst clouds of dust, they extend at times for miles, until one cannot help wondering again and again where all the men and horses come from, and whether enough have not already gone forward to whip the world in arms. They may not satisfy the eye of the fastidious lover of the neat and graceful in military display, but to the eye of the true military man, or to the earnest lover of his country and her glorious institutions, the dust, and clash, and clatter of that sturdy host, their rusty guns and sabres, coats off and sleeves rolled up along embrowned arms, tell the tale of *business*, that they are not filing along there by tens of thousands in grim array to make a mere holiday pageant, but that they are men, freemen, brave, tried, who are to be depended upon by their leaders in the day of battle, who are to be depended upon to save their country.

And if we take up the details of the picture thus in general outlines presented, the effect is not less impressive to the quiet observer. If a body of artillery is before us, we find the gun-carriages laden or strung with almost everything which goes to constitute the absolutely needful in camp life. The "forge" rumbles along, and in the strong jaws of the blacksmith's *vise*, these screwed together so as to clasp it beyond any apparent intention of parting with it, is that indispensable article in military *cuisine*, the frying pan; whilst camp kettles, bags of oats, axes, buckets, shovels, extra saddles, sabres and tin cups alternate with rammers and sponges, nosebags for horses and extra cannon wheels.

But the body moving along may be cavalry; and then another feature presented is the inevitable gang of contrabands that bring up the rear of the respective regiments, the negroes having in charge surplus horses and gangs of pack mules. The pack saddles of these latter are piled to overshadowing dimensions with luggage of every sort known to an army on the march, the contra-bands being usually arrayed in cast off military habiliments, almost always too large for them, whilst the rarely developed *knowingness* of look of these young scamps shows that whatever they may have been once in the way of examples of docility, they will never be of any account in the corn field again.

But they may be infantry who challenge observation, and although this arm of the service is generally less striking in a military pageant, unless viewed under peculiar advantages, since in the entire history of the picturesque in military display the horse is almost always called upon largely to figure, still there is nothing on earth, scarcely, that looks more formidable, more *terrible*, than the long column winding forward, flashing with glittering steel as it waves with the inequalities of the ground. It is true that these same bayonets may be surmounted here and there with a loaf of bread transfixed, perhaps with a tin coffee pot; still, even these do not detract from the business aspect of the march, they even add to the earnest appearance of everything, since men do not thus appear on mere gala-days; but these features tell effectually of the coming camp fire, of the picket posted, or of the bivouac.

We shall never forget the appearance, on a recent occasion, of a long column of infantry which wound along the road on the side of that old battle field, the South Mountain, a fine band at the head of the column playing one of our National airs, as if to challenge the glorious memories of the past connected with that locally. Onward and upward it wound, dark, formidable; the bright morning sun reflected from the thousands upon thousands of shining arms, its tattered flags thrown to the breeze, until every heart beat with proud emotions; every one was obliged to feel that a country defended by men such as moved in solid column through that mountain pass, could never have its liberties and free institutions greatly endangered, could never be made to succumb to the assaults of bad men.

But let us look at these true soldiers whilst in line of battle, let us see them when they are waiting to have the order to charge given which sends them upon lines of batteries; upon the waiting columns of the foe not less determined, and hardly less formidable than their own. Along the crests of the hills, for miles, entrenchments have been thrown up, the work of the past night, and behind these are posted numberless powerful batteries of field artillery, Parrotts, Napoleons, &c., whilst the rifle pits near by swarm with men, ready at an instant's warning to take part in the possibly coming conflict. Behind these again are posted long lines of infantry, their arms stacked, the men lounging about, but within easy distance, whilst here and there the flags of various hues and devices mark the headquarters of the various corps, or division, or brigade commanders, the staff officers and their horses around forming picturesque groups, or galloping hither and thither with needful orders.

And now the enemy must be "felt." Away go long lines of skirmishers across the undulating grass and grain fields to beat up some suspicious looking belts of forest, behind which, in grim and ominous silence, a vast army of the enemy is supposed to be lying. A few rods apart, and followed by their reserves, the skirmishers advance. Suddenly, from out the dark woods, come successive sharp reports of musketry, replied to by our men, and the whole line advances rapidly and disappears in the sombre shadow. Waiting with breathless anxiety to see what will happen next, the heavy boom of a cannon is heard far away on the left, then another and another; and the loungers, the hangers on about the camps, press forward with eager looks. But we hear no more, only the lines of pickets thrown forward move slowly backward, sundry prisoners captured telling of the sharp work which for the evening seems, from various disappointed looks, to have terminated all too soon. A rain storm is gathering, and the general impression is that the "great battle" will not come off today. Let us see the army then *en bivouac*.

A line of stone wall is handy to the spot we have chosen, the better to observe what is going forward, and it is an interesting study to see how old soldiers accommodate themselves to apparently the most uncomfortable or adverse circumstances. Rails are seized upon, placed on the ground to serve instead of mattresses, others are laid endwise on the wall, india rubber blankets are spread on these above, knapsacks are placed against the wall to answer for pillows, and then fires kindled and coffee pots or tin cups in requisition, and directly all immediately off duty are comfortably ensconced under there and similarly improvised shelters nibbling at "hard tack" and any luxury additional procured from the sutlers, telling stories or composing themselves to sleep. Many are not as elaborate in their preparations as in the cases we have attempt to desribe [*sic*]. An armful of wheat straw from the neighboring field serves in a fence corner or under a bush for a bed, the cover for the same being the impervious india-rubber blanket, which, for the time, is roof and bed cover all in one. Here is a drummer-boy who will never lack for resources, full surely. By some accident the head of his drum has been knocked out, and as he cannot help that now, he makes the next best use of his drum that offers in this rainy evening, thus desperate emergency. Covering himself with his rubber blanket, he has placed his drum upon its side upon the ground, and now, sticking his head into it, he is snoozing away with the rain pattering on his tiny roof within ten inches of his nose, oblivious of everything about him.

So much for the rank and file, the *tout ensemble*, of this splendid army. Now for a glance at such of the more distinguished of its leaders as on a casual visit happened to fall within the scope of the observation of a single outsider. And to begin with the General-in-Chief, General Meade:

Tall and spare, with the nose of the bald, First "Caesar", the traditional feature in military heroes from the days of the great Roman to those of Napoleon and Wellington, and with a dark, expressive eye, there is not a more military looking man, soldier or General, in the army; and modest and affable in his deportment, although his quiet, retiring manners may not be calculated to command respect. Added to this, there is an air of earnestness, of sincerity, about him, with an absence of any particular display, calculated to win the best feelings of all with whom he is brought in contact; and thus it is that even those least tinctured with hero worship can say a few words in commendation of one who, in his military career, and especially upon a recent, great occasion, has rendered good service to his country. He stands forth, just now, the one most imposing figure; but there are others we would fain introduce to our readers by a passing glance.

And next of General Sedgwick, one of the best fighters in the army, the man who with desperate valor cut his way out of that close place he occupied in connection with the battle of Chancellorsville. Of medium height and rather stout, he, too, has rather prominent features, and a quiet though determined look. Yet the salient idea conveyed by his countenance is that of thorough good nature; an absence of all pretension, pleasantly refreshing when we find it in contrast with too much of that pomp and military assumption so common amongst many martinets of far lower grades. His close-cropped, curling hair, surmounted by a straw hat, and blue suit of army flannel, would not designate the officer of high rank he is; and when we add that his headquarters of a summer's night are as often selected by the taking possession of a fence corner, with a blanket or two, one may well suppose that he at least is not weakly enamored by the mere "pomp and circumstance of war."

Then there is General Howard, the Commander of the Eleventh Corps at present, one of the heroes of the desperately contested battle of Fair Oaks, where he lost his right arm. He, too, has an unmistakably pleasant face, with hazel eyes and brown hair, and affable in his deportment and neat and soldierly in his appearance, he is one well calculated to win kindly regard. He seldom wears a sword, as why should he?, but he is a hero for all that, a gallant fighter by proxy, and one his country will not forget for the part he has taken in the great drama.

Then there is General Buford, another plainly dressed, a fighting General, as the whole country knows; a man of slight, though nervous build; of good height; young, active, with prominent features, blue eyes and rather light brown hair. We need not commend him as no mere holiday soldier, for his gallant cavalry exploits in conjunction with those of Generals Kilpatrick and Gregg, are fresh in the minds and regards of all.

We have not space nor time to photograph others of the distinguished men, officers of the Army of the Potomac, to whom the country owes a debt or gratitude it can never discharge. Distrusted, and subjected as this army has been to a long series of disasters, such as no such similar splendid military organization ever experienced, it has at last been privileged to vindicate itself; it has achieved distinction enviable in its degree for the humblest private soldier who has shouldered a musket in it ranks. Not broken in spirit by former desperate conflicts and disappointments, it has gallantly borne up against floods of ill-natured comments until now; and called upon still further to stand up for its own honor and the honor and safety of the nation, let the hearts of the loyal go out to it; let them accord to it their highest and most enduring confidence. As was said on another occasion, "the past, at least, is secure;" and for the future, its tried valor will provide for that; its further record be such as will be gratefully cherished by all who love their country.

Appendix 3

Letters

WB included fifteen letters in his memoir: eleven tipped onto pages or into the gutter and four loose in the box containing the bound volume. Two have been included in the text, nine here in the appendix and four have been deleted as repetitious or irrelevant.

Item 1

This letter was tipped into the gutter between text pages 44 and 45. The prefacing text is by WB.

During our stay at Elmira I received many encouraging letters but none were so highly appreciated as the following from your Grandfather, the first received by me on our arrival at Washington.
Chicago Sept 17th 1861
My Dear Son
I have delayed writing you until the present moment hoping that you would be able to visit us before you broke up camp at Elmira but as that is rather a doubtful event I have concluded to drop you a few lines [of] Expression of my Appreciation of your course and to encourage and Strengthen if need be your patriotism and love of Country and in doing this I am conscious that I can offer nothing more congenial to your feelings than to assure you that we all appreciate your position and to renew the assurance before given that we will to the best of our ability kindly and lovingly care for the wife of your heart and the child of your love during your absence. In fulfilling this your arduous and dangerous duty you have the Satisfaction of knowing that your course is just and that the hearts of all good men will be lifted in

prayer to the God of battle for your Success and Safety. Go forth then with a firm heart and a strong arm and Should it be Your fortune to meet on the ensanguent battle field the foe of our glorious flag our prayer Shall be that "Angels will hover in the Sulphur smoke and guard you from the battle stroke" and bring you and the Soldiers of your command back to the homes and hearts of your friends in safety. The eyes of the world and the hope of millions are resting on our noble band of patriot Soldiers, and children yet unborn Shall be taught to Sing grateful poems to their deeds of Valor and Heroism.

Our hearts will follow you wherever you go and with anxious eye Shall we watch the results whether of Victory or defeat as the telegram is flashed along the Electrick wire or chronicled on the daily papers.

Let us hear from you personally as circumstances will allow for your letters will sooth one heart more anxious than all the rest for your personal comfort and safety.

Your little daughter shall be taught to love her Soldier Father and when we give her back to you her endearing kiss shall reward you for all your toil. Be patient, be hopeful, be strong, be courageous and God will protect you and your Country will reward you.

Our Family are in usual health.

Maria is not as Strong as we could desire. The care and excitment consequent on recent events and care of the babe keeps her thin and pale but we hope to see her grow more Strong. The little daughter is doing fine. She improves nicely and is a first rate child. We Shall do all we can to help Maria take care of her.

Lyman is full of patriotism and it is hard for him to deny himself the privalege of doing his duty as a Soldier now in the hour of our Countrys perril but we have as yet in this State more men than arms. Thousands of good hearty young men in this State are now offering themselves and there will be no need of drafting from this State. This being the case and as Lyman is not qualified for an Officer I Shall for the present advise him to hold Still until such time as he is more needed in the field.

Hoping this will find you in good Spirits and sound health and trusting to hear often from you I will close this letter by Subscribing myself as usual Your Affectionate Father.

E A Gage

ITEM 2

While Colonel Stuart was away from the regiment on recruiting duty, he maintained close contact with WB and exerted some influence on the promotions within the 50th. This letter was tipped into the memoir between text pages 164 and 165.

Not to be read or shown to any one but burnt as soon as you read it

Unofficial & profoundly
Confidential
So Colonel says

Geneva 28th Aug.

Dear Captain Brainerd,

Col. Stuart arrived late last night & left home again this morning early to meet engagements to speak in different towns on behalf of the 50th & gave me orders to reply to your letter of the 21st.

Lieut Hoyt was promised the position of Quartermaster Master if on Col. S's return he was found to have filled the place with satisfaction &c. Col. Pettes wrote some time since to Col Stuart that he had best not be in haste to appoint Hoyt without giving any definite reason therefore, but recommending him to secure some good man "outside of the regiment" for the place, etc. I saw the letter myself & we talked the matter over carefully & Col. sat down & wrote you a day or two before the receipt of yours of the 3rd Aug. (which letter I had the honor of answering). Col. also wrote to Pettes by the same mail asking for the specific reasons for his advice which were replied to by P. saying he did not wish to be understood as bringing any charges against Hoyt but wishing the Col to do nothing precipitately. But a letter from Col. Pettes dated Hampton, Aug. 21st speaks in high terms of Hoyt, all you could wish, and the Col. bids me say that the very first leisure hour will be devoted by him to the office of sending to [?] Morgan, Hoyt's name for Quarter Master & that of Sergeant Carroll for the vacant Lieutenancy in your Co.

Nothing is lost to either by a few days delay as the commission will date from the resignation of [?] [?]. Col. has spoken twice at Rome and the recruiting for the old 50th in that place has been improved abundantly, many thanks to your kind father's friendly exertions and your own great popularity.

Col. took great pleasure in giving some account of Wesley Brainerd & in paying him a heart felt trubute of richly merited praise and recommended him in the strongest manner to Mr. [?] as eminently fitted for Lieut. Col. of their regiment but, as he said to me, with a sad fear at his heart that they *would* take you from him, as he well knows he would lose one of his best friends when you go. Your father & Mrs. Brainerd were very kind to him, he was their guest, and your young brother was deeply anxious to see you "home again".

A telegram came yesterday afternoon from Sergeant Drury, "I have 70 men, shall I go on recruiting?" to which I replied, "Yes, get all you can!" Isn't that splendid? Col. has already over 300 men & has no doubt of many more, all splendid fellows, before Sunday night. You have no idea of the immense amount of labor he has performed, as the scramble of aspirants for shoulder straps and other military honors has been so great that the getting men for the regiments in the field already officered has been comparatively up hill work. But Stuart has stumped from Niagara Falls on the west to Onieda Co. on the east & from Lake Ontario to the Pennsylvania line, often speaking twice a day & now, tired & exhausted, he had two engagements today, three tomorrow, Watkins, Millport & Geneva, & leaves at 7 a.m. on Saturday, for Aurora, Cayuga Co., where the Morgans & Wells have gotten up a mass meeting expressly for his regiment with an extra individual bounty of $100 per man who will enlist in the Engineers. Very different from Syracuse. Col. attempted twice to speak in that saline sahara but was peremptorily denied the use of a building or the use of his tongue, indeed, by the War Committee, Elias W. Scavenworth, Chairman, also by the Mayor of the city. "There is as much difference in folks as there is in anybody"! Col. had been obliged to promise a small share of the noncommissioned offices to those who have worked bravely for him & to one splendid fellow he has promised a 1st Lieutancy for 50 men and says he does not intend to promote Dexter, as in the first place, he [has] no earthly claim on the [?] of ifficiency & diligence, & again he had his commission as a *free gift*, while others are trying (for instance Folwell) to book up their number. [?] Folwell having been the means of obtaining about 40 men from Seneca Co. for Co. G. I tell you these things that you may prepare the public mind for the filling of some of the vacancies with new engineer material, as such aids to recruiting were absolutely indispensable.

Charles recommended you strongly to the Committee here for

Major of the 129th & Embick for Lieut. Col., but the War Dept. shut the door on all such things & the poor regiment has gone to Washington without one field, staff or line officer that knows a.b.c. of his duties.

I make no apology for the length of any letter for, as I said in a former letter, I am a woman & must have my full say. If you are going to be long at Alexandria, why cannot I come & stay in Camp with the Colonel? Ever your friend,

F. M. Stuart

Lieutenant Hoyt was indeed promoted to Quartermaster and, as subsequent events proved, fell victim to temptation and was dishonorably discharged. Major Embick moved on to bigger and better things with another regiment and Lieutenant Folwell obtained promotion to major by the end of the war.

Stuart's second wife was Francis Maria Welles. They were married in 1841 and had two daughters and a son.

ITEM 3

This letter was loose in the memoir.

Antietam Battle ground
Sharpsburg Oct 13, 1862
My Dear Niece,

Aside from ordinary considerations of natural relationship & material affection and confidence, there is a special reason why I should denote an hour to the composition of an epistle to you.

That special reason is (keep down envy) that I have seen one dearer to you than all else in the world. Not only have I seen him but am conscious of having received attentions from him for which this acknowledgement to *you* is but an inadequate return. I almost felt myself lionized. And all because I, for the time being, formed a link less than a week long which connected him with you dear [?] & if I could communicate to you one half the pleasure which my intelligence from home served to give him I should fancy myself entitled to be regarded by both of you as benefactor.

Passing on foot from Sandy Hook to Harper's Ferry on the Saturday after I left home, I saw '50' on the cap of an engineer and, on inquiring, found him a member of your husbands company. Of course I returned to his camp where I found him [?] at length with his face turned heavenward & his legs akimbo. His view of the delightful country so precious to a christian soldier was intercepted materially by a heavy canvass "fly" and a volume of Old Homer. I spent the afternoon with him, shared his camp fire and slept by his side all night. It was my first night in camp. We rested on a substratum of Virginia soil compact & firm covered by a government blanket & enjoyed it. A delightful ride next morning amid the most beautiful scenery I ever saw. Up the tow path of the Chesapeake & the bank of the Potomac brot us to the camp of the 8th Ill. cavalry. What if I should say that I like the rough side of your husband exceedingly well. He is a good companion, even for an *old man* like me. Ardent & constant in his attachments, *especially* to his wife. Tenderly solicitous for the welfare of his stranger kin. Being a cavalry man myself you will excuse a reference to one quality which I greatly value, viz. he is a *bold* & *daring* rider as I can attest after the lesson he has given me.

Am I not magnanimous? I have devoted almost a sheet to your husband & have said nothing about my own valorous deeds in putting down the rebellion. All this I leave to posterity. Meanwhile I will endeavor to do the work assigned me, unpleasant and revolting as some of it is. Our men are for the most part without tents & overcoats, having thrown them away in their retreat from the Peninsula. They huddle under low coverings improvised from ponchos. They are necessarily filthy and filth has its attendents which to polite ears should be nameless. The chaplain must visit them when sick and every such visit furnishes occasion for strict *self examination*. (This is private)

My health is excellent. I can ride 40 miles over the mountains as I did day before yesterday, in present of the Rebel raid into Pennsylvania and slept soundly at night. So soundly indeed that while I slept my regiment mounted at 1 O.C am & left me to open my eyes at 6 to find myself alone. *Please don't tell this* but still you know that when marching in column the proper place of the chaplain is in the rear. Well, the most delightful ride on the South Mountain alone amid the most grand & imposing scenery I ever gazed upon for about nine hours, paid me most amply for 5 hours of extra morning slumber.

Now please answer this if you can read it. Much love to your father, mother & brothers, not forgetting the little one. Should be happy to receive letters from all of them. Direct to me as "Chaplain of 8th Illinois Cavalry, Sumner's Division, Washington D.C."

Yours very affectionately
P. Judson

Rev. Philo Judson was Maria Gage Brainerd's Uncle and was the officiating minister at her wedding to WB.

Item 4

This letter was tipped into the memoir between text pages 184/185.

Private
Geneva, November 3d 1862
My Dear Brainerd,

I reached home from New York via Elmira last Saturday and found your letter of the 19th inst. and last evening (Sunday) your welcome & friendly letter of the 29th inst. came to hand.

I surely need not say to you how much I regret your sad disappointment but, I trust not mortification, in the reasonable hope & proper ambition you had for the promotion so long wished for, nor need I add how gladly I should have gratified you had it been in my power to have done it, without manifest injustice to Capt. Spaulding under all the circumstances of the case. This, I am glad to see, you fully understand & properly appreciate and I know give me full credit for doing what I *believe* to be my duty.

As long and intimate as my friendship has been with Capt. Spaulding & much as I esteem him as a friend a gentleman & an Engineer Officer, yet I cannot for a moment believe that he has a warmer personal regard for me than you have for me, so comparatively a stranger, nor would I count upon the friendship of any one I have the pleasure to call my friend, [?], *under all circumstances* of good or evil report, than I would upon you, and this letter now lying before me (showing your great heart) would convince me of it had I ever doubted it before. God grant it may be in my power ere long to gratify your commendabel ambition and also to contrive some plan by which you may visit your home & family after your long separation from them & your faithful service in the camp & field for more than a year.

A month ago I had supposed that I should have been back to the Regiment ere this, but the postponement from time to time of the draft in this State & the impropibility of my filling up the Regiment by Volunteers (after the bounties ceased [and] while the draft *was not dreaded or even anticipated by many*) until *after* the election, has detained me. Now the only chance is to obtain volunteers *immediately after the draft*, under the recent order allowing *conscripts a reasonable time to volunteer in old Regiments & receive the Government bounty*. I have opened offices in Rochester, Buffalo & other cities where the quota is not filled & go to them this next & *hold meetings* to show the advantages of *joining our Regiment*, in [?] to others, so that when the men *who are drafted* may choose to volunteer in it. In this manner I hope to *fill up* some of the Companies at least & thus have the power to nominate the *third* Lieutenant, which, under the order of the War Department of Oct. 11, 1862, requires the companies to be filled to their maximum number before the third Lieutenant can be appointed. I will remember Mr. Jackson & if, when the company is filled, *you think him* the *best* man for the place, will then reccommend him.

Capt. Beers is now in [?] County (having sent from Elmira last week all the recruits who were not sick) looking for recruits there *this* week. He is willing to return to the Regiment the moment he can have his accounts settled with the Pay master in Elmira (not yet having received pay for July & August nor for recruiting expenses for the last *three* months) and have another officer *detailed* to the recruiting Party in his place. I shall endeavor to have Lt. Col. Pettes detail you when Capt. Beers is relieved, as the Lt. Col. wishes to be placed on the *Recruiting Service* when I *return* to give him a chance to visit his Daughters. If he should *not* detail you in place of Capt. Beers I shall, upon my return to the Camp, do all in my power to have you granted leave.

I now hope to return to the Regiment soon after the draft in this State is made & earlier, if my service can be of more value there than in the recruiting service here. I should be glad to see the Regiment *filled*, however, before I return, but am quite *tired* of the *labors of recruiting* (which I perhaps make *harder* than most officers on that duty) and am anxious to return to the gallant boys & friends in the 50th Vol. Engineers. I have travelled the *last* month over a *thousand miles* in *this State* looking for Volunteers & obtained only about *thirty*, after more labor than it required in August to recruit *four hundred*. Oh! if I could only have had the *order* for 150 per company *then* (as *asked* for *repeatedly* in July & Aug.), I could have *enlisted twice* the number *I did* in August.

I have sent down *fifteen good Musicians* (picked from local Bands [and] taken their Instruments with them) who are to be *distributed* among the Companies that are *deficient*, but who are trained (with a very *capable leader* Mr. *Knopp*, who is to be Drum Major) to play as a [?] *Band* provided the Officers & men of the Regiment subscribe a *reasonable* compensation for their services as a Band.

I have told them that while I could not of course promise them anymore than the *wages* of Company *Musicians*, I would subscribe myself & reccommend *all* in the Regiment to do likewise to pay them an *extra* compensation for their services as a Band & doubt not that they would gladly do so. I wish you would see what number you can procure in your Company or [?] (probably never in your charge) open a subscription, about in the following *proportions*—for Line officers—Capts. 75 ct per month & Lieuts. 50 cts., Sergeants 25 cts., Corporals 15 cts., 1st class Privates 10 cents & second class Privates 5 cents per month.

I think the above rates would not be felt onerous & would, if *two thirds* the whole number should subscribe, give from [?] 5 to 15 a month *extra* to the players in the Band, which would compensate them very fairly for their services & instruments.

Do write me fully about your [?] & all things interesting about your duties on the River. Mrs. Stuart & Mary [?] in kind regards to you & remembrance to all engineering friends in the Division. With the [?] of my esteem I am now as ever your friend

Charles B. Stuart

ITEM 5

This letter was loose in the case holding the memoir.

Chicago Dec. 7 1862
Dear Brother

Our formerly pleasant home is today a house of mourning. Sweet little Mary is no more! Nothing left to us but the silent body from which her pure young spirit has gone. Gone to a land of fair skies, where pain and sorrow, sickness and death are not Known. Oh! That we could understand these mysterious dealings of Providence.

If we could, it seems we could more easily submit ourselves to these dreadful afflictions and fully *realize* what our purest Faith teaches us to believe. That the highest good *will* be accomplished! Even through these painful means of sorrow and desolation.

I have written you two short letters previous to this addressing you of Mary's sickness and its probable termination. I have not however given you the details of her sufferings and will not attempt to do so now. Maria can write you these much better than I and I leave it to her. Friday night Mary seemed to be growing rapidly worse and we abandoned all hope of her recovery. She had violent spasms at frequent intervals, followed by short periods of exhaustion. This continued through the night and yesterday, up to one o'clock this morning. It was then evident that she could live but a few moments more. From this time her little remaining strength failed and at ten minutes past one she quietly sank to her final rest.

Maria is almost worn out with anxiety and lack of sleep, though we had a Nurse come Friday evening who held the child almost constantly straight through to the hour of her death. Maria is lying down feeling unable to write. She needs rest and must be kept as quiet as possible.

It is an awful bereavement to Maria and our own hearts all sympathize most deeply with you both. Little Mary, was loved by us all and this sorrow enshrouds itself in all our hearts.

We feel that everything has been done that was possible, sad satisfaction, yet not entirely destitute of comfort. We so much wish that you could be here, you must use all means that are honorable to get leave to come. We should have a short funeral service at the house tomorrow at two PM. We intend to put her body in the "vault" where it can remain for several months without much change in its appearance.

In closing what can I say that will soften the anguish and disappointment this letter will bring to you. Alas! My heart holds such a weary pain that words of consolation seem almost mockery. My earnest love and sympathy is all that I can give you. Should Maria be unable to write I will keep you informed by frequent letters of all matters interesting to you. Please write me if possible.

Truly your friend, *Lyman*

ITEM 6

This letter was tipped into the memoir on text page 240.

Buffalo, Dec. 13th/62
My Dear Friend,

I was in Rome yesterday afternoon when the *sad news* from our brave Boys at Fredericksburgh reached me by Telegram. I was the *first* to hurry to your Father's house & tell him of your wound in the Arm. He received the unwelcome news like a patriot & a christian as he is and hoped you would soon telegraph him or have someone do so, informing him of your condition & whether you wished him to meet you in Washington, provided you should be sent there to recover, it being of course impossible for a *civilian* to reach Fredericksburgh under the present stringent rules.

We hope of course that your wound & that of Capt. McDonald may be *slight*, *as stated* in one of the New York papers of today, but yet fear that it may be *severe*, as telegraphed to another paper this morning, making it all *doubtful*.

But I trust it may be *all for the best* & certainly your Father & your very many friends in Rome & elsewhere have good reasons to feel proud of *your gallantry*, as I certainly am & of that of my brave Nephew, whose life has been gloriously given to his country, and of all the other gallant Spirits of the proud 50th who were in the terrible fight of Thursday. Wish to God they could have all been spared to serve their Country longer, but it could not be & we must be resigned & even thankful that no more victims were sacrificed to rebel treachery & the stern realities of battle.

How much I wish it had been my fortune to be with all my dear friends & shared with you the dangers and the horrors of the victory, however little it might have been in my power to have saved you & others who have suffered death & wounded. Yet my *duty* has kept me here as well as my *orders* from the War Department and the *hope of filling up the Regiment this month*, upon the close of Lake & Canal navigation & the endorsement of the draft, has *alone* kept me from making every effort possible to get returned from recruiting service & being with you at this time. I know how much need there is of *filling the [?]*

the companies to 150 to give us our full compliment of officers and the importance of adding the *two* new companies before my return & also at *this time to be here* to take advantage of the close of navigation this month & obtain boatmen & sailors for *our* Regiment, what need that class of men so much.

But *now*, if your wound should disable you from *active* duty (which I fear it will) and should enable you, after a short time to attend to the *recruiting* service, perhaps I can arrange it so as to have you take Captain Beer's place and allow him & myself also to return to Regiment very soon. Indeed, if you had not changed your mind last month, I should have had the transfer made as just proposed by you in October, but *you* felt anxious to remain for the present movement to Richmond & [?] you had the opportunity, to distinguish your self & do credit to the Regiment.

I saw Mrs. Wilcox at Rome yesterday, who mourns over the loss of her gallant husband in Company C and wishes very much to have *his body* sent home at any expense necessary & she will pay it. I have written to the Chaplain to endeavour to see it done now if possible, if not, then to have such arrangements made as will enable *it to be removed* at some more convenient opportunity. I will, if able, give some directions about this.

Your welcome & friendly letter of the 5th inst came to hand on the 10th inst. I had the letter sent to the Rochester papers with a few extracts omitted, for publication. It was too good to be hid under a bushel.

I telegraphed my wife from Rome of Gus's death & your own & McDonalds wounds from Rome yesterday. I came *direct* from there to this City & have not been home since but hope to go on Monday with some more *Recruits* to Elmira & see her on my return. I am [?] busy in *getting off* all the recruits I can to the Regiment, knowing how much these are needed now, but the *late* close this year of navigation of the canals & the lakes yet open, makes recruiting slow but will doubtless improve next week. I sent off *20* this week to the Regiment & *36* last week & have about 40 more to go *early* next week from here & Elmira. These will help you some & others will soon follow them I trust & myself *before long* I hope. My health has been very good until within the last *ten* days. I have been troubled much with *sore throat* & consequently loss of appetite & strength but I keep moving about at the rate of *several hundred miles a week* for recruits.

With regards to all friends I am [?]

C. B. Stuart.

Write me at Seneca on reciept if *you can* & give particulars about *yourself, Dear Gus* & Capt. McDonald. If you cannot *write* have some-one do it for you.

Capt. Augustus Perkins, killed on the bridge next to the one that WB was wounded on, was Mrs. Stuart's nephew. Private Lewis Wilcox, age thirty-eight, was killed on the same bridge and had been in the army only three months.

ITEM 7

This letter was tipped into the memoir on text page 236.

Battle Field Fredericksburg Va.
Dec 14th 3 PM
My Dear Niece,

I am thankful that I am not compelled to open the fountains of your griefs [?] by communicating any thing unfavorably in regard to the condition of your beloved husband. You have been informed by telegram of his wound. Do not give yourself any [?] concerning it. It is a flesh wound in his left arm and will not prevent his traveling or even attending to business.

[?] can almost see in this affliction a design in our Lord and Father to mitigate the anquish you experience in the loss of your darling Mary. *Now* he can get leave of absense & I hope will be with you in a very few days. But for his *wound* he could not have obtained leave & you would have been compelled to bear your wrought of grief alone. Not that you have no friends to sympathize with you, no Savior upon whose busom you can lay your lacerated heart, no throne of grace from which to draw divine strength. All these you have. But how much you need the presense of your husband & the father of your sainted child in such an hour as this.

I returned from Washington this morning, whither I have been to carry about $15000 of soldier's money. And by a letter from my dear friend Mrs. Ludlam to her husband, learned your bereavement and simultaneously, the wound your husband had received. I rode imme-diately to his tent, found him neatly dressed, sitting on a camp Stool in the warm sun writing to his wife. I met the usual cordial reception, which "Uncle Philo" is accostumed to receive from his much esteemed Nephew. How could I break the sad intelligence of his loss to him, but I did. The effect was such as you might expect. I will not attempt to describe it. Enough to say that the nobility of his nature was evinced as well as his firm belief in the goodness & wisdom of God.

I am suffering too much from a hurt [?] received [by the] falling of my horse to write more. Seeing a family of fugitives, women & children only, flung from their almost demolished City at a ditch they could not [?], I attempted to aid them. My noble horse leaped the ditch but finding no firm footing fell upon the opposite bank & he fell & rolled nearly over me. The injury is not serious.

May God bless you my dear Niece and lay his soothing hand upon your lacerated heart. Love to all the family. I shall write to your father soon. Do not alarm my family by the above. Your Uncle

Philo Judson

Dec. 15. I am better this morning. Armies have been fighting all night. [?] nothing decisive yet. I start for the field of carnage [?].

ITEM 8

This is the full text of the letter written by WB and Major Spaulding on the occasion of Colonel Stuart's resignation. It is a handwritten copy of the original and was tipped into the memoir on text page 307. All of the signatures on this copy were by the same hand.

Camp of the Engineer Brigade
Near Falmouth Va. June 2d 1863
Colonel Charles B. Stuart
50th Regt. N.Y.Vol. Engineers
Colonel,

We have learned with unfeigned regret, that You have been compelled on account of failing health to tender Your resignation as Commanding Officer of this regiment, and that You are soon to leave *us*.

We cannot part with You without first making an acknowledgement of our sincere sorrow that the state of Your health should make it necessary for You to take this step. We have observed the effect of exposure and the climate upon You and though we had hoped and trusted that skillful treatment might again restore You, Yet weeks and months have passed without any visible change, except for the worse, until we are convinced that any further exposure to the hardships of a soldiers life would be doing great injustice to Yourself, Your family and Your friends.

While we are forced to this conclusion we cannot but feel that in parting with You we are loosing an old and a tried friend, one who for nearly two Years has devoted his whole energy, brought to bear with remarkable singleness of purpose superior talents with an enviable character and reputation, sacrificed the comforts & pleasures of a delightful home and family, invested without the possibility of remuneration his private means, risked health, reputation, and life itself for the advancement of the interests of his regiment and for the good of the cause in which we are engaged.

While we express these views as Officers, we are Confident that we represent the feelings & sentiments of every non-Commissioned Officer and private in the regiment.

For the good character the regiment bears, for its dicipline and for its efficency in its peculiar branch of the service, we feel that it is largely indebted to Your care and prudence, and to Your honorable character. While You have required dicipline and implicit obedience to orders, You have treated us with that impartiality, forbearance and Kindness which has won us esteem and affectionate regards.

As Officers who have had the honor to serve under You, we feel that Your name is, and *ever will be*, closely identified with the history of this regiment, that that history has thus far been an honorable one alike creditable to Yourself and to those serving under You. We believe the records of the war department will abundantly testify.

Our associations with You for the past two Years, have made a lasting impression upon our memories. The friendship formed by association in Camp and field in the hour of danger and trial, such as none other but a soldiers life presents, are not easily effaced. We can never forget the energy and ability displayed by You in the perplexing business of organizing this regiment at Elmira. Who of us will ever forget our pleasant winters experience at *Camp Lesley*, and the services You thus rendered the regiment during that session of Congress. Or the Peninsular Campaign, Yorktown, White House, the Chickahominy and the seven days battle and retreat. Our Camp at Harisons Landing where You were over come by desiase and during Your absence north. Your extraordinary labors and success in recruiting the ranks of Your regiment. Your return to us in February and participation in the exposure and dangers of the recent battles on the Rappahannock. Your valuable and conspicuous services at "Franklins Crossing" and "Banks Ford," are familliar to us *all*, though we were at times seperated. But why attempt to enter into particulars? A thousand incidents of duty and of pleasure which we have shared with You will be treasured in our hearts. We will cherish, with You, the memory of those who have fallen in the line of their duty. And now, we must part with You.

You will bear with You to Your Northern home our respect and best wishes for Your prosperity and our prayers that You may be restored to health. For ourselves, we trust we shall never disgrace our old Commander, And when, after this rebellion is Completely subdued, Stuarts Engineers shall return to their homes, we hope You will be there to welcome us with the Kindness and affection with which we now bid You *farewell*.
Signed

Wm H. Pettes Lt Colonel
I Spaulding Maj.
Wesley Brainerd Maj
E.O. Beers Maj
Wm W. Folwell Capt. Co I
Chas N. Hewitt Surgeon
Geo. W. Ford Capt. Co. H
M. H. McGrath Capt. Co. F.
Asa C. Palmer Capt. Co. D
A. B. Dolan 1st Lieut.
Lewis V. Beers Asst. Surgeon
Henry O. Hoyt 1st Lt. & A.Q.M.
Edward C. Pritchett Chaplain
A Clarke Barnes Asst Surgn
Wm. B. Johnson 1st Lieut.
M. Van Brocklin 1st Lieut
Edwin Y. Lansing 1st Lieut
William V. Van Reusselaer 1st Lieut
Fred. W. Pettes 1st Lieut
Mahlon B. Folwell 1st Lieut
John M. Libbalds 2nd Lieut

Daniel M. Hulse 2nd Lieut
C. Q. Newcombe 1st Lieut
Wm H. Whitney 2nd Lieut
Thos J. Langden 2nd Lieut
John T. Davidson 1st Lieut
Geo W. Nares 1st Lieut
John L. Roosa 1st Lieut
George Templeton 2nd Lieut
Sidney Geo Grwynne
Lieut & Actg. Adjt.

ITEM 9

This letter was tipped into the memoir on text page 283.

Chicago April 23 1863
Maj Wesley Brainerd

My Dear Son

Your letter and package by express enclosing Seventy Dollars came safe to hand and is more than Satisfactory. The sentiments expressed in your letter are such as might be anticipated from a Soldier who loves his Country and is willing if need be to give his life to preserve the honor and maintain its integrity. Our prayers shall follow you through the strife and din of battle that God will in his own good time bring our Country safely through this firey ordeal, purged from all that has caused this conflict of arms, that She may be a bright and a shining light, a refuge for the oppressed of other lands, the land of the *Free* and home of the brave. Language is inadequate and words are too feeble to express the pride we feel in our brave Soldiers or the Sorrow that fills our heart when they fall. God grant to us all that manly courage, that Trusting Confidence, that the right will prevail which alone can Sustain us during these scenes of carnage and strife. I see by the papers that Hookers Command are in motion and work will follow. We shall watch you with much anxiety.

We received a letter from Maria yesterday. She is well but undecided as to where she will spend the summer. Lloyd had gone to Rome, left here yesterday [and] will be absent about 6 weeks, when he returns will go in again with Messrs *Preston Willard & Keen*. Lyman is well and remains as usual cashier of the M L & T Co.

The Photograph Gallery is laid on the shelf for the present. If you were here I would advise that enterprise. The weather [here] is brisk, of [course] we are wide awake, let us hope to hear from you again at no distant day.

Truly Yours

E A Gage

Preston Willard & Keene was a banking firm on Clark Street, Chicago. Principals were David Preston, Jofiah Willard and Samuel B. Keene. Lloyd Glover Gage (1846–1885) was only seventeen when this letter was written and his position with the firm is unknown; however, at that age it can be assumed he was little more than a messenger.

M L & T Co was the Merchant's Savings Loan and Trust Company at the corner of Lake and Dearborn, Chicago. Lyman Judson Gage (1836–1927) was listed as a cashier for the firm in 1863.

Lyman's first wife, Sarah, died in 1874. Lyman married (for the second time) his brother's wife, Cornelia, in 1887 after Lloyd passed away in 1885. Cornelia died in 1901 and Lyman was married (for the third time) in 1909 to Francis Ada Ballou.

Appendix 4

Orders, Miscellaneous

WB included twenty-seven orders or copies of orders, four official documents, and eighteen miscellaneous items in his memoir. The orders and official documents that were redundant with WB's narrative have been deleted. Only three of the miscellaneous items have been retained.

Item 1

This printed poem was pasted onto text page 135 in the memoir. The preface is in WB's words as they introduced the item.

To show you that we had Poets as well as workmen among us I have preserved this specimen of poetry which you will no doubt pronounce very rare. The author of this choice piece of poetry I have never been able to discover, though that it originated somewhere in the Quartermasters Department was established beyond a doubt. The name of "Stuarts Engineers," quite popular at that time, soon fell into disuse and a few months afterwards it was dropped entireley.

STUART'S ENGINEERS

BY A VOLUNTEER

'Twas on a hot September day, beneath a burning sun
We left our camp at Elmira, to start for Washington.
They called us grey back Engineers, not one word could we say;
For as we marched along the street we wore a suit of grey.

With measured tread we marched to where the iron horse did stand,
With waving scarfs and noisy cheers, started for Dixie's land!
With headlong speed we rushed along—with shouts and
noisy cheers—
A band of volunteers, my boy, to serve as Engineers!

At New York City each man got a Harper's Ferry gun;
The Fire Zouaves there to us said, Grey backs, we'll make you run.
We'll drive you in the river deep, you grey back volunteers.
But they could not their promise keep with Stuart's Engineers.

For we with muskets in our hands, just stop'd to see the fun;
But while we camped upon the sand, the bloody zouaves run.
Our Quaker friends were kind to us in Philadelphia street;
They sat before us viands choice, and furnished us with meat.

And very feeling hearts they had—the hungry they did feed—
A long long life to them, my boys, they gave in time of need.
'Twas Sabbath day when we came near the streets of Baltimore,
Where Massachusetts boys had passed nearly six months before;

Although they scowled on us, my boys, we saw no cause for fear,
For we were armed with muskets, boys, and we were Engineers.
Our Colonel is an Engineer, and unto us did say,
"I want you to keep sober, boys, I'll watch you night and day;

"You enlisted for the present war to work—if needed fight—
And if the guard house you would shun, you never will get tight;
But if you're bound to run the guard, vile whiskey to obtain,
And then I have to send for you, you'll wear the ball and chain.

"And gambling is forbidden, too, by night, also by day;
And if you swear too much, my boys, you forfeit all your pay."
Three cheers for Colonel Stuart, boys we are his volunteers.
And he is death on whiskey, boys, among his Engineers.

And we can work or we can fight—we're promised extra pay—
If we work from morn till night, 'tis forty cents a day.
We have long range rifles true, our hearts are always light,
Our weapons they are always new, because we keep them bright.

And when they want to build a fort, or bridge that has no piers,
They call on us—we like the sport—for we are Engineers.
This war will not last long, my boys, and soon we'll take the train,
Then we will have a happy time when we get back again.

But some of us are dead, my boys; we cannot all return;
Some never will go back again—for them we oft will mourn.
Some of our boys have gone to rest on famed Virginia's soil;
And they are sleeping peacefully, they are free from every toil.

Disease came watching 'round our camp, and death was at the door,
And bore them to eternity on yonder misty shore.
Now when the war is over, boys, home go the volunteers,
Then there will be a mighty noise from Stuart's Engineers.

Our Uncle Sam is kind to us, he finds us all in meat
He says the sutler robs you, boys, when you go there to eat.
But save your money—send it home and buy your wife a dress;
And buy bread for your little ones, and clothe their nakedness.

Yes, Uncle Sam is very kind, and he beats all the rest.
He says you'll have a fame, boys, by going South or West,
A hundred dollars in hard gold is what he'll give to you
As bounty money, if you're bold and put these rebels through.

Now here we are near Richmond, boys, we've built some roads you see;
And many a bridge we've laid across the Chickahominy,
War is a trade, and we have learned a few things since we came
Down to this land of cotton, boys, where grows the sugar cane.

We understand what hard bread is; we've seen a knapsack drill;
We used to double quick it, boys, when we were on Hall's Hill.
We've marched for hours through mud and rain, through storm
and driving sleet,
With not a drop of whiskey, boys, or anything to eat.

But now we're near McClellan, boys, we work both day and night.
He says, "drink whiskey all you want, but mind and don't get tight.
For whiskey is your enemy, be sure and keep him bound,
Or he will shoot you in the neck, and throw you on the ground."

Now, when the war is over, boys, we'll have a holliday;
And we would make a joyful noise to get our extra pay;
If we don't get it, all the same, for we were volunteers;
It is worth something for the name of Stuart's Engineers.

ITEM 2

This was a printed piece and tipped onto text page 67 in the memoir. The prefacing words are WB's.

My Son, look to that order of the great and good General Washington and also to the order of General McClellan, who, whatever may be said by his enemies, was undeniably a Christian Soldier and Gentleman.

GEN. WASHINGTON'S GENERAL ORDER,
AUG. 3, 1776.

"That the troops may have an opportunity of attending public worship, as well as to take some rest after the great fatigue they have gone through, the General, in future, excuses them from fatigue duty on Sundays, except at the ship-yards, or on special occasions, until further orders. The General is sorry to be informed, that the foolish and wicked practice of profane cursing and swearing, a vice hitherto little known in an American army, is growing into fashion. He hopes the officers will, by example as well as influence, endeavor to check it, and that both they and the men will reflect that we can have little hope of the blessing of Heaven on our arms, if we insult it by our impiety and folly. Added to this, it is a vice so mean and low, without any temptation, that every man of sense and character detests and despises it."
Sparks' Writings of Washington, Vol. iv., p. 28.

GEN. MCCLELLAN'S GENERAL ORDER,
SEPTEMBER 6, 1861.
HEADQUARTERS, ARMY OF THE POTOMAC,
WASHINGTON, SEPT. 6, 1861.

General Orders, No. 7.

The Major-general Commanding desires and requests that in future there may be more perfect respect for the Sabbath on the part of his command. We are fighting in a holy cause, and should endeavor to deserve the benign favor of the Creator. Unless in the case of an attack by the enemy, or some other extreme military necessity, it is commended to commanding officers, that all work shall be suspended on the Sabbath; that no unnecessary movements shall be made on that day; that the men shall, so far as possible, be permitted to rest from their labors; that they shall attend divine service after the customary Sunday morning inspection, and that officers and men shall alike use their influence to insure the utmost decorum and quiet on that day. The General Commanding regards this as no idle form; one day's rest in seven is necessary to men and animals: more than this, the observance of the holy day of the God of mercy and of battles is our sacred duty.

George B. McClellan,
Major-general Commanding.

Official: A.V.COLBURN, Assistant Adj. Gen.

ITEM 3

This printed extract order was tipped onto text page 106 in the memoir.

GENERAL ORDER #1
EXTRACTS

The Trenches.

Head-Quarters, Army of the Potomac.

Camp Winfield Scott, near Yorktown, Va., April 23d, 1862.
General Orders.

The following orders for the construction of batteries and trenches during the operations before Yorktown, will be strictly observed, viz:—

I. The senior Brigadier General upon the detail for the day and night will act as General of the Trenches, whose duty it will be to superintend the operations and arrange the guards of the trenches to repel sorties and protect the works.

The General of the Trenches will be designated in orders from Head-Quarters and will report to the Chief of Staff for instructions, at noon of the day before he goes upon duty. He will also confer with the Engineer or Artillery officers in charge of the trenches, and receive information from them as to the manner of performing the work and visit the localities with these officers, before dark, so as to make himself familiar with the ground.

Field Officers will be assigned to duty with each working party in the batteries and trenches and with each guard. They will be under the immediate orders of the General of the Trenches and report to him for orders and instructions one hour before their tour of service commences.

The General of the Trenches will see that the working parties and guards are properly posted and directed, and will be responsible to the General Commanding for the faithful performance of all duty during his tour of service.

He will, at the termination of his tour of duty, make a report, in writing, to the Chief of Staff of all the operations that have been under his charge. He will also make hourly reports of all that transpires in the trenches.

Night working parties will be double the size of day parties—one-half forming a support to the guard.

All officers must take care to give and to ask for all information necessary to carry on promptly, skillfully and successfully all duties confided to them.

At night there will be two reliefs of working parties.

The officer in charge of the first relief will explain in detail, to the officer in charge of the second relief, the manner of posting his men and how the work is to be done, so that there may be no delay in posting the second relief at its work.

The workmen will march to the trenches with their firearms and cartridge boxes, which they will place near them while at work, and always carry their overcoats, to cover them while not at work.

The troops for guarding the trenches and working parties will be detailed at least twelve hours in advance, and they will furnish no other details during this tour.

Twenty-four hours (or twelve, at least,) before mounting guard in the trenches, the battalions detailed for guard do not furnish workmen; and the companies of those battalions whose tour it would have been to work in the trenches do not go there for twenty-four hours after guard, if possible, or at least, twelve.

The battalions first for detail or guard of the trenches and the companies first for detail for work in the trenches, furnish no other detail and are held on picket ready to march at the call of the General of the Trenches. He will be notified what battalions are on picket.—[See Par. 593, and following, Army Regulations.]

II. The guard of the trenches will be divided into three entirely distinct parts, viz:

1st. Those designed to man the parapets and banquets. They are deployed behind the loopholes of the places of arms and the advance parallels, particularly at those places most exposed to attack. These men are stationed in pairs, that one may have his piece constantly loaded.

2d. Those designed to act as supports, (corresponding to the supports of the advance guards.) Their stations are in the portions of the parallels or places of arms specially entrusted to their care. These stations must be large enough to admit of free and rapid passage in and out, to enable them to reinforce at a moment's notice any places attacked. The supports will be made up of entire companies, never less than one company each, and will be strengthened according to circumstances.

3d. Reserves, intended to act with great promptness and force, to repulse the enemy, should he have pierced our lines. They must be placed where they are protected from the enemy's fire, but where they can be used at a moment's warning.

III. The boyaux of communication must be left as free as possible, and the men never allowed to collect in them. They will be occupied only by a few sentinels posted at considerable intervals, with orders to keep the boyaux clear.

Before night all officers commanding guards of the trenches will assure themselves by personal examination that the connection is perfect with the guards on their right and left. Any neglect of this may be attended by the most serious consequences.

At nightfall, and whenever it is practicable during the day, sentinels selected for their coolness, intelligence and quick vision, must be placed in pairs, as well concealed as possible, a little in front of the parallels and places of arms.

Should the enemy make a sortie these sentinels fire the instant they see them, and fall back into the parallel, giving the alarm. The buglers and drummers (of whom there should always be some in the parallels with the officers commanding the firing parties and supports) immediately sound "to arms," or beat the "long roll," which will be repeated all along the lines, and as far back as the reserves. All rush to arms at once, the firing parties man the loopholes, and as soon as they see the enemy within range, open a sharp but deliberate fire upon them; the workmen, if there are any, drop their tools, grasp their arms, are formed by their officers, and either take part in the firing or join the supports according to circumstances. The enemy, approaching entirely exposed and unable to injure our men, who are sheltered by the parapet, must suffer enormously from the cross fire they will meet with. If, nevertheless, they succeed in reaching the parapet, the guards must coolly step back upon the reverse of the trench and as the enemy jump down into the trench, attack them at close quarters; this manoeuvre will always cause them to hesitate and will embarrass them. This is the moment for the action of the supports, who will attack with the bayonet, and with little or no risk to themselves inflict great loss upon the enemy, who in retiring will be obliged to turn their backs and use both hands in climbing the parapet. The pursuit should not be continued more than 50 yards beyond the parallel, into which the guards will then quietly return to resume their original positions and be ready to repeat the same manoeuvre if necessary.

The Reserves must act with the greatest energy—throw themselves headlong upon the enemy should they force the trenches, fire upon them at the closest range and finish the work with the bayonet. The Reserves should also be careful not to pursue the enemy too far, in order not to expose themselves to the artillery fire of the fortress.

Sorties will be followed up to the enemy's works only by the express order of the General Commanding the Army.

Unfinished trenches need not be defended with so much obstinacy; but the guards and working parties taking with them their arms and tools, may retire to the nearest parallel, the fire from which will cause the enemy to pay dearly for the few gabions they may be able to overturn.

IV. In defending trenches against sorties, the guards must not move forward to meet the enemy in front of the trenches; this would only be giving up the advantage of position and would mask the fire from the trenches: on the contrary the enemy must be left fully exposed to this fire. If he reaches the trenches, the guard will form on the reserve, and aided by the working parties and supports resist him obstinately. It is only when the enemy is broken and retires, that the guards will follow him out of the trenches with the bayonet. Finally, no officer detailed for duty in the trenches will on any account leave them during his tour of duty, except in case of wounds or sickness, which fact will immediately be reported to Head-Quarters through the General of the Trenches.

The utmost silence and order must be preserved in the trenches and in marching to and from them.

All working parties for the trenches will go equiped for action.

To prevent the glistening of the bayonets and arms betraying the movements of the troops, bayonets will be carried in the scabbard and the gun slung, while going to and from the trenches.

An officer will be sent with each relief of the guards and working parties to the batteries and trenches, who will return to his camp to escort the next relief to their proper positions.

No officer, soldier or citizen shall be allowed to enter the trenches, or to approach their vicinity unless specially detailed on duty there. The only exceptions to this rule will be in the case of General Officers, the Staff of the Major General Commanding, and the Staff Officers of the General near the trenches. All persons violating this rule will be arrested and sent to the nearest Provost Marshal.

All officers on duty in and near the trenches are required to keep the men under cover of the parapets, and to arrest and send to the nearest Provost Marshal all officers or men with the exceptions heretofore mentioned who, unnecessarily, expose themselves to the view of the enemy.

By Command of Major General McClellan:

S. Williams,

Assistant Adjutant General.

ITEM 4

This printed order was issued during the Peninsula Campaign and was tipped into the memoir in chapter six. The date on the order would place it two days before the engagement between elements of Porter's V Corps and Branch's North Carolina Brigade at Hanover Court House.

GENERAL ORDER #2

Head-Quarters, Army of the Potomac,
Camp near Coal Harbor, Va., May 25th, 1862.
General Orders No. 128.

I. Upon advancing beyond the Chickahominy the troops will go prepared for battle at a moment's notice, and will be entirely unencumbered. With the exception of ambulances, all vehicles will be left on the eastern side of the Chickahominy and carefully parked. The men will leave their knapsacks packed with the wagons and will carry three days rations in their havresacks. The arms will be put in perfect order before the troops march, and a careful inspection made of them, as well as of the cartridge boxes, which in all cases will contain at least forty rounds. Twenty additional rounds will be carried by the men in their pockets. Commanders of batteries will see that their limber and caisson boxes are filled to their utmost capacity.Commanders of Army Corps will devote their personal attention to the fulfilment of these orders, and will personally see that the proper arrangements are made for parking and properly guarding the trains and surplus baggage, taking all the steps necessary to insure their being brought promptly to the front when needed. They will also take steps to prevent the ambulances from interfering with the movement of any troops. These vehicles must follow in rear of all the troops moving by the same road. Sufficient guards and staff officers will be detailed to carry out these orders.The ammunition wagons will be held in readiness to march to their respective brigades and batteries at a moment's warning, but will not cross the Chickahominy until they are sent for. All Quartermasters and Ordnance officers are to remain with their trains.
II. In the approaching battle, the General Commanding trusts that the troops will preserve the discipline which he has been so anxious to enforce and which they have so generally observed. He calls upon all officers and soldiers to obey promptly and intelligently all orders they may receive.Let them bear in mind that the Army of the Potomac has never yet been checked: Let them preserve in battle perfect coolness and confidence, the sure forerunners of success. They must keep well together; throw away no shots, but aim carefully and low; and above all things rely upon the bayonet. Commanders of regiments are reminded of the great responsiblity that rests upon them;— upon their coolness, judgment and discretion the destinies of their regiments and the success of the day will depend.

By command of Major General McClellan:
S. Williams, Assistant Adjutant
General Official: Aide-de-Camp.

ITEM 5

This is the full text of General Order No. 7 that WB pointed out caused so much resentment in the regiment. It was tipped into the memoir in the middle of chapter 12.

Head-Quarters, Engineer Brigade,
Camp near Falmouth, Va., April 17, 1863
General Orders No. 7
Order of March of the Ponton Trains of the Engineer Brigade

The ponton trains will be arranged and designated as follows:

The train in charge of the 50th regiment New York volunteer engineers, (now with Major Ira Spaulding,) and having the mule teams broken in, will be called the "First Train of the 50th Regiment." The Captains of these trains especially responsible for the details of equipment and march will be, for the "first train," Captain Geo. W. Ford, and the "second train," Captain W. W. Folwell, Major Spaulding being in general charge of both trains.

Major E. O. Beers, 50th regiment, (with Captain Asa C. Palmer, company "H", captain of the train,) will take charge of the train recently repaired by him, which will be called the "Third Train of the 50th Regiment."

The train now under the charge of Captain H. V. Slosson of the 15th regiment, and having the horses broken in, will continue under his command and be styled the "First Train of the 15th Regiment." The other train recently parked with this will be under the command of Captain Jos. Wood, Jr., 15th regiment, as captain of the train, which will be called the "Second Train of the 15th Regiment." Both trains will be commanded by Major Walter L. Cassin, 15th regiment.

The train at these Head-Quarters will be commanded by Captain Timothy Lubey and will be called the "Third Train of the 15th Regiment" and be under the general charge of Major Thomas Bogan.

In addition to these, there will be assigned to each train another company to be under the general command of the captain of the train as already designated. To the 1st, 2d, and 3d trains of the 50th regiment, companies "C," "F," "D," "K" of Captains Geo. N. Falley, M. H. McGrath, Asa C. Palmer, and James H. McDonald, respectively, and to the 1st, 2d, and 3d trains of the 15th regiment, companies "A," "I," and "K" of Captain A. P. Green, Lieutenant Thomas Sanford, and Captain Sewall Sergeant, respectively.

Each train will be divided by its captain into four sections, comprising "first section" and "second section of pontons," and "first section" and "second section of equipage." The first and second platoons of the captains of the trains, will be assigned to the first and second sections of pontons, and the 1st and 2d platoons of the second company to the first and second sections of camp equipage.

The remaining companies of each regiment will form a *road party*, the largest company and any force beyond one company to each train, preceeding [sic] the first train, the second in size the second train, and so on, and if there are but two companies, the second in size will be divided so that about one-half shall preceed [sic] each of the last two trains.

The working parties of the 50th regiment will be under the charge of Major Wesley Brainerd, and of the 15th, under Major E. C. Perry. Each working party will be supplied with tools in the proportion of one to each two men, who on a march will relieve each other in carrying these tools, at equal intervals, by the order of the officer. The kind of tool will be, not over one-tenth picks, though with two at least to each party, and the balance about equally, shovels and axes. The whole command being necessarily available as a working party at all times will have their knapsacks and extra rations carried on the trains if not overloading the teams, but so placed as to be readily unloaded in case of emergency. The men will habitually only carry their arms, 40 rounds of ammunition and rations in their haversacks, blankets and overcoats.

The ambulance trains will be about equally divided for the rear of each train until otherwise directed. The largest proportion of ambulances with the rear train, the officer in charge to have his position in the rear of 2d train of 15th regiment, and one non-commissioned officer in rear of 2d train of 50th regiment. The instruments of the band will be carried in the trains to which they belong and the musicians will be prepared to assist at the ambulances.

The colonels of the regiments will take the general direction and charge of the trains and see to their proper closure, &c. on the march, taking care to expedite their marches to the utmost degree practicable and being each responsible for the same, directly to the General commanding, from whom, or through his staff, all necessary orders will be given.

The place of the colonel will be generally (when not called elsewhere by duty) between the 1st and 2d trains of his regiment. The colonel will be careful to send an officer to the General commanding each evening on the parking of the trains of his regiment and when practicable between 10 and 12 a.m. of each day to impart full information as to the condition of their trains or to receive orders.

The colonel will be assisted in his general supervision by the lieutenant colonel, whose position will be habitually between the 2d and 3d trains and who will be especially charged with selecting positions for parking the trains as well as with their prompt starting on marches at the hour ordered, with a view to which their mode of parking will be arranged, and trains 1 and 2 of each regiment will be parked as near together as practicable. These marches, will, unless otherwise directed, usually begin at the earliest light that the roads can be plainly visible, the breakfast of the men, the striking and packing of tents and harnessing and hitching up of teams being completed before that hour.

The respective field officers commanding trains will select a careful, reliable non-commissioned officer with a guard of five men for each company under their command, to bring up the rear of each train, to prevent straggling, marauding, skulking, &c., and persistent stragglers will be summarily dealt with. The officers in charge of the different sections of the train are held responsible, that no one rides on the wagons without the written authority of a medical officer thereof, for which one medical officer, if so many are available, will accompany each train, the senior of the regiment with each 2d train. Each section of a train will start off when ready at the hour ordered, to follow the movement of its next advance section, and without being delayed for a rear section. And in all cases when a wagon is delayed by an accident or any cause whatever, those in the rear must if it is possible pass on, leaving such wagon to take its proper place, whenever practicable, and without delay to the trains, unless contrary orders shall be given by the colonels, in writing, to the captains of the train.

Should wagons become fixed or "set" in bad places, it may be expedient to unload them at once, to drag the wagons out separately, as the balks, chess, &c., can be readily carried for short distances on the shoulders of the men, and it may even be advisable to dismount a ponton to move it by rollers, or even directly on soft muddy ground like the stone "boat" or "drag."

As it is desirable to give as much freedom of passage as possible to the trains; when two roads are known to be practicable to the point aimed at, unless otherwise specially directed, the trains of the 50th regiment will habitually take the right hand roads in such cases, [while] those of the 15th regiment follow the left roads. One wagon load of ammunition, with the regimental wagons, forage, &c., with the beef on the hoof, will follow in rear of the 2d train of each regiment, in charge of its regimental quartermaster and assistant commissary of subsistence.

The Head-Quarters wagons under the brigade quartermaster will take the head of the column of the 50th regiment, near which will be the habitual position of the commanding General.

The wagons of the *trains* will not be halted for locking the wheels or for watering the animals, except under the direct orders of the captains of the trains or other superior officers, who, in giving such orders, will, at the same time, station reliable officers or non-commissioned officers at the points for locking and for unlocking wheels and watering the animals, who will be responsible that these operations are performed with the utmost expedition possible. A supply of extra wagon poles and kingbolts will be carried with each train to avoid delays from these sources.

The road parties, will, whenever it is at all practicable, in very bad, especially miry places, construct double road ways, that the train may pass, if any wagons should break down at these points, and in making "corduroy" ways, especial care will be taken to lay as near as may be the logs or poles of the same diameter next each other, the buts on the same side, and to place above these the smaller branches of limbs well straightened before laying on the earth of the road-way. The stumps of trees in road ways must be cut close to the ground. *These rules are to be general for all roads* worked by this brigade.

All orders to any of the officers of the regiments in relation to the halts or movements of the trains given in the name of the General commanding by the assistant adjutant general, aides and quartermaster will be at once complied with.

As a prompt obedience to orders, especially as to hours for details or for marches, is indispensable to the successful movement of this brigade (it may be said—in its present situation—for this whole army,) the officers of the command may rely upon it that any delinquency in this respect will be inexorably followed up and traced to the offender to the end that, even in a first offence discovered in any person not of a fair character as an officer, or for *any* second offence of this kind, such delinquency may be reported to the General commanding this army for his action.

By order of Brigadier General H. W. Benham:
Commanding Engineer Brigade.
Official:
Channing Clapp
Capt. and A. Ass't Adj't Gen'l
Major Wesley Brainerd.

ITEM 6

This order was issued by WB to his battalion on its movement from Rappahannock Station to the II Corps encampment prior to the opening of the Wilderness Campaign. It is indicative of the detailed management required of battalion commanders in the engineering regiment. The movements of such a complicated unit, encumbered with their special material and wagons had to be carefully worked out. This was tipped into the gutter between text pages 417 and 418.

Order of march for 1st Battalion 50th Regt N.Y.V.E.

No of men to Wagons	No of Wagons	Kind of Wagons		
4	1	Head Quarter Wagon		
4 = 12	2	Commissary Wagons		
4	1	Company	"	Co G
4	1	Tool	"	
6	1	Abutment	"	
6 = 42	7	Pontons	"	
6 = 18	3	Chess	"	
11 = 12	3	Forage	"	
102				

4 = 12	3	Forage Wagons	
4	1	Quartermaster Wagon	
4	1	Teamsters "	
4	1	Company "	Co F
4	1	Tool "	
6	1	Extra Balk "	
6 = 42	7	Ponton "	
6 = 18	3	Chess "	
4	1	Forage "	
98			
4	1	Company Wagon	Co B
4 = 24	6	Entrenching Tools	
4 = 20	5	Forage for Train	
8	2	Forage for Entrenching S.Teams	
	2	Hospital	
		Rear Guard	
	2	Ambulances	
56			

Remarks:

1st Section—Co "G" in charge. 102 men will march [with the wagons] and their knapsacks will be carried within [the] wagons. The remainder of the company will march in advance as "Advance Guard."
2nd Section—Co "F" in charge. 98 Men march with [the] wagons, the ballance in rear of Section.
3d Section—Co "B" in charge. 56 men [march] with the Wagons, the ballance in rear of train & in front of ambulances as *"Rear Guard"*.

Officers in charge of Sections will allow no man to ride in the wagons. They will attend personally to the proper arrangement of the Knapsacks in the wagons at the rate of not more than 4 to the Army Wagon and 8 to the Pontons & 6 to the Chess and Balk wagons. No man will be allowed to place his gun upon any of the wagons.

The proper arrangements of the trains gives less transportation for the knapsacks of Co "B" than could be wished for. The only way to remedy the difficulty as far as possible is to relieve those who are to carry their knapsacks every morning. By a judicious packing of the company wagon a few more knapsacks may be carried than is laid down for it. The same arrangement will be made with the other companies "F" & "G" so far as the surplus knapsacks are concerned.

Should anything happen to any of the wagons to prevent its moving, the commander of the Section will immediately send a report of the fact to the commanding Officer.

When a halt is made for the night the train will park as in the order of march unless instructions are given to the contrary.

The commander of the 1st Section will detail 1 Sergeant and 10 men as a "Pioneer Party." They will be furnished with 2 Picks, 2 Axes and 6 Shovels.

A copy of this order is furnished to every Company Commander and to the Quartermaster, all officers will see that its provisions are strictly complied with.

By Order of
Wesley Brainerd
Major Comdg 1st Battalion
50th Regt N.Y.V. Eng.

Memorandum of Baggage to be carried in Head Quarters Wagon:

2 Wall Tents
Comdg. Officer's Mess Chest, Surgeons Mess Chest
Quartermasters Mess chest & Pails for both messes
Comdg Officers Desk
Adjutants Desk
Quartermasters Desk
Baggage for 4 officers & Q.M's clerk
Servants tents & Blankets
16 Days rations for 4 officers, 1 clerk and Servants
3 Days forage for 6 officers Horses
3 " " " 6 Mule Teams

Memorandum for Teamsters Wagon:
Camp Kettles, Mess pans & Pails
Wagon Masters and Assistants baggage
16 Days rations for Wagon Masters and Teamsters
Picket rope
3 Days forage for 6 Mule Team

ITEM 7

This is the full text of the circular order for Grant's review of the II Corps.

Head Quarters, Second Army Corps
Cole's Hill, Culpepper Co. Va., April 8, 1864
Circular

The troops will be reviewed by Lieut. Gen'l. Grant to-morrow at 12 o'clock P.M., if it does not rain, on the plain near Stevensburg.

Divisions and Independent Brigades will be assigned their positions by Lieut. Col. Morgan, Chief of Staff, as follows:

The First and Second Divisions in the first line, First Division on the right, Third and Fourth Divisions in the second line. Third Division on the right. The cavalry on the right of the infantry on the prolongation of the first line. The Artillery Brigade in rear of the Cavalry on the prolongation of the second line.

The Infantry will be formed in lines of battalions in mass, doubled on the centre, with the least intervals between battalions necessary for a change of direction by the flank, (not more than 20 paces,) see Pars. 394, 395, 396, and 397, vol. ii. "Casey's Tactics."

The front of the divisions should not be less than twenty (20) files, nor much more than thirty (30) files. The intervals between the two lines of battle will be one hundred (100) paces.

The first call of the bugle, sounded by direction of the Corps commander, will be the signal that the reviewing officer is approaching and for the troops to be called to attention. The infantry will then be brought to a "parade rest," by order of the Division Commanders. All general, field and staff officers will continue mounted.

At the second call of the bugle, the colors of each Regt. of infantry, with the front rank of the color guard, will advance six paces to the front of the centre of their battalions, the rear rank taking place in the front rank of the Divisions. The Colonels will take their places eight paces in front of their colors; the Lieut. Colonel and Major on the line of the colors, the Lieut. Colonel opposite the right, the Major opposite the left of the battalion. The adjutant on the right of the first division, the regimental quartermaster on the left. All other regimental staff officers in one line in rear of the battalion.

Brigade Commanders will take their places opposite the centre of their Brigades. Brigade colors being dressed to the right, Brigade Staff Officers in one line in rear of the Brigade commanders. Brigade orderlies in one line in rear of the Brigade Staff.

Division commanders on the right of their Divisions in advance of the line occupied by Brigade commanders. Division Staff officers in one line in rear of Division Commanders. Division orderlies in one line in rear of the Staff.

The Division Generals will join the reviewing officer and the Corps Commander as their Divisions are reached and will continue with them during the inspection of the Corps. The salutes will be given by successive brigades. After the salute each brigade will remain at shouldered arms until it is inspected and also while the reviewing officer is passing in rear of it. At all other times it will be at ordered arms and parade rest.

All officers accompanying the reviewing personage will allow sufficient space for their seniors to be in advance of them.

After the inspection and the reviewing officer has passed to the right and at the third call of the bugle, Division Generals will put their commands, by battalion to the right, into column, the colors and officers taking their posts. Commanders of Divisions which do not move immediately after performing this movement will cause their men to stand at an ordered arms and parade rest.

The march will commence at the fourth call of the bugle, sounded by direction of the Corps Commander. The troops will pass in review by division front at half distance.

Each Division Commander, when he has passed the reviewing officer, will take his place on the right of the Corps Commander. His staff forming on the right of the staff of the Commander of the Corps, orderlies in the rear, on a line with the orderlies of the Corps Commander. Each Brigade commander will take his place on the right of his Division commander while his brigade is passing; his staff on the right of the staff of his Division commander, orderlies on the right of division orderlies.

The pioneers will march by brigade. The Major General commanding the corps would suggest that the music be consolidated, if practicable, as much as possible, either by division or brigade.

In passing in review, the Infantry will precede and the Artillery follow the Cavalry. After each brigade has passed the reviewing officer, it will move rapidly so as under no circumstances to obstruct the column in the rear. The greatest attention will be paid to this matter by every commander. The troops will return to their camps immediately after passing the reviewing officer.

The troops will be on the ground promptly at 12 o'clock, and will parade with knapsacks, haversacks and canteens and with great coats neatly rolled on top of the knapsacks.

The commander of the Artillery Brigade will designate the battery to fire the appropriate salute.

For any other information concerning the forms to be observed in this review, reference is made to Art. III. form I. for a review of a Corps of Infantry, by Brig. Gen'l. Torbert, which may be followed, except when it conflicts with this order. Not more than the strength on one Regiment will be left in the camps of the brigades at Stony Mountain and on the Germania Ford Road.

A field return, showing the actual aggregate present on review, will be furnished on the field by a close investigation. Each orderly sergeant, adjutant, and Brigade and Division Staff officer should be provided with a blank form for this purpose. The form will also embrace a column for the aggregate of the armed men absent by virtue of being on picquet or on duty at Stony Mountain and on the Germania Ford Road or other detachments of such nature.

Each Division Commander will send an officer and sufficient number of pioneers to prepare its own ground in advance of its arrival.

By Command of Major General Hancock
Francis A. Walker.
Assistant Adjutant General.
Official

Notes

Abbreviations

Works frequently cited have been identified by the following abbreviations:

ARAGNY	*Annual Report of the Adjutant General of the State of New York for the Year 1898: Registers of the First, Fifteenth, and Fiftieth Engineers and the First Battalion of Sharpshooters.*
OR	*The War of the Rebellion: A Compilation of the Official Records of the Union and Confederate Armies.*
RUS	*Register of Officers and Agents, Civil, Military and Naval in the Service of the United States on the Thirtieth of September, 1861.*
USMA	*Register of Graduates and Former Cadets of the United States Military Academy.*

Introduction

1. Lucy Abigail Brainard, *The Genealogy of the Brainerd-Brainard Family in America 1649–1908*, vol. 1 (Hartford Press: The Case, Lockwood and Brainard Co., 1908), 130.
2. *Portrait and Biographical Record of the State of Colorado* (Chicago: Chapman Publishing Company, 1899), 304–5.
3. John H. White Jr., "Once the Greatest of Builders: The Norris Locomotive Works," *Railroad History* 150 (Spring 1984): 17–28.
4. *Portrait and Biographical Record*, 306.

CHAPTER ONE

1. WB is misquoting Alexander Pope (1688–1744). It should read: "Honour and shame from no Condition rise;/Act well your part, there all the honour lies" (*An Essay on Man*, Epistle IV, ll. 193–94). See William K. Wimsatt, ed., *Alexander Pope: Selected Poetry and Prose* (New York: Holt, Rinehart and Winston, 1972), 226. Whether WB learned it wrong, misremembered it, or simply changed it to suit his own meaning we cannot know.
2. In June 1840, prominent Whigs moved the Vernon Compass newspaper to Rome and renamed it the *Roman Citizen*. Alfred Sandford (b. 1823, d. Rome, 1898) came with the paper as errand boy. In 1854, Alfred Sandford became sole proprietor of the *Roman Citizen* and remained so until February 1884, when he sold it and retired. The paper ceased publication in 1903. The other paper in Rome in 1860 was the *Sentinel*, controlled by Democrats. Daniel W. Wagner, "Part IV, History of Journalism in Rome," *Daily Sentinel*, June 8, 1964.
3. This advertisement appeared in the *Roman Citizen*, April 20, 1859:

    "Having recently purchased the premises, formerly occupied by M. Burns as a Sash & Blind Factory, and having added a quantity of New Machinery, we are now prepared to execute in the neatest manner, all kinds of *Turning and Planing*, *Plain and Scroll Sawing*, *Dovetailing*, and every variety of Jobbing, requiring the use of machinery. We have constantly on hand and manufacture to order, a great variety of *Plain and Ornamental Cottage Bedsteads*, also, a stock of *Common Bedsteads*, *Table Legs*, *Ballasters*, *&c*.

    A.H. Brainerd & Son,
    Rome, N.Y."

    In the same paper appeared an article describing the testing of two "Fanning Mills," devices used to separate grain and seeds from their hulls and other debris by using fan blades in an enclosed space.
4. Col. Peter Gansevoort (1749–1812) commanded 750 men at Fort Stanwix (later renamed Fort Schuyler) on the Mohawk River at the present site of Rome, New York, in 1777. He was besieged by Col. Barry St. Leger with a force of eighteen hundred Regulars, Indians, and Loyalists on August 3, 1777. Gansevoort refused St. Leger's demand to surrender and was finally relieved by a column of one thousand under Gen. Benedict Arnold on August 22. Henry Steele Commager and Robert B. Morris, eds., *The Spirit of 'Seventy Six* (New York: Bonanza Books, 1983), 561–68. The following two articles are from WB's memoir:

    Gansevoort Light Guard.

    A military Company under the above name was formed in this village about three months ago, and is now duly organized.—The Company numbers at present some 40 members.

An election of civil officers for the ensuing year, was held last Thursday evening; the following persons being duly elected:

F.W. Oliver, President.
Wesley Brainard, Vice President.
C.F. Bissell, Treasurer.
S.B. Wallworth, Recording Secretary.
F.A. Mallinson, Assistant Secretary.
The following are the Commissioned officers.
C.H. Skillin, Captain.
W. Brainard, 1st Lieutenant.
B.H. Wright, 2d Lieutenant.
M.W. Rowe, 3d Lieutenant.
NON COMMISSIONED.
George Merrill, 1st Sergeant.
H.J. Gifford, 2d Sergeant.
Edward Thompson, 3d Sergeant.
C.F. Bissell, 4th Sergeant.
A.H.S. Curtiss, Corporal.
Wm. H. Phillips, 2d Corporal.
Charles Hemmens, 3d Corporal.
E.M. Moore, 4th Corporal.

This military company we learn is in a prosperous condition and promises to be one of the permanent "institutions" of our village.

With the number of military companies now organized in our village, we don't see as Rome need feel any fear from invasion. She is able to defend herself and protect her citizens.

[newspaper and date unknown]

GANSEVOORT LIGHT GUARD—This favorite corps made its fourth public appearance yesterday afternoon, accompanied by the Rome Brass Band. The weather was magnificent—a favor the "Guard" had not enjoyed at any parade previous to the one of yesterday.

After a short and well-executed field drill in the Arsenal Yard, the "Guard" took up its line of march through our principal streets, to the enlivening notes of the Rome Brass Band, and its own much improved martial band, closing with a series of beautiful and difficult manoeuvres in Dominick and James streets.

The G.L.G.'s are a credit to themselves and no small addition to the public "institutions" of our place. We wish them abundant success.

[newspaper and date unknown]

5. George N. Hollins was the captain of the U.S. warship *Cyane* in 1855 when he got involved in a tawdry imbroglio in Nicaragua. The town of Greytown, a self-

governing metropolis of four to five hundred, sat astride the San Juan River and commanded that route across the isthmus. Greytown was a town that lived by commerce and got caught up in a battle of competing business interests. There was a riot of sorts during which Solon Borlund, the American minister to Central America, got cut by a flying bottle and the resident state department official called in the navy to assert U.S. presence and calm things down. Hollins steamed in, demanded reparations, and then let loose with a two-hour bombardment of the defenseless town, this against all entreaties, then landed a party and destroyed the town, causing three million dollars in damages but no loss of life. The incident proved a political embarrassment as well as an internationally condemned act of barbarism. Allan Nevins, *A House Dividing, 1852–1857*, vol. 2 of *The Ordeal of the Union* (New York: Charles Scribner's Sons, 1947), 365–67.

Hollins was a captain in the USN when war broke out, and, after resigning his commission, he was made a commander in the CSN. He captured the Chesapeake Bay steamer *St. Nicholas* shortly thereafter. He participated in the defense of New Orleans, but was transferred to courts duties in April 1862 and remained on various courts and boards to the end of the war. Mark M. Boatner III, *The Civil War Dictionary* (New York: David McKay Co., 1959), 405; Stewart Sifakis, *Who Was Who in the Confederacy*, vol. 2 of *Who Was Who in the Civil War* (New York: Facts on File, 1988), 132–33.

6. The Prince of Wales (later to become King Edward VII) visited the United States and Canada in 1860 and was at Kingston, Canada, on September 4 and 5. The incident that WB relates had its seeds in an address the Prince made in Quebec at the French-speaking University of Laval. In addressing the Roman Catholic bishops as "Gentlemen" rather than "My Lords," the Prince offended the Catholics, a strong contingent in Quebec. The Duke of Newcastle, in a published apology, then angered the Protestant Orangemen in Toronto, causing widespread resentment, demonstrations, and arch building. It all came to a head in Kingston with WB and the Gansevoort Light Guard right in the middle of a near riot of two thousand very upset Irish Protestants. Philip Magnus, *King Edward the Seventh* (New York: E. P. Dutton, 1964), 32–36.

   Although WB has clearly written "peaches" in the memoir, it seems a strange reference.

7. John Benson Lossing (1813–1891) wrote two books on the Civil War. WB can be referring only to the first, as the second was not published until 1912. Lossing's *Pictorial History of the Civil War in the United States of America*, published in 1866–68 in three volumes, was illustrated with woodcuts from sketches by the author. Nevins, in *Civil War Books: A Critical Bibliography*, comments: "Since Lossing was both an eyewitness and an historian, his study has value in the extent of his coverage and in his unusual first-hand impressions." Allan Nevins, James I. Robertson Jr., and Bell Irvin Wiley, eds., *Civil War Books: A Critical Bibliography*, vol. 2 (Baton Rouge, La.: Louisiana State Univ. Press, 1967), 20.

8. President-elect Lincoln left Springfield on February 11, 1861, passed through Rome on February 18, and arrived in Washington on February 23. Hundreds of citizens of Rome went to Utica, fifteen miles east, to hear and see Lincoln speak

on the eighteenth. John S. Bowman, executive ed., *The Civil War Almanac* (New York: Gallery Books, 1983), 45–46; "The President Elect," *The Roman Citizen* February 20, 1861.

9. WB is referring to the Seventh Regiment, New York State Militia, an elite state unit stationed in New York City and filled with many of the cream of New York Society. Richard Wheeler, *A Rising Thunder: From Lincoln's Election to the Battle of Bull Run: An Eyewitness History* (New York: HarperCollins Publishers, 1994), 139.
10. On May 19–20, 1856, Massachusetts Senator Charles Sumner, then a Democrat and ardent abolitionist, made a vituperative speech entitled "The Crime Against Kansas," during which he attacked the South, slavery, South Carolina, and Sen. Andrew P. Butler among others. Representative Preston S. Brooks, Butler's nephew, was infuriated, and on May 22 he assaulted Sumner on the U.S. Senate floor with a cane, beating him senseless. It took Sumner years to fully recover, and he was reelected as a Republican. Brooks was hailed as a hero in the South. Nevins, *House Dividing*, 437–48.
11. WB was employed as draughtsman and locomotive builder at the Norris Locomotive Company in Philadelphia, Pennsylvania, from 1850 to 1858 and traveled widely for the company as a locomotive starter. He then became a master mechanic on a railroad in Georgia sometime in the late 1850s, staying there until war became inevitable, when he returned to Rome. *Portrait and Biographical Record of the State of Colorado* (Chicago: Chapman Publishing Company, 1899), 306.
12. Maj. Robert Anderson surrendered Fort Sumter on April 13, 1861, after a bombardment by southern batteries that began at 4:30 A.M. on April 12. Bowman, *Civil War Almanac*, 51.
13. Although WB states that this article is from the *New York Herald*, I found the exact duplicate in the *New York Times* of Saturday, April 20, 1861. The Sixth Massachusetts Regiment lost four killed, thirty-nine wounded. There were twelve civilian deaths and dozens wounded. Patricia L. Faust, *Historical Times Illustrated Encyclopedia of the Civil War* (New York: Harper and Row, 1986), 37.
14. President Lincoln's call for seventy-five thousand volunteers was dated April 15, 1861. Roy P. Basler, ed., *The Collected Works of Abraham Lincoln*, vol. 4 (New Brunswick, N.J.: Rutgers Univ. Press, 1953), 331–32.
15. The Seventh New York State Militia Regiment left New York City on April 19, 1861, and arrived in Washington, D.C., on April 25, 1861, helping to secure the capital from attack. The Seventh was there until June 3, 1861, when it was mustered out of Federal service. The unit was called out again and served from May to September 1862 during the Peninsula/Jackson's Valley Campaigns, when Lincoln and Stanton feared for the safety of Washington, and again from June to July 1863, during the Gettysburg Campaign, for the same reason. Frederick Phisterer, *New York in the War of the Rebellion: 1861 to 1865*, 3rd ed., vol. 1 (Albany, N.Y.: State of New York, 1912), 546–47.
16. On April 19, 1775, Massachusetts Minutemen fired the first shots of the American Revolution on Lexington Green and at Concord Bridge. Commager and Morris, *Spirit*, 69–76.

17. Company G of the 14th New York Infantry was raised in Rome to serve two years. There were six officers in the regiment who enrolled in Rome, and three of them eventually became captains of Company G. The Regiment lost 104 men in two years and was with the Army of the Potomac for all its battles through Chancellorsville. Phisterer, *New York in the War of the Rebellion*, 3: 1901–11.
18. Eli A. Gage was WB's father-in-law and Lyman Gage was Eli's son, Maria's brother.
19. Ephraim Elmer Ellsworth, the twenty-four-year-old colonel of the 11th New York Fire Zouaves, was killed on May 24, 1861, after removing a Confederate flag from atop the Marshall House Hotel in Alexandria, Virginia. His former militia unit, the U.S. Zouave Cadets of Chicago, toured the country in 1860. After that, Ellsworth worked in the law office of Lincoln and Herndon in Springfield, Illinois, served in the presidential campaign, and went with Lincoln to Washington. Ellsworth raised the 11th New York from among New York's fire departments. Faust, *Historical Times*, 240.
20. The First Bull Run Campaign was on July 16–22, 1861, and climaxed in a confused battle between two armed mobs of green soldiers on July 21. It ended in a rout of the Union troops and led to overconfidence in the South and grim determination in the North. Six days later, President Lincoln brought Maj. Gen. George McClellan to Washington to make an army. Faust, *Historical Times*, 92.

    On August 2, 1861, Congress passed the National Income Tax Bill to raise five hundred million dollars to provide for Union support. Bowman, *Civil War Almanac*, 61.

    Lincoln requested four hundred thousand men to aid the Union before a special session of Congress on July 4, 1861. Basler, *Abraham Lincoln*, 4: 431–32.
21. Charles Beebe Stuart (1814–1881) had a long and illustrious career as an engineer. He conceived the railway suspension bridge over the Niagara River two miles below the falls and, despite widespread disbelief that it could be done, contracted and supervised its completion in 1855 by John A. Roebling. He was state engineer for New York, engineer in charge of the Brooklyn dry docks, and engineer-in-chief of the U.S. Navy. He was chief partner of Stuart, Serrell & Company, president of the Iowa Land Company, and had numerous interests in building several railroads. He authored three books: *The Naval Dry Docks of the United States* (1852), *The Naval and Mail Steamers of the United States* (1853), and *Lives and Works of Civil and Military Engineers of America* (1871). After his Civil War service, he served in a number of engineering capacities and was widely respected in his field. Dumas Malone, ed., *Dictionary of American Biography*, vol. 9 (New York: Charles Scribner's Sons, 1928), 163.

CHAPTER TWO

1. The training depot at Elmira was one of three in New York State. From July 30, 1861, to May 1864, 20,796 officers and men passed through the facility. In May 1864, the depot was turned into a Federal prison camp and 12,123 Confederate prisoners were kept there, of which 2,963 died. Faust, *Historical Times*, 241;

Matthew S. Walls, "Northern Hell on Earth," *America's Civil War* 3 (6) (Mar. 1991): 25–29.

2. The lake connecting Geneva with Elmira is Seneca Lake.
3. The trip from Rome to Elmira is approximately 135 miles.
4. Shoddy was either an inferior woolen yarn made from fibers taken from used fabrics and then reprocessed or the cheap woolen cloth made from this material.
5. This is a reference to a book in WB's library which, because of the abbreviated nature of the reference, is impossible to specifically identify. In the *Army Officer's Pocket Companion*, which WB most probably carried, there is a diagram and description of the layout of the typical regimental camp for infantry. Photographs of the day show clearly that this layout was followed to the letter by infantry regiments, often regardless of the topography of the campsite. William P. Craighill, *The Army Officer's Pocket Companion and Manual for Staff Officers in the Field* (New York: D. Van Nostrand, 1863), 119. [See fig. on p. 190.]
6. Charles B. Norton was relieved of duty on January 6, 1863, and resigned from the service on January 26, 1863 [see Appendix 1], but there is no clue to the reason for his resignation. He married Fannie Parker on January 18, 1863, and they moved to Europe, living seven years in Paris. While there, Norton was appointed commissioner general to the Paris Exhibition by the U.S. Congress and served in other positions of management over other expositions. Norton was wounded at Second Manassas and breveted brigadier general of volunteers for meritorious service. He died at the Palmer House in Chicago on January 29, 1891, age sixty-six, and his wife Fannie was present at his death. He was buried at Woodlawn Cemetery in New York. Charles B. Norton Pension Records, National Archives and Records Administration.
7. Edward Christopher James (1841–1901) came from an illustrious family that traced back to Roger Williams on one side and the Plymouth Colony on the other. James was a well-educated son of a justice of the New York State Supreme Court. His military career was meteoric. He served as adjutant of the 50th, assistant adjutant general and aide-de-camp to General Woodbury, major of the 60th New York, lieutenant colonel and colonel of the 106th New York. He commanded brigades at the age of twenty-two. He retired from military service in the Spring of 1863 for disabilities incurred while on duty. He returned to the practice of law and distinguished himself in Ogdensburg and later in New York City as senior partner in James, Schell & Elkus. Malone, *American Biography*, 5: 575.
8. Bolton W. De Courcey-O'Grady claimed to be a descendant of one of the noble houses of Ireland, "The O'Grady." He married Helen Coffin on September 2, 1852, in Canada, and they had two daughters, Annie Gertrude (b. 1856) and Amelia Geraldine (b. 1858). O'Grady lived with his family in Washington, D.C., until his disappearance in late August 1864. His wife filed for a widow's pension in May 1888, stating that O'Grady had been killed in Virginia while on a tour of inspection as a staff officer under General Barnard. This could not have been the case, as O'Grady had resigned from the army in April 1863. [See Appendix 1.]

   The O'Grady's story does not end there. He disappeared but was far from dead. It appears that he was maintaining another family in Washington, living with a Mrs. Chappelle and her two sons, and evidently leading his wife to understand that he was still in the army and therefore had to be gone for long periods of time. He then abandoned his legal wife, Helen, and his daughters, somehow fabricating his death and making them believe that he had been killed in Virginia. Then he moved to Chicago with Mrs. Chappelle. He took the name Walter De Courcey, and it was probably in Chicago that WB ran into him, and where O'Grady (De Courcey) denied his former identity, fearing that WB would somehow get word to his legal wife, who thought him dead.

From there O'Grady moved to Hobart, Indiana, with Mrs. Chappelle and her sons, but abandoned her sometime in the 1870s. He then married Amelia Isabella Shannon on June 22, 1881, and they had a daughter, Geraldine Montefiore Isabella Caroline, in 1884. O'Grady died on April 1, 1900, at Olympia, Washington. The second Mrs. O'Grady (Mrs. De Courcey) then filed for her widow's pension and the ensuing investigation turned up the whole story.

O'Grady himself had filed an invalid pension claim in 1897 and told a special examiner that he had divorced Helen O'Grady in Chicago in 1866, but Helen, when questioned, stated that she did not recall a divorce, and the pension examiners could find no proof of one. The examiners did turn up evidence of O'Grady's involvement with women in Covington, Kentucky, and in Louisiana. Bolton W. O'Grady Pension File, National Archives and Records Administration.

9. Ira Spaulding's health was broken by the war, and he spent ten years trying to recuperate. He married his wife, Clara, on April 24, 1867, after a sojourn in Florida in an attempt to cure his chronic cough and attacks of pleurisy. In her deposition to the pension board, his wife stated that his physical problems stemmed from a wound in the side Spaulding received while directing the erection of earthworks but that he rarely complained of his condition and insisted on treating himself. His health did not improve while working for the Northern Pacific Railroad in Minnesota, and he died suddenly in Philadelphia on October 2, 1875, with an autopsy stating cause of death to be "Valvular Heart Disease." Ira Spaulding Pension Records, National Archives and Records Administration.
10. Sadly, the *Register*, the *Adjutant General's Report*, and the photo album are not in our possession, nor do we know if they exist. WB mentions many items that he preserved from his war service, but he moved many times prior to his death, outlived all of his immediate family, and left no hint as to the disposition of his effects.

Chapter Three

1. The Ellsworth Fire Zouaves, later the 11th New York Volunteers, left New York on April 29, 1861, and were mustered into service on May 7, 1861, at Washington, D.C., for two years. They lost forty men at First Bull Run and were stationed in New York Harbor and Westchester County in September 1861, before being shipped out to Fortress Monroe. The regiment served at various

posts during its enlistment, which ended when its members were mustered out on June 2, 1862. Phisterer, *New York in the War of the Rebellion*, 1: 19, 3: 1861.

2. *New York Times*, September 20, 1861: "Yesterday a most disgraceful riot took place among the Fire Zouaves who were encamped on the Battery. After having an altercation with the Elmira regiment, also stationed there, they assaulted unarmed inoffensive citizens, tossing them in blankets, and cruelly beating one man whom they thought was a reporter for the New York Times. Finally a number of them were got aboard a steamer that was to take them to Amboy, enroute for Fortress Monroe."

   Although WB thought that the 11th was being discharged and awaiting final pay, it is more likely that they had just been paid and were having a last fling fueled by strong drink before facing a tour of duty at Fortress Monroe.

3. Although WB states that the man recovered, there is a record of Pvt. John L. Newcomb, age forty-four, who was killed on September 21, 1861, at New York City by the accidental discharge of his gun. *ARAGNY*, 1064.

4. The Cooper's Shop in Philadelphia, Pennsylvania, was the Civil War's most famous volunteer way station. In one year, 87,518 transient soldiers were fed and refreshed there. There were dining rooms, sleeping quarters, and other facilities that operated twenty-four hours a day. Bell Irvin Wiley, *The Life of Billy Yank: The Common Soldier of the Union* (Baton Rouge: Louisiana State Univ. Press, 1986), 35.

5. The city of Washington found it expedient to exercise control over the thousands of soldiers that traveled through that town on the way to the various fronts in the East. Two large buildings near the depot, Soldiers Rest and Soldiers Retreat, were set aside to house, feed, police, and forward large numbers of soldiers in an efficient manner. The commissaries were notified of incoming trains, and soldiers would sit down to a hot meal upon arrival and would take cooked rations with them when they left. Margaret Leech, *Reveille in Washington: 1860–1865* (Alexandria, Va.: Time-Life Books, 1980), 231.

6. While McClellan was training the green Union army in the late summer and fall of 1861, Confederate outposts were pushed in as close as Munson's Hill, an elevation in Virginia seven miles southwest of Washington. Leech, *Reveille*, 139, 143; Maj. George B. Davis, Leslie J. Perry, and Joseph W. Kirkley, *The Official Military Atlas of the Civil War* (New York: Fairfax Press, 1983), plate 7.1.

7. Fort Corcoran was an earthwork enclosure constructed at the Virginia end of the Aqueduct, which carried the Chesapeake and Ohio canal over the Potomac and also served as a bridge. It was named after Col. Michael Corcoran of the 69th New York Militia and hero of First Bull Run. William C. Davis, *First Blood: Fort Sumter to Bull Run*, The Civil War Series (Alexandria, Va.: Time-Life Books, 1983), 60–61; Leech, *Reveille*, 136.

8. The De Kalb Regiment, designated the 41st New York Infantry, was composed of Germans and raised mostly in New York City. It was initially commanded by Col. Leopold Von Gilsa, who later commanded a brigade in the XI Corps at Chancellorsville and Gettysburg. Phisterer, *New York in the War of the Rebellion*, 3: 2237; Stewart Sifakis, *Who Was Who in the Union*, vol. 1 of *Who Was Who in the Civil War* (New York: Facts on File, 1988), 429.

9. Hall's Hill was three and one-half miles due west of Ft. Corcoran. Davis, *Military Atlas*, 89.1.
10. At this time the 50th was in the 3rd Brigade, Fitz-John Porter's Division, V Corps, Army of the Potomac. Phisterer, *New York in the War of the Rebellion*, 2: 1670. [See also note 18, p. 347.]
11. This training regimen was planned and instituted throughout the Army of the Potomac by General McClellan to bring the volunteer regimental and company officers to a high state of competence. Stephen W. Sears, *George B. McClellan: The Young Napoleon* (New York: Ticknor & Fields, 1988), 112.

    Gen. Daniel Butterfield enjoyed a meteoric rise in his military career, despite not having any background in the military. A businessman with strong political connections, he was made a colonel and commanded the 12th New York Militia during the Bull Run Campaign. He quickly ascended the ranks to major general and served well on the Peninsula. He rewrote a bugle call to become the haunting "Taps" and designed the army's corps badges. He became General Hooker's chief of staff, then served in that same capacity for General Meade and was wounded at Gettysburg. He commanded XX Corps at Atlanta but went home because of sickness. After the war he served as superintendent of the army recruiting service and colonel of the 5th Infantry. He died in 1901 and was buried at West Point by special order. Faust, *Historical Times*, 100–101.
12. On July 14, 1861, Horace Greeley, publisher of the *New York Tribune*, ran a masthead that announced "Forward to Richmond," echoing the cries for action pressuring Lincoln from all sides. These cries were silenced after First Bull Run, but as soon as McClellan came on the scene and the Army of the Potomac grew to impressive numbers, the demands for action were again taken up. E. B. Long and Barbara Long, *The Civil War Day by Day: An Almanac 1861–1865* (New York: Doubleday, 1971), 94.
13. Ball's Bluff, October 21, 1861, was a tactical misadventure across the Potomac fords twenty-six miles northwest of Washington against Confederate forces near Leesburg, Virginia. Led by Col. Edward D. Baker, the 15th and 20th Massachusetts Infantry were ambushed, suffering 921 killed, wounded, and missing. Baker, a personal friend of President Lincoln's, was among the dead. The debacle led directly to the formation of the Legislative Committee on the Conduct of the War. Boatner, *Civil War Dictionary*, 41; Davis, *Military Atlas*, plate 7.1; Faust, *Historical Times*, 154–55.
14. In the Union army, 29,336 men died from typhoid fever, a disease caused by insanitary conditions. Philip Katcher, *The Civil War Source Book* (New York: Facts on File, 1992), 120.
15. Isaac F. Seamans, Byron R. Seamans's brother, was the fifer in Company C. *ARAGNY*, 1125.
16. WB is here quoting the last two lines of "Ode on a Distant Prospect of Eton College," by Thomas Gray (1716–1771), written in 1742, published 1747: "Since sorrow never comes too late,/And happiness too swiftly flies,/Thought would destroy their paradise./No more; where ignorance is bliss,/'Tis folly to be

wise." Roger Lonsdale, ed., *The New Oxford Book of Eighteenth Century Verse* (New York: Oxford Univ. Press, 1989), 352.

17. The chaplain of the 17th New York was Thomas G. Carver. Phisterer, *New York in the War of the Rebellion*, 3:1922.

18. According to Col. Henry Royer in an address on June 21, 1888, at the dedication of the monument to the 96th Pennsylvania at Gettysburg, the 96th was in camp at Pottsville, Pennsylvania, from September 9, 1861, to October 30, 1861. It did not arrive in Washington until sometime in early November, which would make it difficult for the men to have participated in this incident, which seems to have taken place in early to mid-October. John P. Nicholson, ed. and comp., *Pennsylvania at Gettysburg: Ceremonies at the Dedication of the Monuments*, vol. 1 (Harrisburg, Pa.: Wm. Stanley Ray, 1914), 519–20.

    A more likely explanation is that WB simply got the unit designation wrong. The 50th New York was brigaded with the 17th New York, the 83rd Pennsylvania, the 25th New York, and the 16th Michigan. The 83rd Pennsylvania was commanded by Col. John W. McLane, which seems close enough to explain the mistake. So, instead of the 96th Pennsylvania, it was actually the 83rd Pennsylvania and Colonel McLane instead of Colonel McLean. *OR*, ser. 1, vol. 11, pt. 2: 345; ser. 1, vol. 5: 16.

    Bruce Catton relates the same story, attributing it to *Recollections of a Private* by Warren Lee Goss, and mentioning the regiments as being from Brooklyn and Manhattan. Goss mentions the story as perhaps being a "myth" and refers to the two regiments involved as the ——th Brooklyn and the ——th New York. Could these mystery regiments have been the 17th New York and the 25th New York? Goss served in Company A, 1st U.S. Engineers, and as an engineer might very well have had conversations with WB or others in the 50th who related the story of the baptism under orders. Some confusion over the exact units involved may be expected. Catton, *The Army of the Potomac: Mr. Lincoln's Army* (New York: Doubleday, 1962), 177–78; Warren Lee Goss, *Recollections of a Private* (New York: Thomas Y. Crowell, 1890), 256.

    The 17th New York Infantry was called the Westchester Chasseurs and was raised in New York City and surrounding counties, but none came from Brooklyn. The 25th was raised entirely in Manhattan. Phisterer, *New York in the War of the Rebellion*, 3: 1922, 2012.

19. On September 10, 1861, President Lincoln, Governor Curtin (Pennsylvania), Secretary of War Cameron, and Gen. George McClellan attended a presentation of colors to Pennsylvania regiments in McCall's Division. After being introduced, General McClellan was surrounded by troops anxious to shake his hand. The escort officer requested the crowd to move back, saying that McClellan would say a few words if they did. With that, the General said: "Soldiers: We have had our last retreat. We have seen our last defeat. You stand by me, and I will stand by you, and henceforth victory will crown our efforts." Frank Moore, ed., *The Rebellion Record: A Diary of American Events*, vol. 3 (1861–68; reprint, New York: Arno Press, 1977), 22.

CHAPTER FOUR

1. Paper money was authorized by the Legal Tender Act of February 25, 1862, and began circulating in early April. The original issue was $150 million and later issues totaled $300 million more before the end of the war, by which time the paper money had become both necessary and popular. Government obligations had been issued periodically between 1812 and 1861 in the form of interest-bearing Treasury notes, which bore interest at 5 or 6 percent and were payable after two years. The wartime demand notes were authorized by congressional acts of July 17 and August 5, 1861. Numismatists claim that it was these demand notes that were nicknamed "Greenbacks," but other authorities disagree, stating that it was not until the Legal Tender Act in 1862 that paper money acquired this term. It is unclear which issue WB is referring to here. Greenback value fluctuated with the Union's success or failure on the battlefields of the war. Faust, *Historical Times*, 323, 432, 433; Bradley P. Reed, *Coin World Almanac*, 6th ed. (Sidney, Ohio: Amos Press, 1990), 250.
2. The only soldier to fit the description WB gives is Sgt. Nathan Teall, age twenty-five, who had been sergeant major from September 1, 1861, to March 1, 1862, and died in Geneva of undisclosed reasons on April 20, 1862. Teall had been reduced in rank but then promoted again to sergeant. *ARAGNY*, 1898, 1162.
3. Bough houses were gates at the entrances to each company street constructed of tree branches interwoven to make a decorative arbor.
4. The 44th New York Volunteer Infantry, nicknamed Ellsworth Avengers and People's Ellsworth Regiment, was raised from every ward and town in the state and served in the Army of the Potomac through all its campaigns and battles, losing 188 men killed or wounded, 335 to all causes, during the war. Phisterer, *New York in the War of the Rebellion*, 3: 2289, 2290.
5. The Long Bridge was approximately one mile long and connected southwest Washington to Virginia. The Navy Yard was located in southeast Washington on the right bank of the eastern branch of the Potomac River (Anacostia) across from Uniontown and Poplar Pt. Davis, *Military Atlas*, plate 89.1.

   The Washington Engineer Depot was located one-half mile north of the Navy Yard, near the foot of East 14th and/or East 15th Streets. Dale E. Floyd, ed., introduction to *"Dear Friends at Home . . ." : The Letters and Diary of Thomas James Owen, Fiftieth New York Volunteer Engineer Regiment, During the Civil War*, Corps of Engineers Historical Studies, no. 4 (Washington, D.C., 1985).
6. On November 1–3, 1861, a violent storm lashed the eastern seaboard, scattering the ships headed for the Port Royal expedition as they passed south of Cape Hatteras. One troop transport and three supply ships of the seventy-seven-ship armada were lost. Robert Underwood Johnson and Clarence Clough Buel, eds., *The Opening Battles*, vol. 1 of *Battles and Leaders of the Civil War* (n.p.: Castle, n.d.), 675–76. The *New York Times* reported on November 3, 1861: "The severe easterly rain storm which prevailed yesterday [Nov. 2] and last night seriously interfered with military and telegraphic operations. The rise in the Potomac . . .

owing to the storm, is unequaled within the memory of the oldest inhabitant. The wharves at the Navy-yard were all under water. Part of the drilling ground and the practice batteries were in like condition."

7. The 15th New York Engineers were originally mustered in on June 17, 1861, in New York City for two years as an infantry unit. Raised principally in New York City, the 15th arrived in Washington on June 30 and was officially attached to the Engineer Brigade in March 1862. Phisterer, *New York in the War of the Rebellion*, 2:1651.

   The Engineer Brigade of the Army of the Potomac was organized as follows when it moved to the Peninsula:

   > Brig. Gen. J.G. Barnard, Chief Engineer, commanding.
   > 1st Lt. H.L. Abbot, Topographical Engineers, aide-de-camp
   > Brig. Gen. [Volunteers] D.P. Woodbury, Major, U.S. Engineers, commanding Brigade volunteer engineers:
   > Col. J. McLeod Murphy, 15th N.Y.Volunteers
   > Col. C.B. Stuart, 50th N.Y.Volunteers
   > Capt. J.C. Duane, commanding U.S. Engineer Battalion [regular army] consisting of three companies commanded by 1st Lieutenants Reese, Cross and Babcock, U.S. Engineers
   > Staff: [all U.S. Engineers]
   > Lt. Col. B.S. Alexander
   > Capt. C.S. Stewart
   > 1st Lieutenants Comstock, McAlester and Merrill
   > 2nd Lieutenant Farquhar
   >
   > *OR*, ser. 1, vol. 5: 24–25

   Regular army engineers formed the nucleus of the Engineer Brigade and retained their regular army rank. When directly in command of volunteer troops, they would also carry a volunteer rank, hence some officers carried both regular army and volunteer designations. The regular army engineers were considered the elite of the army.

8. James Lesley Jr. is listed in the *Register of 1861* as chief clerk, War Department. *RUS*, 101.

9. The 50th New York Volunteer Infantry was transferred by orders of General McClellan to engineer duty and assigned to General Woodbury's brigade on October 22, 1861, to receive instruction as engineers. The Volunteer Engineer Brigade was designated in March 1862, and contained the 50th and the 15th. The regiments were officially recognized by the 20th Section of the Act of Congress, passed July 17, 1862. This act authorized a volunteer engineer regiment to have 12 companies with 154 men each and a headquarters staff of 18, a total strength of 1,866 men per regiment. The act also recognized the men in the brigade "on the same footing in all respects in regard to its organization, pay, and emoluments, as the corps of engineers in the regular army." Acceptance by the regular engineers and their officers did not come until after the battle of Fredericksburg, where the volunteer engineers distinguished themselves.

Phisterer, *New York in the War of the Rebellion*, 2: 1650–51, 1669–70; William F. Fox, *New York at Gettysburg* (Albany, N.Y.: State of New York, 1900), 1081–95.

The "castle" WB refers to is the emblem of the Army Corps of Engineers and was worn on the hat and stamped on the uniform buttons.

10. The book WB is referring to is *A Complete Treatise on Field Fortification*, by Dennis Hart Mahan (1836) or a revised edition that was printed in 1861. Mahan was an influential instructor at the United States Military Academy from 1830 to 1871. Faust, *Historical Times*, 468–69.
11. The only recorded instances I could find of President Lincoln attending ponton drills were on November 21, 1861, and December 21, 1861, when he rode his carriage across the completed span. Earl Schenck Miers, ed., *Lincoln Day By Day: A Chronology 1809–1865*, 3 vols. (Dayton, Ohio: Morningside, 1991), 3: 78, 84.
12. Gabions were cylindrical baskets woven of twigs with open ends, filled with dirt and used in making field fortifications. J. C. Duane, *Manual for Engineer Troops* (New York: D. Van Nostrand, 1862), 67; Boatner, *Civil War Dictionary*, 320. Fascines were bundles of closely bound wood stacked around to protect infantrymen and artillerymen in field works. Duane, *Manual for Engineer Troops*, 55; Boatner, *Civil War Dictionary*, 276.
13. James C. Duane's *Manual for Engineer Troops* was published in 1862 and was the basic text for engineering operations. Duane, who graduated from the United States Military Academy third in the class of 1848, taught at West Point from 1852 to 1854 and from 1858 to 1861, and was chief engineer of the Army of the Potomac from 1863 to 1865. *USMA*, 240.
14. Col. H. Naylor lived one mile east of Uniontown, D.C., on the left bank of the eastern branch of the Potomac River on Eastern Branch Road. Davis, *Military Atlas*, plate 89.1.
15. Arlington House was built by George Washington's stepson, John Parke Custis. It became the family estate of Robert E. Lee through his father-in-law, George Washington Custis. It was seized by the Federal government upon Lee's resignation from the Union army and is now the site of the National Cemetery. Boatner, *Civil War Dictionary*, 24.
16. Col. Henry Naylor was born on September 26, 1799, in Prince George County, Maryland, and died on January 24, 1871. Colonel Naylor built the house that WB refers to in 1853 and named it Mount Henry. It was a sprawling, two-story mansion that housed three generations of Naylors. Naylor inherited land and slaves, which he used to build one of the largest and most valuable properties in the District of Columbia. He was described as "a wise and energetic man, . . . adequately schooled, ambitious in the reasonable meaning of the word, entertained well, dressed well, observed the proprieties of his period, and practiced that manner which many plain people of his time called 'courtly'." He was married in 1838 in Washington, D.C., to Susan Matilda Smith and served as a justice of the peace, trustee under deeds of trust, administrator and executor of estates, and member of the Levy Court for over fifty years. "History of Naylor Family is reviewed by Rambler," *Washington Evening Star*, November 16, 1924, and November 23, 1924. Frederick I. Ordway, *Register of the Society of Colonial Wars in the District of Columbia* (Washington, D.C., 1967), 155.

17. John Purdy was listed as a "painter and glazier" and lived on the north side of Pennsylvania Avenue, west of and near the Capitol. In 1847 his friend Benjamin B. French mentioned that he operated a coal and wood yard on 7th Street near the Washington canal. In 1861, Purdy was surety for French's bonds when French was appointed commissioner of Public Buildings in the District of Columbia. Purdy died in 1881. Benjamin Brown French, *Witness to the Young Republic: A Yankee's Journal, 1828–1870*, ed. Donald B. Cole and John J. McDonough (Hanover: Univ. Press of New England, 1989), 192, 374; *The Washington Directory Showing the Name, Occupation, and Residence of Each Head of a Family and Person in Business Together With Other Useful Information* (Washington, D.C.: S. A. Elliot, 1827), 63.
18. Spiritualism was the vogue in Washington and elsewhere during the war. Mrs. Lincoln was especially addicted to talking to the dead through mediums at seances after the death of her twelve-year-old son, Willie, of typhoid fever on February 20, 1862. Mrs. Laury of Georgetown, Father Beeson, and Mr. Colchester were all well-known mediums of the time. Leech, *Reveille*, 376–77; Bowman, *Civil War Almanac*, 86.
19. "Zoo zoos" is a reference to Zouaves, a term used to denote regiments which adopted a style of uniform patterned after those worn by famous units in the French army. Colorful baggy red pants, vests, and tasseled fez hats made these units very conspicuous. The Zouave uniform included leggings or gaiters. Boatner, *Civil War Dictionary*, 954; Faust, *Historical Times*, 850.

    WB mentioned previously that the Gansevoort Light Guard was a "fancy dress Company." A photograph of a member of the 46th Regiment (of which the GLG was part of) clearly shows a semi-Zouave uniform, with full trousers and leggings. WB probably carried over his clothing from his prewar experience in the militia (see fig. on p. 49). Michael J. McAfee, curator of history at West Point Museum, United States Military Academy, letter to editor, December 24, 1991.
20. A government issue, .58-caliber, 1861 Springfield rifled musket barrel holds 6.2 ounces of liquid. It would have taken five or six men to smuggle a quart of whiskey into the camp using this method.
21. Francis Preston Blair Jr. (1821–1875) was a politician (founder of the Free Soil Party in Missouri), a newspaper editor (*The Barnburner*), a member of the U.S. House of Representatives (1856–58, 1860–62), and a major general commanding volunteers serving in the Western Theater. After the war Blair was nominated as the Democratic candidate for vice president with Horatio Seymour, war governor of New York, in the 1868 election. Seymour and Blair were defeated by Republican candidates U.S. Grant and Schuyler Colfax. Boatner, *Civil War Dictionary*, 67.

    His older brother, Montgomery Blair, was Lincoln's postmaster general. A prewar lawyer, he had been counsel for Scott in the Dred Scott case before the Supreme Court in 1857. Boatner, *Civil War Dictionary*, 67.

    Their father, Francis Preston Blair Sr. had served President Lincoln in a number of informal capacities. His home in Silver Springs, Maryland, was destroyed by Early's troops during their raid on Washington in 1864. Sifakis, *Who Was Who in the Union*, 36.

22. "For dancing was the rage, . . . the ladies tossed their cataract curls in the mazes of the polka and the lancers." Leech, *Reveille*, 351.

CHAPTER FIVE

1. Maj. Albert J. Myer was the aggressive leader of the U.S. Army Signal Corps and developed many of the methods used to communicate when military telegraph was not practical, including signal pistols, flags, rockets, and torches. Faust, *Historical Times*, 688–89.
2. Charles C. Coe of Rome, New York, constructed the largest balloon ever at that time; it had more than one million cubic feet of capacity. Frederick Stansbury Haydon, *Aeronautics in the Union and Confederate Armies*, vol. 1 (New York: Arno Press, 1980), 35.

   That capacity would have been necessary to get large amounts of explosives into the air. Balloons were used for little more than reconnaissance during the war, as free ballooning—ascending without a tether—was not well understood and the problem of getting back to one's own lines after dropping the bombs was never quite worked out.
3. On November 20, 1861, McClellan and Lincoln, along with numerous cabinet members and other dignitaries, reviewed seventy thousand troops near Munson's Hill, Virginia. The event was described by John Nicolay, Lincoln's secretary, as "the largest and most magnificent military review ever held on this continent." Miers, *Lincoln Day By Day*, 3: 78.

   Although the date seems to be out of sequence with WB's narrative, I could find no other reference to as large a review as this.
4. On March 10, 1862, the Army of the Potomac moved against Confederate forces at Manassas, which had fallen back to a position near Rappahannock Station. The Federals moved into vacated Confederate camps but soon returned to the Alexandria vicinity without engaging the enemy. Quaker guns were logs painted to look like artillery when seen from a distance. Stephen W. Sears, *To the Gates of Richmond: The Peninsula Campaign* (New York: Tichnor and Fields, 1992), 16–17.
5. On March 17, 1862, the Army of the Potomac began the move to the James and York Rivers, beginning the Peninsula Campaign. Bowman, *Civil War Almanac*, 91.
6. General Irvin McDowell (1818–1885) was an 1838 graduate of the United States Military Academy, ranked twenty-third in a class of forty-five. He served on border duty, as instructor at West Point, in the Mexican War, on the frontier, and in Washington, where he was given a brigadier's commission and command of the principal Union army south of the Potomac River. After First Bull Run, McDowell served as a division and corps commander until he was relieved of command and severely criticized after Second Bull Run. He was exonerated but did not command in combat again. He retired a major general in the regular army in 1882. Boatner, *Civil War Dictionary*, 531.
7. William Howard Russell (1820–1907), a reporter for the *London Times*, was the

most prominent English journalist of the day. His unrestrained, penetrating, and critical dispatches about both the North and the South earned him deep resentment and led to his virtual expulsion from the United States in the spring of 1862. Faust, *Historical Times*, 649.

8. Fort Ward was one-half mile south of the Leesburg Turnpike and three miles northwest of Alexandria. Davis, *Military Atlas*, plate 6.1.
9. Bristoe Station lay along the line of the Orange & Alexandria Railroad. The line ran for 170 miles between Alexandria and Lynchburg and was a vital link during the Virginia campaigns for both sides. Faust, *Historical Times*, 547.
10. Invented by Dr. Richard Gatling, the Gatling gun saw only limited service late in the war due to its proclivity for breaking down in the field. A handcranked machine gun with six barrels capable of firing six hundred rounds a minute, the gun's problems outweighed its gains, and it was not purchased in any numbers until after the war when an improved version was produced. Faust, *Historical Times*, 302.

    Mitrailleuse is French for "Machine Gun."

    It is unclear how WB would have managed to see Gatling guns on this march as there are no records of the guns in the field at this time. It is much more likely that WB saw a "Coffee Mill" gun, which was also handcranked but only had one barrel and was in the field at that time in limited numbers.
11. New Hampshire native Daniel P. Woodbury graduated from the United States Military Academy in 1836, married into a southern family, and became one of the leading engineer officers in the Union army. He took command of the volunteer engineer brigade in March 1862. Boatner, *Civil War Dictionary*, 947; Sifakis, *Who Was Who in the Union*, 463.
12. Cheeseman's Landing does not appear on contemporary area maps. There is a Cheeseman's Creek, but a landing there, fourteen miles above Fort Monroe, would not have afforded the 50th an opportunity to pass near the battlefield at Big Bethel on its way to the Yorktown lines. As Cheeseman is the name of a local family, the landing could have been any place where the family owned property at the water's edge, perhaps on the banks of the Back River. A landing there would have placed them at the requisite distance from Fort Monroe, "a few miles north," and properly placed for a march from there to Yorktown, passing right through the Big Bethel battlefield. Cheeseman's Creek empties into the Poquosin River and thence to Chesapeake Bay and is approximately thirteen miles north of Fortress Monroe. Davis, *Military Atlas*, plate 18.1.

    The Battle of Big Bethel, June 10, 1861, was an early defeat for raw Union troops. A confusing fight filled with mistakes cost the Federals 18 dead, 53 wounded. Confederate losses were 1 dead, 7 wounded. Faust, *Historical Times*, 59.
13. WB is referring to the *Report of the Engineer and Artillery Operations of the Army of the Potomac from its Organization to the Close of the Peninsular Campaign* by Brig. Gen. J. G. Barnard and Brig. Gen. W. F. Barry (New York: D. Van Nostrand, 1863). John Gross Barnard designed the Washington defenses and during the Peninsula Campaign was McClellan's chief engineer. William Farquhar Barry organized the ordnance of the Army of the Potomac and served as the chief of artillery during the Peninsular Campaign.

14. Battery No. 4 was in a ravine under the plateau of Moore's house on a branch of Wormley's Creek approximately 2,300 yards northwest of WB's camp and 2,400 yards from Yorktown's defensive works. Davis, *Military Atlas*, plate 19.2.
15. WB reported two hundred men of the 74th New York Infantry for insubordination on April 22, 1862, and the report was forwarded by General Woodbury through the adjutant general's office. *OR*, ser. 1, vol. 11, pt. 1: 327. The records do not show what, if anything, was done to discipline the troops.

    The 74th New York served with the Army of the Potomac throughout the war and lost 124 killed and 301 wounded during that time. Phisterer, *New York in the War of the Rebellion*, 3: 2768.
16. Maj. Alexander S. Webb is mentioned in General Barry's report as follows: "The conduct of Major Webb in running the 13-inch sea-coast mortars, with their material and ammunition, into the mouth of Wormley's Creek, under the fire of the enemy, was particularly conspicuous for perseverance and great coolness and gallantry." *OR*, ser. 1, vol. 11, pt. 1: 349.

    Major Webb, formerly of McClellan's staff, wrote *The Peninsula: McClellan's Campaign of 1862*, as part of the *Campaigns of the Civil War* series in 1881. During the war Webb rose to the rank of Major General.
17. The 44th New York at Yorktown lost 1 enlisted man killed in action and 2 men wounded. The regiment lost 147 men to disease during its entire career. Phisterer, *New York in the War of the Rebellion*, 3: 2290.

Chapter Six

1. On May 3, 1862, the only battery not turned over to the artillery was No. 13. However, many of the batteries were under final stages of construction and not all guns were mounted in all batteries. Work parties were underway in over half of the siege batteries on May 3. *OR*, ser. 1, vol. 11, pt. 1: 336, 347.
2. Torpedoes, land mines made up of artillery shells with very sensitive fuses, were reported by Generals Locke, Porter, and Barry. *OR*, ser. 1, vol. 11, pt. 1: 313, 349–50; Robert Underwood Johnson, and Clarence Clough Buel, eds. *The Struggle Intensifies*, vol. 2 of *Battles and Leaders of the Civil War* (n.p.: Castle, n.d.), 201.
3. Confederate Gen. John Bankhead Magruder (1810–1871) was an 1830 graduate of the United States Military Academy, ranked fifteenth in a class of forty-two. He served in garrison, on the frontier, in the Seminole Wars, and fought in Mexico, where he was cited for bravery. He was nicknamed "Prince John" for his courtly manner and lavish entertainments. He resigned his Federal commission at the start of the war and served well in various command capacities for the Confederates during the Peninsula Campaign, including the fight at Big Bethel and the defense at Yorktown. However, his performance during the Seven Days was found wanting, and he was excoriated by the Richmond press. He was later sent to command in Texas, Arizona, and New Mexico. As the war ended, Magruder refused parole, served Maximilian in Mexico as a major general, and finally returned home to the lecture circuit. Boatner, *Civil War Dictionary*, 501.

Richard N. Current, editor in chief, *Encyclopedia of the Confederacy*, vol. 3 (New York: Simon and Schuster, 1993), 988–89.

4. In the memoir is a piece of heavy cotton canvas twelve inches wide by seven and one-quarter inches high with handwriting in pencil: "Follow us and we'll give you what you need. Just come out a few miles all we want is a fair Showing."
5. *OR*, ser. 1, vol. 12, pt. 1: 400–401.
6. Gen. William Buel Franklin (1823–1903) was an 1843 graduate of the United States Military Academy and ranked first in a class of thirty-nine. He served on the frontier, on western surveying expeditions, as instructor at West Point, in the Mexican War, and on construction and harbor-improvement projects. He rose in rank rapidly during the war, through brigade, division, and corps command with the Army of the Potomac. At Fredericksburg he commanded the Left Grand Division and was subsequently blamed for defeat by Burnside and Congress. He later commanded a corps in the Red River Campaign, was wounded there, and finished out his service on boards and "awaiting orders." He left the army in 1866 and became vice president and general manager of Colt Fire Arms Company. Boatner, *Civil War Dictionary*, 303–4.
7. Williamsburg was a rear-guard action by Longstreet's and D. H. Hill's divisions in trenches prepared earlier by Magruder. They were attacked unsuccessfully on May 5 by Union divisions under Hooker, Smith, and Kearney. Losses were 2,239 out of 40,768 engaged for the Union and 1,603 out of 31,823 engaged for the Confederates. Faust, *Historical Times*, 829.
8. This property was in the Washington-Custis family for generations and passed down eventually to William Henry Fitzhugh "Rooney" Lee, General Lee's son. Mark Nesbitt, *Rebel Rivers: A Guide to Civil War Sites on the Potomac, Rappahannock, York, and James* (Pennsylvania: Stackpole Books, 1993), 72.
9. The records of the 50th show two lieutenants with the last name of Perkins—Augustus S. and Henry W.—but do not state which of these men was sick aboard the steamer. Augustus S. Perkins was a captain when he was killed at Fredericksburg and Henry W. Perkins evidently left the 50th to accept higher rank in another unit. WB does not mention whether Perkins returned to the 50th after his bout with sickness, but, as both men show up in the subsequent records, we can assume that whichever one was sick did indeed return.
10. The white dwelling from which the landing got its name was on property once owned by Martha Custis, wife of George Washington. The house that stood on the property was only some thirty years old, the original having burned. The property was passed down through the generations and when WB saw it, it was owned by Confederate Gen. William Henry Fitzhugh "Rooney" Lee, Gen. Robert E. Lee's second son. This house was burned on June 27–28, 1862, during the Seven Days battles. Nesbitt, *Rebel Rivers*, 71–74.
11. The Battle of Hanover Court House (also called Kinney's Farm and Slash Church) took place on May 27, 1862. McClellan ordered Maj. Gen. Fitz-John Porter to take elements of the V Corps and move against Confederate forces near the Union right flank in the vicinity of Hanover Court House. This was done to protect against incursions against the Union flank, rear, and supply lines. Porter drove off the Confederates after a stiff fight that resulted in 355

Union casualties and 265 Confederate casualties, although Porter claimed to have scooped up 731 prisoners that the Confederates did not acknowledge. McDowell was never supposed to cooperate with this movement, but one of the results McClellan hoped to achieve was to secure his right flank so that McDowell could join the Army of the Potomac by an overland route. Boatner, *Civil War Dictionary*, 373; William F. Fox, *Regimental Losses in the American Civil War 1861–1865* (1989, 1974; reprint, Dayton, Ohio: Morningside Bookshop, 1985), 543, 549.

12. General Sumner put lumberjacks of the 1st Minnesota and the 5th New Hampshire to work building two bridges across the Chickahominy River to link his corps with Union forces south of the river. The boggy bottomlands necessitated a span of a quarter of a mile. Although "rickety," the bridges served their purpose. Casey's Division was 2nd Division, IV Corps. The events that WB narrates actually happened on Saturday, May 31, 1862, during the Battle of Seven Pines (Fair Oaks). Stephen W. Sears, *To the Gates of Richmond: The Peninsula Campaign* (New York: Ticknor and Fields, 1992), 113.
13. Lt. Walter L. Cassin was actually the adjutant of the 15th New York Engineers who was perhaps acting as a temporary aide to Woodbury during this time. *ARAGNY*, 424.
14. In the memoir WB again refers his son to the Barnard/Barry book on the Peninsula campaign with this statement: "You will see a condensed report of this transaction on page 95, also Page 221, also page 23, also page 50 of the same book I have mentioned."

    On page 95 is WB's report on the attempted building of the bridge over the Chickahominy and is essentially the same as in his narrative. WB was relieved by Captain Ketchum of the 15th Engineers, who resumed the work, changed the direction of the bridge, had his entire structure collapse in the current, rebuilt it, and had an operational bridge ready by 2 A.M. the next day.

    Page 221 is the part of General Woodbury's report to General Barnard that deals with the part WB played in building the bridge. This report states that the bridge as completed was wide enough for two ranks of marching infantry or one width of artillery. Woodbury said, in part, "No fault can be found with the officers and men of the engineer brigade. I have never seen officers work with more zeal, or men work harder, than they have done during the last two days."

    Page 23 is that part of General Barnard's report to General Marcy, army chief of staff, that covers the bridge-building episode, explaining in detail the reasons for the delays in building the bridge.

    Page 50 is the end of Barnard's report. Barnard hands out "mentions" and commends WB for "untiring energy and fidelity." Barnard and Barry, *Report of the Engineer*, 23, 50, 95, 221.
15. The Battle of Seven Pines (or Fair Oaks) was fought May 31 and June 1. When heavy rains turned the Chickahominy into a raging torrent, Confederate Gen. Joseph Johnston took advantage of the situation and attacked the isolated III and IV Corps south of the river. Confederate division commanders botched General Johnston's plan and attacked piecemeal. Sumner was able to cross some of his troops over the two bridges he had built and the Confederates were

repulsed. General Johnston was himself wounded in the fighting. General McClellan became timid after viewing the battle's carnage and decided on a siege. Confederate Gen. Robert E. Lee was elevated to army command and ordered a return to original lines, giving the tactical victory to the Union. The Confederates did not flee back to Richmond. Faust, *Historical Times*, 668.

CHAPTER SEVEN

1. Lt. Gen. Don Juan Prim y Prats, the Count of Reus and Marquis de los Castillejos, visited the Army of the Potomac during June 1862. Prim had commanded the troops that entered Mexico in December 1861, causing a good deal of consternation in Washington, but, as Spain had since pulled out of Mexico, Prim and his glittering staff and retinue were welcomed at the Army of the Potomac. *New York Herald*, June 12, 13, and 17, 1862.

   McClellan referred to the general and his retinue as "a large dose of Spaniards." In return, the general was greatly impressed with the feats of engineering he saw. Sears, *To the Gates*, 160.

   On his return to Spain, Prim led a revolution, was defeated, and lived in exile in Paris, where he led the Progressives in their continuing desire to overthrow the Spanish government. Prim returned to Spain late in 1866 at the head of another revolution, this one successful, and he was named premier, the most important political figure in Spain. During the unrest that followed in a country that had a monarchy without a king, Prim was an unwitting accomplice to Bismarck's machinations that led to the Franco-Prussian War. Prim was killed by an assassin on December 30, 1870. Rhea Marsh Smith, *Spain: A Modern History* (Ann Arbor, : Univ. of Michigan Press, 1965), 336–41, 354–57.
2. Henry C. Vogel (b. 1807) was the pastor of the Rome First Baptist Church from 1838 to 1859. His "notoriety" is attributable to the fact that he was very well known and his picture hangs in the church today. Virginia V. Herrmann and Fritz Updike, letter to editor, February 12, 1995.

   Vogel served in the 61st New York Infantry, first as chaplain and then concurrently as assistant surgeon. The 61st was a hard-fighting regiment, suffering 110 casualties out of 432 engaged at Fair Oaks and 769 during the entire war. Its career included every battle of the Army of the Potomac except Second Manassas. Phisterer, *New York in the War of the Rebellion*, 3: 2554–55, 2572; Frederick H. Dyer, *A Compendium of the War of the Rebellion*, vol. 1 (Dayton, Ohio: Morningside Press, 1994), 201.
3. There has always been some confusion over the naming of the various temporary and permanent bridges over the Chickahominy. No two sources seem to agree and Confederate reports add to the confusion by using the same names for bridges that the Union used but applying those names to different bridges.

   From WB's description it would appear that what he calls the Woodbury and Alexander Bridge was actually named simply the Woodbury bridge on the maps and crossed the Chickahominy where Boatswain's Swamp enters the river. There was an Alexander Bridge about a mile downstream from the Woodbury Bridge, but they were definitely two separate structures. Sears, *To the Gates*, 231; Davis, *Military Atlas*, plate 19.1; Alexander S. Webb, LL.D., *The Peninsula: McClellan's Campaign of 1862* of *Campaigns of the Civil War*, introduction by William F. Howard (Charles Scribner's Sons, 1881; reprint, Wilmington, N.C.: Broadfoot Publishing, 1989), 131.

4. Chickahominy fever was a term the soldiers applied to a variety of sicknesses like typhoid and malaria. Federal army returns for June 20 listed 11,000 men sick enough to be unfit for duty and during the month of June, 705 soldiers died of disease. McClellan suffered a bout of malaria and Union Gen. William H. Keim died of Chickahominy fever. Sears, *To the Gates*, 163–64.
5. Cavalry Bridge was evidently not a name used by anyone other than WB. Just above the Woodbury Bridge there is a structure on the maps named Duane's Bridge and this fits WB's description. It crosses the Chickahominy where River Road meets the river. Davis, *Military Atlas*, 19.1.

   General Franklin ordered the destruction of this bridge on June 27 after Confederates were seen on the north side of the river close to it. Sears, *To the Gates*, 227.
6. Just as Johnston had attacked an isolated corps of the Army of the Potomac south of the Chickahominy on June 1 at Seven Pines, now Lee pounced on the isolated V Corps north of the river at Mechanicsville on June 26. During the intervening 25 days, McClellan had shifted 41,000 men, Sumner's II Corps, and Franklin's VI Corps south of the Chickahominy, leaving Porter with 16,800 men to face Robert E. Lee's attack force of 47,000, although only 16,300 Confederates were engaged at Mechanicsville. Faust, *Historical Times*, 483–84; Boatner, *Civil War Dictionary*, 540–41.
7. Jackson's Valley Campaign unhinged McClellan's plan on the Peninsula by bluffing Lincoln into withholding McDowell's Corps to guard Washington from Jackson. Jackson scattered three separate Armies while outnumbered two to one in five separate battles in thirty days. Jackson's campaign tied up sixty thousand Union soldiers when McClellan was demanding more men with which to attack Richmond. Faust, *Historical Times*, 677.
8. The reinforcing troops were from Slocum's division (1st Division, VI Corps) and they arrived on the battle lines just in time to bolster Porter's defense.

   The Confederates skirmished with Porter's rear guard at Beaver Dam Creek on the morning of June 27, and this was the firing that WB heard off in that direction. The attacking troops could not have been Jackson's Corps, as his troops were opposite the Union right flank, almost two miles from WB's position on the Union left flank near the river. The troops that drove the engineers off were probably from one of Longstreet's brigades sent to probe the Union positions near the river.

   Soon after the engineers left, General Franklin destroyed the bridge before the Confederates could get possession. Sears, *To the Gates*, 227.
9. The Battle of Gaines Mill (also called Cold Harbor and Chickahominy) was the third of the Seven Days battles and was fought on June 27, 1862, as part of Lee's offensive against Porter's isolated V Corps. Initial Confederate attacks against a

strong Union position on a plateau behind Boatswain's Swamp were piecemeal and met with stiff resistance until 7 P.M., when a concerted assault cracked Porter's line in the center, forcing his orderly retreat across the bridges over the Chickahominy. The Union had losses of 6,837 out of 34,214 engaged. The Confederates lost 8,751 out of 57,018. Faust, *Historical Times*, 296–97; Boatner, *Civil War Dictionary*, 321.

10. Almost 2,500 wounded were abandoned by the precipitate retreat from Savage's Station. John Newton and Gerald Simons, *Lee Takes Command: From Seven Days to Second Bull Run*, The Civil War Series (Alexandria, Va.: Time-Life Books, 1984), 52.

    Livermore reports 3,067 Union missing during June 29–July 1. Overall missing (presumed captured) during the Seven Days was 6,053. Thomas L. Livermore, *Numbers and Losses in the Civil War in America: 1861–65* (1900; reprint, with an introduction by Edward E. Barthell Jr., Bloomington, Ind.: Indiana Univ. Press, 1957), 84–87.

11. During the Army of the Potomac's retreat from its positions on the Chickahominy, the 14th New York Infantry, along with the rest of the 2nd Brigade, Morell's division, was ordered to fall back quickly to the east side of Gaines' Creek, losing all their commissary and equipment. They were closely pursued by Confederate troops under Gen. James Longstreet. It was during a stiff rear-guard action near Gaines' Mill on June 27 that Lieutenant Colonel Skillen was killed during a bayonet charge. The 14th lost 11 killed, 73 wounded, and 16 missing in this action on June 27. The brigade lost 96 killed, 354 wounded, and 136 missing. They crossed White Oak Bridge on June 28. *OR*, ser. 1, vol. 11, pt. 2: 313; Phisterer, *New York in the War of the Rebellion*, 1: 163.

12. The White Oak Swamp Bridge was completed on June 27 by engineers and work details from Morell's brigade working under General Woodbury. The Volunteer Engineer Brigade reported twelve men missing during operations from June 25 to July 2, eight from the 50th and four from the 15th. *OR*, ser. 1, vol. 11, pt. 2: 37; Phisterer, *New York in the War of the Rebellion*, 1: 161.

13. The engagement at White Oak Swamp Bridge, June 30, was between General Jackson's Confederate command and General Franklin's Union VI Corps. Union troops of French's brigade fired the trestle bridge at 10 A.M. as soon as the last of the rear guard was over and just as Confederate troops arrived. Colonel Crutchfield, Jackson's chief of artillery, cut a concealed path for his cannon through the woods and opened fire at 1:45 P.M., running off Union artillery placed on the ridge. Troops of D. H. Hill's division along with Munford's cavalry then attacked across the swamp but were driven back by Union artillery firing from masked positions guarding the approach roads. The artillery then settled into a vicious fire fight that lasted until nightfall, when the Union retreated to Malvern Hill, and the Confederates rebuilt the bridge and followed. *OR*, ser. 1, vol. 11, pt. 2: 55, 57, 75, 77, 216, 464, 556, 557, 561, 566, 627, 653.

14. Haxhall's House was a landing on the James River just below Malvern Hill, and Turkey Bend was the name of the horseshoe bend in the James at that point. The present-day James follows a slightly different course—the inner part of the

bend is now an island—but it retains a variation of the prewar name, Turkey Island. Davis, *Military Atlas*, plates 16.1, 17.1, 19.1.

15. General McClellan inspected the Union lines on Malvern Hill early in the morning and very possibly WB saw him ride out for that inspection. But by 9:15 A.M. McClellan was aboard the gunboat USS *Galena* and at 10:00 A.M. was steaming downriver and away from the fighting on Malvern Hill without leaving anyone in command of the army in his absence. McClellan inspected the ground at Harrison's Landing, his proposed new base for the army, and returned in mid-afternoon, appearing on the battlefield by 3:30. After he returned, he posted himself far to the right flank and far from the fighting. Sears, *To the Gates*, 309, 330.
16. The Seven Days Campaign lasted from the Battle of Oak Grove on June 25 to the Battle of Malvern Hill on July 1. Gen. Robert E. Lee forced Gen. George McClellan from his positions before Richmond to Harrison's Landing on the James River. In the process, Lee demoralized McClellan and destroyed his effectiveness to command in the eyes of the Lincoln administration. Malvern Hill was a rear-guard action during the retreat to the James and a Union victory, with Lee suffering 5,355 casualties while McClellan lost only 3,214. However, the entire Seven Days was a defeat for the Union and changed the course of the war in Virginia. An advance toward Richmond was the last thing McClellan had on his mind after Malvern Hill. Faust, *Historical Times*, 667; Sears, *To the Gates*, 337–56.
17. Harrison's Landing on the James River was the site of Berkeley Plantation, first settled in 1619 and home to the Harrison family since 1726. Benjamin Harrison V was a signer of the Declaration of Independence, and President William Henry Harrison had been born there. The Harrisons, who grew and exported tobacco, sold the property in the 1840s, and in 1862 it was owned by Powhatan Starke, who grew grain with the help of thirty-six slaves. John M. Coski, *The Army of the Potomac at Berkeley Plantation: The Harrison's Landing Occupation of 1862* (n.p., 1989), 1.
18. In order to deflect criticism after the retreat to the James River, McClellan characterized the movement as a "change of base" from White House on the York River to Harrison's Landing on the James; he claimed that he had planned the move all along. The fact that he had accomplished this movement while under attack convinced McClellan that he had done a great thing. Few were taken in by this subterfuge. It was a Union defeat, and little that McClellan said could change that fact, although he steadfastly blamed the administration for not supporting him with more men and supplies. Sears, *George B. McClellan*, 193–247.
19. The Sanitary Commission was a largely voluntary organization that provided ambulance services, medical attention, nursing care, convalescent facilities, supplies, food, lodging, medicine, clothing, and personal items to Union soldiers. These were all needs that overwhelmed the government agencies. The commission raised seven million dollars in cash and over fifteen million dollars in donated goods to provide for soldiers' comfort, to aid their dependent families,

and to help them with pensions and wages. It did all this through seven thousand aid societies and major branches in ten large cities. The commission conducted Sanitary Fairs to raise money and to promote public awareness of the needs of the soldiers. Faust, *Historical Times*, 656.

Scurvy is a disease caused by Vitamin C deficiency and characterized by weakness, anemia, and bleeding gums. Most soldiers revived quickly when given fresh vegetables. Wiley, *Billy Yank*, 230–31.

20. Each regiment was allowed one official sutler, a businessman who supplied the soldiers with necessary and, many times, unnecessary items. A sutler sold his wares from a temporary store or a wagon and had a monopoly to provide a long list of approved stores, including food, newspapers, playing cards, books, tobacco, razors, cutlery, and sometimes illegal alcohol. Sutlers would extend credit and inflate prices beyond officially sanctioned limits, angering the soldiers and sometimes inviting retaliation. Faust, *Historical Times*, 738.
21. On August 10 Lieutenant Comstock, U.S. Engineers, was ordered by Army of the Potomac Headquarters to make a reconnaissance in the area of the Chickahominy River near Barrett's Ferry for the purpose of determining a bridge location suitable for the retreat of the army to Ft. Monroe. The 1,980-foot bridge was completed at 9:30 A.M. on August 14, and all of the army except Heintzelman's Corps had crossed over by 2:30 P.M. on August 18. *OR*, ser. 1, vol. 11, pt. 1: 123.
22. Philip Kearny was one of the best-known and most-respected officers in the army. Well educated, widely traveled, and a millionaire, he had been in the prewar army, had studied cavalry tactics at the French Cavalry School at Saumur, had fought in Algiers with the Chasseurs d'Afrique and in Napoleon III's Imperial Guard at Solferino, where he earned the French Legion of Honor. He had been an aide to Scott in Mexico, where he lost an arm at Churubusco. During the Civil War, Kearny rose to major general and commanded the 1st Division in Heintzelman's III Corps. Faust, *Historical Times*, 408–9; Boatner, *Civil War Dictionary*, 449.

General Kearny was an outspoken critic of McClellan, and when he received the order to retreat to the Landing from Malvern Hill Kearny blurted out to aides that the order could "only be prompted by cowardice or treason." He blasted McClellan for "mismanagement" of the campaign and was highly critical of the position at Harrison's Landing as inviting the destruction of the army. Coski, *Berkeley Plantation*, 4, 13.

CHAPTER EIGHT

1. John Pope had been a friend of Lincoln's before the war when Pope was a captain in the prewar army. Pope was made a major general after his victory at Island No. 10 in the Mississippi River and was brought east to manage affairs in northern Virginia as commander of the Union Army of Virginia. He took over on June 26 and by September 2 had been soundly defeated by Robert E. Lee at

the Battle of Second Manassas. He was relieved and sent West to fight Indians. Boatner, *Civil War Dictionary*, 658–59; Faust, *Historical Times*, 593.

2. The Battle of Second Manassas was fought August 29–30 over much of the same ground as the first battle at that place. General Pope did well against the Confederates until he had to contend with Robert E. Lee's strategy and Stonewall Jackson's tactics. The Union defeat was followed by a retreat back into the defenses around Washington. Pope blamed his defeat on animosity toward him from the McClellanite faction in the Army of the Potomac. Boatner, *Civil War Dictionary*, 101–5; Faust, *Historical Times*, 92–95.
3. At the Battle of Chantilly on September 1, General Kearny accidentally rode into Confederate lines during a thunderstorm and was shot dead. Faust, *Historical Times*, 408–9; Boatner, *Civil War Dictionary*, 449.

   Gen. Louis Desaix was a division commander under Napoleon at Marengo on June 14, 1800, where he made a forced march in time to come in on the Austrian right and save the day. He was killed in this attack. Michael Glover, *The Napoleonic Wars: An Illustrated History 1792–1815* (New York: Hippocrene Books, 1978), 73.
4. Lincoln had few good choices after the Battle of Second Manassas. Pope had to go as he did not have the confidence of the army, especially the officer corps. The only real alternative was McClellan, whom the army did have confidence in, and Lincoln gave him his army back. This angered the Republicans in Congress as well as Lincoln's Cabinet as they thought McClellan, a Democrat, was not fit for the job after his failures on the Peninsula. There were also accusations that McClellan deliberately refused to do all he could have done to help Pope and that there was a conspiracy among the McClellanite faction in the Army of the Potomac against Pope's leadership. General McClellan took the beaten and dispirited army and, in a few days, had it on the road chasing Lee into Maryland, a feat that certainly displayed both McClellan's abilities as an organizer and motivator and the army's belief in him as a commander. Stephen W. Sears, "Lincoln and McClellan," in *Lincoln's Generals*, ed. Gabor S. Boritt (New York: Oxford Univ. Press, 1994), 3–50.
5. Willard's Hotel on Fourteenth Street and Pennsylvania Avenue was the foremost establishment of its kind in Washington during the war. Its large dining rooms offered varied and hearty fare with robust servings. It was said that much of the business of government was done in the bar and public rooms. The Willard brothers had fashioned a large, rambling place that attracted the high, the famous, and the hangers-on to its gregarious atmosphere. Leech, *Reveille*, 10.
6. John Reese Kenly (1818–1891) was a well-known Maryland lawyer and supporter of the Union cause. He fought in Mexico as a major of Maryland and D.C. volunteers. In 1861 he again donned his uniform, led a Maryland regiment, and was made a brigadier on August 22, 1862. He primarily commanded a Maryland brigade in the 8th Corps in the Middle Department, moving back and forth between that command and 1st Corps, Army of Potomac, until the end of the war. It was Kenly's command that was defeated by Jackson at Front Royal during the Valley Campaign, and Kenly was wounded and captured. Kenly also

commanded the ineffective pursuit of Early after Early's attack on Washington. Boatner, *Civil War Dictionary*, 453–54; Faust, *Historical Times*, 412.

7. Stephen W. Sears, author of a noted biography of General McClellan and books on the battles of Antietam and the Peninsula Campaign as well as the editor of McClellan's papers, points out that General McClellan was faced with an opportunity at Antietam that few Union commanders ever had: he had almost whipped Robert E. Lee and the Army of Northern Virginia. Sears states that the problem was that McClellan did not know what he had done and was too afraid of losing the battle to take the final chance to win it. In this assessment, McClellan felt that he had to save the Army of the Potomac to save the republic for which it fought and would not risk defeat, so he lost the battle. Stephen W. Sears, *Landscape Turned Red: The Battle of Antietam* (New Haven: Ticknor and Fields, 1992), 303–11.

8. Harper's Ferry was indeed a trap to any army foolish enough to get caught there. General White was ordered to Harper's Ferry on September 3, but then ordered to Martinsburg, not arriving back at the Ferry until September 12, where he allowed Col. Dixon S. Miles to retain command of the local tactical situation as White felt that Miles outranked him by virtue of his regular army commission and knew more about the immediate situation. Miles had graduated from the United States Military Academy in 1824 and was breveted twice during the Mexican War.

   Maryland Heights was occupied on September 13 by Gen. Lafayette McLaws and, with the rest of the Confederate force surrounding the Ferry, compelled its surrender. Miles, under a cloud after continued charges of drunkenness on duty, was killed on September 15 by an exploding artillery shell just as the 12,000 troops were being surrendered along with 13,000 small arms, 73 cannon, and a prize of other equipment.

   Gen. Julius White (1816–1890) was a prewar lawyer and legislator in Illinois. He had fought at Pea Ridge, was promoted to brigadier general, and commanded at Harper's Ferry. General White was arrested after the debacle at Harper's Ferry but eventually exonerated of any blame. He later became a division commander in the IX Corps. He lived in Evanston, Illinois, after the war, but it is not known if WB knew him while he was there. Faust, *Historical Times*, 340–41, 493, 820–21. Sifakis, *Who Was Who in the Union*, 272–73, 450.

9. General Sumner's II Corps was ordered to Harper's Ferry on September 22, 1862, and upon arriving during the next few days, the corps took up positions on Bolivar Heights. Sumner commanded all troops at the Ferry until he went on a leave of absence on October 7, 1862. Frank J. Welcher, *The Eastern Theater*, vol. 1 of *The Union Army 1861–1865: Organization and Operations* (Bloomington, Ind.: Indiana Univ. Press, 1989), 317.

10. The area between Maryland Heights and South Mountain where McClellan had his headquarters is appropriately named Pleasant Valley. While the army was camped in this area, Confederate Gen. Jeb Stuart and eighteen hundred cavalrymen rode on a raid completely around the Union army via

Chambersburg, Pennsylvania, from October 9 to October 12, embarrassing McClellan and infuriating Lincoln. Davis, *Military Atlas*, plates 42.1, 25.6.

11. Tom-Big-Bee River was written by S. S. Steele, date unknown.

    On Tom-big-bee river so bright I was born,
    In a hut made ob husks ob de tall yaller corn,
    And dar I fust meet wid my Jula so true,
    An I row'd her about in my gum-tree canoe.

    Chorus:
    Singing row away row,
    Oe'r de waters so blue,
    Like a feather we'll float,
    In my gumtree canoe.

    All de day in de field de soft cotton I hoe,
    I tink ob my Jula an sing as I go;
    Oh, I catch her a bird, wid a wing ob true blue,
    An at night sail her round in my gumtree canoe.

    Wid my hands on de banjo and toe on de oar,
    I sing to de sound ob de river's soft roar;
    While de stars dey look down at my Jula so true,
    An' dance in her eye in my gum-tree canoe.

    One night de stream bore us so far away,
    Dat we couldn't cum back, so we thought we'd jis stay,
    Oh, we spied a tall ship wid a flag ob true blue,
    An' it took us in tow wid my gum-tree canoe.

    From Joe Mitchell Chapple, comp., *Heart Songs* (Boston, Mass.: Chapple Publishing, 1909), 250–51.

12. Philo Judson served in the 8th Illinois Cavalry as chaplain from September 4, 1862, to June 24, 1863. Abner Hard, *History of the Eighth Cavalry Regiment, Illinois Volunteers* (n.p., 1868).
13. President Lincoln visited the Army of the Potomac October 1–4, 1862. On October 1, he reviewed troops on Bolivar Heights with McClellan. Miers, *Lincoln Day By Day*, 3:143.
14. The eleven-hundred-foot bridge at Berlin, Maryland, was completed on October 25. Two divisions of IX Corps crossed on October 26, and I, VI, and the rest of IX Corps crossed from October 27 to November 2. *OR*, ser. 1, vol. 19, pt. 1: 86.
15. The Army of the Potomac moved to the vicinity of Warrenton, Virginia, on October 26 and stayed there until November 15. At its closest point, the army was thirty-five miles from Berlin. Bowman, *Civil War Almanac*, 118–19; James M. McPherson, ed., *The Atlas of the Civil War* (New York: Macmillan, 1994), 24.

Anyone trying to get from Berlin, Maryland, to Warrenton, Virginia, would have to pass directly through a part of Virginia that later became known as "Mosby's Confederacy," and included parts of Loudoun, Prince William, Fairfax, and Fauquier Counties that were dominated by stubborn Confederate resistance. These guerrillas were later organized as partisan rangers led by John Singleton Mosby and designated the 43rd Battalion of Virginia Cavalry. Jeffry D. Wert, *Mosby's Rangers* (New York: Simon and Schuster, 1990), 25–71.

16. John E. Jones was not enrolled in the regiment and does not appear on the roster, nor does he appear anywhere else in any of the references to the regiment or to WB. It would appear then that the only memorial to Jones is the praise for faithfulness and service that WB makes in his memoir.
17. Gen. George McClellan was removed by President Lincoln on November 5, one day after the midterm elections. The IX Corps commander, Gen. Ambrose Burnside, was promoted in his place the same day. Bowman, *Civil War Almanac*, 118. The best discussion of why McClellan was replaced can be found in Sears, *Landscape Turned Red*, 298–335.
18. The orders to move the Berlin bridge to Washington were dated November 6 but were sent to the engineers via the mails instead of over the telegraph by a functionary at general headquarters and arrived on November 12, six days later. On that same day, Burnside submitted his plan of campaign for approval, and Lincoln approved the plans on November 14, the day that the first of the tardy pontons were arriving in Washington. Bruce Catton, *The Army of the Potomac: Glory Road* (New York: Doubleday, 1962), 21–25.

The quote is from the New Testament, Epistle of James 3:5. The King James Version reads: "Behold, how great a matter a little fire kindleth." The Revised Standard Version reads: "How great a forest is set ablaze by a small fire!"

CHAPTER NINE

1. On an 1862 map of the area there is a Friends Meeting House and a Methodist Church, both on side roads and approximately three and one-third miles from Mount Vernon. Pohick Church is on Telegraph Road, the main road connecting Alexandria and Occoquan, and is approximately six and one-quarter miles from Mount Vernon. Davis, *Military Atlas*, plate 8.1.

The Reverend Lee Massey was the rector at Pohick Church, Truro Parish, prior to the Revolution and said, "I never knew so constant an attendant in church as Washington." Paul Leicester Ford, *The True George Washington* (1896; reprint, Freeport, N.Y.: Books for Libraries Press, 1971), 77.

2. General Sickles commanded the second division, III Corps (General Stoneman), Left Grand Division (General Hooker), and had about three thousand men and some artillery. He acted as rear guard during the movement to Fredericksburg. On November 16 he was still at Manassas Junction. On the nineteenth Hooker expected him to be up to Hooker's camp at Hartwood, Virginia, within two days, but on November 22, Sickles was twenty-five miles

away at Wolf Run Ford on the Occoquan River. He was in camp by at least December 2, as he was ordered on that date to furnish a regiment of "ax-men" to report to General Woodbury, evidently to clear the roads leading to the Snicker's Neck crossing point. *OR*, ser. 1, vol. 21: 54, 102, 760–61, 773, 789, 816.

3. Dumfries is on Quantico Creek where the Telegraph Road crosses it and approximately halfway between Washington and Fredericksburg in Prince William County. Davis, *Military Atlas*, plate 8.1.
4. Years later WB wrote an article in *Battles and Leaders* in which he clearly wrote that he and the pontons arrived at Falmouth on November 27. All indications in the *OR* back up an arrival date of November 25. There does not appear to be any explanation of this discrepancy. *OR*, ser. 1, vol. 21: 82–97, 103, 148–51, 790–91, 800; Robert Underwood Johnson and Clarence Clough Buel, eds., *The Tide Shifts*, vol. 3 of *Battles and Leaders of the Civil War* (n.p.: Castle, n.d.), 121–22.
5. The *New York Times*, December 2, 1862, reported: "Gen. Woodbury is said to have been placed under arrest by order of the Secretary of War on account of negligence in forwarding the pontoon bridges for the use of the Army of the Potomac in the late movement."

   The *New York Times*, December 3, 1862, ran the following article: "The [Washington] *Star* has the following: 'The War Department has caused the arrest of Maj. Spaulding, of the Fiftieth New York Engineer Volunteers, and the other officers who were with him in the recent transportation of the pontoon train from this point to Falmouth, Va. It is alleged that Maj. Spaulding started with the train two and a half days later than he should have started, and was five days on the road longer than was necessary.'" There is no attribution as to when this article ran in the *Star*; however, WB states that he read about it on November 26 and attributes it to the "NY papers."

   General Woodbury wrote a detailed explanation of his side of the delay to General Burnside late on November 24, the day he had arrived at Falmouth with a small number of pontons and had been placed under arrest. He was careful to make it clear that no one had told him of the importance of the movement of the pontons to Fredericksburg and that the delay was due to lack of communications from army headquarters, not from any lack of zeal on the part of his men. *OR*, ser. 1, vol. 21: 793–95.
6. The Joint Committee on the Conduct of the War took depositions from Generals Burnside, Sumner, Franklin, Hooker, Halleck, Woodbury, Haupt, and Meigs.

   In his testimony, General Burnside stated that it was General McClellan who had given the order to take up the bridge at Berlin and move the pontons from there to Washington. The order was signed November 6 by Major Duane, chief engineer of the Army of the Potomac. But Burnside then blamed General Halleck for neglecting to make sure that the pontons were pulled up at Berlin, moved to Washington, refitted, and forwarded to Fredericksburg in a timely fashion. Burnside stated that he had enough to do to become familiar with a new command and move it to Fredericksburg and assumed that anything that had to be done in Washington would be taken care of by Halleck, as Burnside had been led to believe in a meeting with Halleck and Meigs on November 12.

General Woodbury stated that the first he had heard of the movement of the army or of the necessity of the pontons accompanying the army was in a telegram he received from General Halleck on November 13. After looking into the state of things, including the late arrival of the pontons from Berlin, Woodbury went to Halleck and begged him to delay the start of the campaign, as pontons could not get to Fredericksburg on time. Halleck refused. Halleck also refused to give Woodbury warrant to subvert the forms of service and act with emergency powers to requisition wagons, teams, drivers, and equipment to get the pontons to Fredericksburg any quicker than was then possible.

General Halleck stated that all troops, including those immediately at Washington, were under the express direction and command of first McClellan and then Burnside, so the responsibility for movements of troops, including engineers and pontons, was solely theirs, as was any responsibility in issuing orders and requisitions. Halleck then testified that, at his meeting with Burnside on November 12, he and Meigs had drafted the telegram that Woodbury received on November 13. Further, Halleck denied that he had a conversation with Woodbury wherein the latter asked for a delay in the movement of the army because the pontons could not be got down in time. He testified that the delay, as far as he knew, was due to weather and inexperience on the part of the engineers, but backed off the inexperience charge under questioning.

The sum total of the testimony was that each officer thought the other responsible and each thought that he had done everything required of him. Therefore, no one was responsible. The issue of the six-day delay from the time the original order was cut on November 6 until it was received by the engineers at Berlin on November 12 was never addressed in the hearings. The *New York Times*, December 24 and 25, 1862, carried the testimony verbatim from the Joint Committee report.

7. Burnside had reached the Rappahannock opposite Fredericksburg with Sumner's VI Corps on November 17. It had begun to rain, and Burnside, fearing that the rain would cause the river to rise above a point where the fords could be used, refused to allow Sumner to cross because he did not want him stranded on the south bank and liable to defeat.

   The engineers did not reach army headquarters with enough pontons to cross the river until November 25. By that time Robert E. Lee had assembled Longstreet's Corps, and it had dug in on the heights behind the town. During most of this week, Burnside had a much larger force than Lee but did not have the flexibility to change his plans to meet this delay. By the time the engineers did arrive, Burnside was no longer so determined to cross at Fredericksburg and was delaying again.

   Jackson's Corps arrived on November 29 in a snowstorm, and when it deployed, Lee's line of defense was complete and very sturdy. During this four-day delay Burnside was considering his options. Catton, *Glory Road*, 26–34; Bradley Finfrock, *Across the Rappahannock: From Fredericksburg to the Mud March* (Maryland: Heritage Books, 1994), 32–44.

8. On a map prepared on May 16, 1863, there is a residence marked "Alexr Walker" on the main road between Fredericksburg and King George Court

House, approximately twelve miles from the former. This is the only house marked "Walker" but is over three miles from the river, so does not fit WB's description.

Comparing WB's drawing of the first reconnaissance area with this same map and making allowances for the passage of time and the imperfect mapmaking skills of the colonel, there is a bend in the Rappahannock below Fredericksburg that roughly coincides with the drawing. Houses and roads are where he says they were, but the residence that WB marks as "Walker" is marked on the official map as "Hollywood," which may be the name of the place rather than the name of the resident. Davis, *Military Atlas*, plate 39.2.

9. One of the meanders that the Rappahannock takes on its way to the Chesapeake Bay is a northern bulge that forms an irregular horseshoe bend with a broad opening. The land within the bend on the south bank forms two bulges; the larger one is named Skinker's Neck and the other, smaller one, wholly contained within the larger, is called Buckner's Neck. From downtown Fredericksburg to the center of Skinker's Neck is twelve miles. Davis, *Military Atlas*, plate 39.2.
10. It is obvious when comparing WB's map of the second reconnaissance that the portion of Skinker's Neck that he drew is actually Buckner's Neck. The crossing place is therefore at the center of the Buckner's Neck bulge in the Rappahannock. Port Royal is just ten river miles below Berry Plain, which is opposite the narrow part of the neck and one and one-half river miles below the crossing place. Davis, *Military Atlas*, plate 39.2.
11. There is no house marked "Gaines" on period maps, but there are two houses marked "Taylor." One is on the main river road and is marked as "Col. Taylor." The other is almost a mile distant and on the river, marked "Taylor." This second house fits WB's description and, as the Gaines family rented from Taylor, perhaps the house carried his name on the maps. These two houses are opposite Buckner's Neck and near Berry Plain. Davis, *Military Atlas*, plate 39.2.
12. WB made his second reconnaissance on November 28, which would put him back there with Woodbury and Hunt on the twenty-ninth, and at that time there were no Confederates at the Neck. But Jackson's Corps of the Army of Northern Virginia arrived in the Fredericksburg area on the twenty-ninth, and Robert E. Lee wasted no time in ordering Jackson's Divisions southward to cover the lower Rappahannock crossings on his exposed right flank. Jackson sent D. H. Hill with his division to Port Royal and Jubal Early and his division to Skinker's Neck. These forces were in position in a matter of days. This movement by the Confederates was coincidental with WB's reconnaissance and probably had little to do with Colonel Taylor's spying. Edward J. Stackpole, *The Fredericksburg Campaign: Drama on the Rappahannock* (New York: Bonanza Books, 1957), 110–13.

   Burnside's engineers had actually recommended the crossing at Skinker's Neck as an alternate site, but noted some serious limitations: the roads approaching the crossing were in bad shape and the ground was frozen or muddy, meaning that extended time would be necessary to make the needed improvements to the approaches to the crossing points. On top of that, alternating days of rain, sleet, and snow, freezing nights and days of warmth had turned all of the roads between Fredericksburg and Skinker's Neck into quagmires, making a large-scale movement by the army problematic. *OR*, ser. 1, vol. 21: 87–89; Finfrock, *Across the Rappahannock*, 42; William Marvel, *Burnside* (Chapel Hill, N.C.: Univ. of North Carolina Press, 1991), 169.

The Federals also put two balloons up in the area just in time to observe Early's division bivouacking on the Skinker's Neck ground. Stackpole, *Fredericksburg Campaign*, 112; Finfrock, *Across the Rappahannock*, 46.

Then Federal gunboats steaming upriver to secure the crossing site at Berry Plain were fired on by John Pelham's Confederate artillery, and, after a lengthy duel, the gunboats retired. All these factors added up to defeat the Skinker's Neck proposal. Finfrock, *Across the Rappahannock*, 42, 45, 47.

CHAPTER TEN

1. Detachments from the 4th Maine (made up of a large number of lumberjacks) and the 120th New York were sent down to the vicinity of Skinker's Neck to build corduroy roads on the approaches to the designated crossing points. As all the reports were in and the Skinker's Neck crossing points had been eliminated from consideration, this was probably a ruse by Burnside to attempt to convince Lee that the real point of attack was at Skinker's Neck so that the southern commander would weaken his center and left to reinforce his right. Lee was not convinced and kept his dispositions as they were. Finfrock, *Across the Rappahannock*, 51.
2. WB is quoting an old British soldier's song, "Why, Soldiers, Why?" A drinking tune, it is also known as "The Duke of Berwick's March" or "How Stands the Glass Around?" and dates to at least 1729, its first appearance in print. Called "Wolfe's Song" by some, it was supposedly sung for General Wolfe the night before Quebec, September 13, 1759. The quote is from the second verse.

   Why, soldiers, why,
   Should we be melancholy, boys?
   Why, soldiers, why,
   Whose business 'tis to die!
   What, sighing? Fie!
   Damn fear, drink on, be jolly, boys,
   'Tis he, you or I!
   Cold, hot, wet or dry,
   We're always bound to follow, boys,
   and scorn to fly!

   Irwin Silber, compiler and editor, *Songs of Independence* (Harrisburg, Pa.: Stackpole, 1973), 143–44. Lewis Winstock, *Songs & Music of the Redcoats: A History of the War Music of the British Army 1642–1902* (Harrisburg, Pa.: Stackpole, 1970), 58–60.

3. The Phillips House stood approximately one mile east of the Rappahannock River and on the main road connecting Falmouth with King George Court House. Davis, *Military Atlas*, plate 33.1.

   Maj. Gen. Edwin Sumner was the oldest active corps commander to serve in the Civil War. At the time of Fredericksburg he was sixty-five, a veteran of forty-four years' service, and he was intensely patriotic and courageous. Sumner came under criticism after Antietam for handling his lead division poorly and getting it chewed up in the West Woods. After Fredericksburg he was exhausted physically and requested relief from duty. He was transferred to Missouri, but contracted pneumonia and died in Syracuse, New York, in March 1863. Faust, *Historical Times*, 733.
4. Col. Charles Tompkins was in command of the right center group of artillery stationed around the Lacy House. In his group he had thirty-eight guns, a mixture of light rifled fieldpieces and Napoleons. Tompkins added thirty-six more cannon about 9 A.M., and some long-range heavy weapons joined in shortly after noon. L. Van Loan Naisawald, *Grape and Canister: The Story of the Field Artillery of the Army of the Potomac* (1960; reprint, Washington, D.C.: Zenger Publishing, 1983), 236–43.
5. Nine men from the 50th were killed on the bridges at Fredericksburg: Private John Barber, 22; Pvt. Charles Beswick, 27; Pvt. William Blakesley, 21; Pvt. Hanson Champlin, 18; Cpl. Philip Comfort, 25; Capt. Augustus Perkins, 23; Cpl. Robert Pettre, 26; Pvt. John Sturges, 40; and Pvt. Lewis Wilcox, 38. Forty men were wounded, including Pvt. Richard Bascomb, 39; Pvt. Philander Dunlap, 21; Capt. James H. McDonald, 21; Pvt. Peter McKenna, 25; and Pvt. Warren Watson, 42. Ages are those given at time of enlistment. *ARAGNY*, 817, 831, 834, 871, 880, 1083, 1086, 1156, 1198, 820, 919, 1040, 1043, 1185.
6. The first regiment to cross the river in ponton boats was the 7th Michigan, Lieutenant Colonel Baxter commanding. They were followed by the 19th and 20th Massachusetts. Finfrock, *Across the Rappahannock*, 58–59.
7. The fighting that ensued when the Union troops landed was both bitter and protracted, much of it house to house, yard to yard, hand to hand. Each street became a battleground, and the fight lasted until dark, when the Confederates withdrew to the heights above Fredericksburg. Finfrock, *Across the Rappahannock*, 59–62.

CHAPTER ELEVEN

1. After the lower bridges had been completed, WB thought that Burnside had an excellent opportunity to march a couple of divisions across the Rappahannock and into the town, flanking the Confederate defenders out of their positions. What WB was not aware of was the Confederate cannon on the heights behind Fredericksburg, which had remained silent up to this time in order not to draw counterbattery fire from Union guns on Stafford Heights. Any Union troops marching toward Fredericksburg from the south would have been excellent targets for the Confederate cannon. Burnside chose to cross the upper and lower

bridges simultaneously, and, as it turned out, at night. Finfrock, *Across the Rappahannock*, 62.

2. Union losses at Fredericksburg were 1,284 killed, 9,600 wounded, 1,769 missing out of 106,007 effectives available for duty. Confederate casualties were less than half that. Livermore, *Numbers and Losses*, 96.
3. Capt. Barton Stone Alexander was a regular army engineer and never held volunteer rank during the Civil War. He instructed engineer troops early in the war and then served as assistant engineer on the Washington defenses. He was in a good position to know the truth about the ponton controversy and wrote the article in the *New York Herald* defending the engineers and their pontons.

   Although WB refers to Alexander as "Col.," at the time of the article, Alexander was still only a captain in the regular army, and he was not promoted to major until March 1863. Alexander died in 1878 on active duty as a lieutenant colonel. A discrepancy appears to exist, as Alexander signs his article as a "Lt. Col.," but this was a brevet rank, not his actual rank. It was acceptable to refer to an officer by his brevet rank. Boatner, *Civil War Dictionary*, 6.

   The lengthy article, "Our Pontoons in Virginia," is an exercise in military sidestepping. Alexander defends the use of the particular type of ponton chosen by himself, Captain Duane of the engineers, and General McClellan, general in chief, as the best available at the time. Then he goes into a description of their exemplary use up to the Fredericksburg campaign but then stops short, ending his article by saying "merely because, for some reason or other, they were not found at the Rappahannock when most wanted, it is no more fair to condemn the boats because they are heavy than it would be to condemn our cannon when the roads in Virginia will not permit them to be moved." Alexander carefully avoided placing any blame on anyone for the tardy arrival of the boats. The full text of the article can be found in Appendix 2, Item 8.
4. Burnside wrote to General in Chief Halleck immediately after Fredericksburg: "For the failure of the attack I am responsible." *OR*, ser. 1, vol. 21: 67.
5. In General Order Number 8, written in late January 1863, Burnside accused Gen. William B. Franklin of being partly responsible for the Fredericksburg debacle. Burnside said Franklin did not do all that he could have done while in command of the Left Grand Division during the attack on the Confederate right. The same order also accused Generals Hooker, Brooks, Newton, Cochrane, and Smith of culpability in the defeat. Partly as a result of this order and helped by the disastrous Mud March, January 20–23, 1863, Burnside was relieved of command of the Army of the Potomac on January 25 and replaced by General Joe Hooker. Catton, *Glory Road*, 94–96, 109–10; Faust, *Historical Times*, 285; *OR*, ser. 1, vol. 21: 998–99.

   President Lincoln was assassinated in 1865, Secretary of War Edwin Stanton died in December 1869, and General in Chief Henry Halleck died in January 1872. Faust, *Historical Times*, 332, 440, 713.
6. A "receiving vault" was a place where coffins were kept temporarily until a monument or grave marker could be prepared or the grave dug. In Chicago in the middle of the winter, a coffin would be placed in the receiving vault until the ground thawed enough to dig the grave. It was also a place where the coffin

would be kept until WB arrived home so that he could see his daughter one last time before her burial.

7. The Rev. Dr. Thomas Brainerd (1804–1866, D.D. Amherst College, 1848) was pastor of the Old Pine Street Church in Philadelphia from 1836 to his death. It was one of the oldest congregations in America and during his tenure, he admitted 1,200 new members and married 800 couples. WB and Rev. Brainerd were distant cousins. Lucy Abigail Brainerd reported: "During the Civil War he was earnest in support of the Government, both in the pulpit and elsewhere, so that 130 young men of his congregation entered the army or navy. One of his last acts was to erect a tablet in the vestibule of the church to 18 who lost their lives in the struggle. When Lee retired from Gettysburg, and on his final surrender to Grant, Dr. Brainerd conducted the religious services at the spontaneous meetings of citizens, which completely filled Independence Square. A history of his old church, published in 1905, called him 'the great patriot pastor of Philadelphia.'" *Genealogy of the Brainerd Family*, 164–68.

CHAPTER TWELVE

1. The infamous Mud March was undertaken January 20–23, 1862, as an attempt by General Burnside to carry on a winter offensive. A torrential two-day rainstorm destroyed the movement before it got started. Burnside was relieved a few days later. Boatner, *Civil War Dictionary*, 573.
2. In Lossing's *Pictorial History of the Civil War*, there is a sympathetic version of the replacement of Burnside. "The arrangement made at that time, whereby the country might be best served, was highly creditable to the President and to General Burnside." Benson J. Lossing, *Pictorial History of the Civil War in the United States of America*, vol. 2 (Philadelphia, Pa.: David McKay Publisher, 1894), 496–97.

   Burnside, the reluctant commander of the Army of the Potomac, had, after two disasters in which he accused everyone of conspiracy against him, finally convinced the Lincoln administration that he had been right when he said he did not think himself up to the task.
3. Three men from Company H died of disease during March 1863, all of them new recruits to the 50th: Private John Hazard, March 13; Private William Jennison, March 20; and Private Asa Sweet, March 15. They were the only deaths from disease reported during that period. Hazard and Sweet both enlisted at Willet, New York, on November 7, 1862, and were probably friends. Jennison enlisted on November 8, 1862, at Lapeer, New York. *ARAGNY*, 970, 996, 1158.
4. General Woodbury was transferred to command the District of Key West and Tortugas, Department of the Gulf, in March 1863. He died of yellow fever on August 15, 1864. Boatner, *Civil War Dictionary*, 947; Sifakis, *Who Was Who in the Union*, 463.
5. Henry Washington Benham graduated from West Point in 1837, one year after Daniel Woodbury, and first in his class. Posted to the Engineer Corps, Benham served in Mexico and on prewar engineering duties. He was given a promotion

to brigadier general of volunteers after success in West Virginia late in 1861, but this was revoked after the disaster at Secessionville in June of 1862, when he was also relieved of command and arrested for disobedience of orders. He was reinstated to brigadier general by President Lincoln early in 1863 and commanded the Volunteer Engineer Brigade from March 1863 to the end of the war. He received brevets to major general in the U.S. Army and U.S. Volunteers. Sifakis, *Who Was Who in the Union*, 29–30; Boatner, *Civil War Dictionary*, 58–59.

6. On April 6 President Lincoln arrived for a visit with the Army of the Potomac and reviewed General Stoneman's cavalry corps near Falmouth, Virginia. On April 8, Lincoln reviewed II, III, V, and VI Corps. On the ninth, Lincoln reviewed I Corps, and on the tenth he reviewed XI and XII Corps, after which he returned to Washington. Miers, *Lincoln Day By Day*, 3:178.
7. A division of Couch's II Corps was ordered to create a diversion at Bank's Ford to help deceive the Confederates as to the army's true crossing point. While this diversion was being accomplished, the V, XI, and XII Corps crossed at Kelly's Ford farther up the Rappahannock. These corps then split up, with Meade's V Corps crossing the Rapidan at Ely's Ford and Howard's XI Corps and Slocum's XII Corps crossing at Germanna Ford. Ernest B. Furgurson, *Chancellorsville 1863: The Souls of the Brave* (New York: Alfred A. Knopf, 1992), 67–94.

   Bank's Ford was well guarded by elements of Wilcox's Confederate brigade and had been since Burnside's Mud March of a few months before. *OR*, ser. 1, vol. 25, pt. 1: 195–96; Jay Luvaas and Harold W. Nelson, eds. *The U.S. Army War College Guide to the Battles of Chancellorsville and Fredericksburg* (Carlisle, Pa.: South Mountain Press, 1988), 137–38.

CHAPTER THIRTEEN

1. Hooker's opening moves in the Chancellorsville campaign were masterful and professionally carried out. With his simultaneous crossings of the upper fords over the Rapidan and the lower crossings below Fredericksburg, Hooker was able to hold Lee in place for a crucial two days and place the greater part of the Union army in Lee's unprotected rear. Catton, *Glory Road*, 160–65; Furgurson, *Chancellorsville*, 87–88.
2. Shortly after he took over the command of the Army of the Potomac, General Hooker appointed his provost marshall, Gen. Marsena Patrick, to create an intelligence-gathering system. Patrick made Col. George Sharpe, 120th New York Infantry, his deputy in charge of the Bureau of Military Information, and in short order Sharpe had a network of spies operating behind Confederate lines. On May 1, one of Sharpe's operatives in Richmond made it into Union lines carrying the intelligence that WB refers to. Furgurson, *Chancellorsville*, 33–34, 136.
3. Although WB clearly wrote "Banks" ford in his memoir, the action he described took place at Franklin's Crossing below Fredericksburg.

   General Benham misunderstood his orders for laying the ponton bridges across the Rappahannock at Franklin's Crossing and thought that he was the ranking officer in charge of the entire crossing operation. Because of this, he got into a heated argument with Generals Brooks and Wadsworth of the VI Corps and placed General Russell under arrest, although Russell ignored this. Benham then got into a fight a mile farther downstream at Fitzhugh's Crossing, where I Corps was supposed to cross. He was reported to be so intoxicated that he fell off his horse and lacerated his face. Hooker demanded an explanation for all this, and Benham wrote a long letter that evidently won him a reprieve from Hooker's wrath. Furgurson, *Chancellorsville*, 97–98.

In his letter to Hooker through General Williams, assistant adjutant general, Benham pointed out that he felt the main cause of delay in laying ponton bridges was the confusion of command at the crossing point. Therefore, he requested that all troops assigned to protecting and assisting the engineers be placed under his direct command. This would, Benham stated, lower casualties among the engineers and speed up the work. The problems at Franklin's and Reynolds's Crossing on June 5 were blamed on this confusion of command. *OR*, ser. 1, vol. 27, pt. 3: 62–64.

Gen. Horatio G. Wright graduated from the United States Military Academy in 1841, ranked second in a class of fifty-two. He taught at West Point and supervised numerous construction projects. He served at First Bull Run, Port Royal, Secessionville, Hilton Head, was made brigadier general, and led a division at Gettysburg, Rappahannock Station, and Mine Run. He later served on engineering boards before once again commanding in the field as a major general during the summer campaigns of 1864. He retired as chief of engineers in 1884. Wright was not at Chancellorsville, but took command of the 1st Division, VI Corps, shortly after and took command of the VI Corps in April 1864. Boatner, *Civil War Dictionary*, 949.

4. At approximately 5 P.M. on May 2, 1863, Confederate Gen. "Stonewall" Jackson led twenty-six thousand men of his II Corps in an attack against the vulnerable right flank of the Union army, held by the hapless XI Corps commanded by Maj. Gen. O. O. Howard. This attack routed the XI Corps and shattered the Union right flank, endangering the entire Army of the Potomac. Only nightfall and a last-ditch defense saved the day for the Union.

   After darkness fell and while the two opposing forces nervously held their lines in close proximity, General Jackson rode out to reconnoiter between the lines to gauge the possibility of a nighttime attack. While returning from this mission, Jackson's party was fired on by the 18th North Carolina, which had just come up to the front lines, and Jackson was wounded in the hand and arm, which necessitated the amputation of his left arm above the elbow. Jackson died of pneumonia on May 10. Faust, *Historical Times*, 126–29, 391–92; Furgurson, *Chancellorsville*, 151–95, 201–6, 250–51, 329.

5. Hooker's staff members were extremely sensitive to the possibility that Lee might be reinforced by Longstreet's divisions and at one point were convinced that such a reinforcement had been accomplished. Furgurson, *Chancellorsville*, 136, 296.

   However, no additional troops in any numbers joined Lee's forces at Chancellorsville, and Hooker enjoyed enormous numerical superiority: 97,382 to 44,588. Livermore, *Numbers and Losses*, 98–99. It was not Hooker's plan that was faulty, but Hooker's leadership that failed the Army of the Potomac. Hooker simply was not up to the task of facing the aggressive and pugnacious Lee at Chancellorsville.

6. This skirmish was between Confederates under General Colston and Federals of General Slocum's II Corps. Colston was ordered by Lee to move up River Road, feel out the Union lines, and prevent any elements of General Hooker's army from moving to link up with General Sedgwick's force then moving up from Fredericksburg. Colston's force was turned back by Union cannon. This small fight late on May 3 would have been under full view of Union troops on the heights across the river. Furgurson, *Chancellorsville*, 269–70.
7. The Confederate battery of ten guns was under the command of Maj. Robert Hardaway and had been run out the River Road to Hayden's farm, near Anderson's division. Hardaway blasted away at the large bivouac of Federal troops, cavalry, and wagons, firing fifteen rounds per gun, and then withdrew under fire from skirmishers from Union General Slocum's corps. Confederate General Anderson countered with troops of his own, and the fight threatened to escalate but Anderson received orders to move to Salem Church and broke off the engagement. Furgurson, *Chancellorsville*, 293.
8. Contrary to WB's assertion, there was little more than skirmish firing on May 5. General Sedgwick had retreated back across the Rappahannock the night before, and General Lee spent the day concentrating his forces for an assault on General Hooker's position early on the sixth. A thunderstorm broke on the marching troops late in the afternoon and delayed the concentration until the next morning. Furgurson, *Chancellorsville*, 311–13.
9. The withdrawal of the Army of the Potomac across the Rappahannock was a confusing and disorderly affair, marred by conflicting orders, darkness, and rain. The last unit to cross the bridges was Barnes's brigade (1st Brigade, 1st Division, V Corps), and at 9 A.M. the bridge cables were cut. Furgurson, *Chancellorsville*, 313–18.
10. During the Chancellorsville campaign, the Union army suffered 16,792 killed, wounded, and missing, the Confederates, 12,764. Livermore, *Numbers and Losses*, 98–99.
11. Folley had replaced WB as captain of Company C on December 27, 1862, and resigned on May 23, 1863. *ARAGNY*, 934.
12. General Hooker called a conference of his corps commanders at midnight, May 4, and after much discussion a vote was taken that tallied three to two for advancing against Lee. Hooker, however, had already made up his mind to retreat back across the river, and the orders were cut early on the fifth. Hooker was one of the first to cross. Furgurson, *Chancellorsville*, 304–5; Johnson and Buel, *The Tide Shifts*, 171.
13. General Hooker was standing on the porch of the Chancellor House about 9 A.M. on May 3 when he was wounded in the head by a piece of wooden pillar blasted apart by a Confederate shell. He was knocked senseless but then, after being given some brandy, rode about his lines to show himself and quash a rumor that he had been killed. Later, from a tent a half mile behind the lines, Hooker

turned over command to General Couch but ordered Couch to withdraw the army to a defensive position around U.S. Ford. Furgurson, *Chancellorsville*, 240–42.

Whether General Hooker was drinking during the campaign or not remains a point of argument. Hooker stated that he did not drink, but others said they witnessed his drinking, and he certainly was incapacitated during much of the battle, his moods swinging from braggadocio to timidity. Hooker was cleared by a congressional committee after the war, which placed the blame for defeat on Generals Howard, Stoneman, and Sedgwick. Furgurson, *Chancellorsville*, 240–42, 286–87, 333–35, 339–40.

Chapter Fourteen

1. Gen. Robert E. Lee began the movement away from the Fredericksburg area on June 3, 1863, and by the eighth had concentrated two corps and his cavalry at Culpeper, leaving one corps behind at Fredericksburg. Faust, *Historical Times*, 307–8.
2. The V Corps of the Army of the Potomac was the only corps that had left the Fredericksburg area; the rest of the corps were still in their camps but not in the immediate area. On May 28 Barnes's division, V Corps, was sent to cover the upper reaches of the Rappahannock River, and on June 4 Sykes's division, V Corps, was ordered there to extend the right flank of the army to cover the river crossings as far as Richard's Ford. Welcher, *The Eastern Theater*, 1:264, 372.
3. Frank Welcher in *The Eastern Theater*, 1: 400, identifies this unit as Grant's 2nd Brigade of Howe's 2nd Division of VI Corps.
4. Charles E. Cross was born in Massachusetts and graduated from the United States Military Academy class of 1861 second in his class of forty-five. Assigned to the Army Corps of Engineers, Cross was breveted to colonel posthumously. He was twenty-six when he was killed at Franklin's Crossing on June 5, 1863. *USMA*, 254.
5. The 50th reported losses at Deep Run, Virginia (Franklin's Crossing), on June 5 as one enlisted man killed, one enlisted man mortally wounded, and nine enlisted men wounded. The 15th reported two enlisted men killed and four enlisted men wounded. The regular engineers lost Captain Cross, who was killed. Phisterer, *New York in the War of the Rebellion*, 2: 1651, 1670.
6. The regiments that crossed in the ponton boats were the 5th Vermont and the 26th New Jersey. The 2nd, 3rd, 4th, and 6th Vermont followed and established the bridgehead on the south bank. These regiments comprised the 2nd brigade, 2nd division, VI Corps. The captured Confederates were from A. P. Hill's Corps and led Sedgwick to report that Lee's army was still in force in Fredericksburg. Welcher, *The Eastern Theater*, 1:401.
7. WB is overestimating the importance of the skirmish at Franklin's Crossing. Lee had left A. P. Hill's fifteen-thousand-man corps behind at Fredericksburg from the beginning of his movements to guard the crossing there in case Hooker might try to move on Richmond or on Lee's flank and rear. Hooker proposed a move against Hill's Corps but this was disapproved by Lincoln

and Halleck, underlining their sensitivity to the security of Washington. Edwin B. Coddington, *The Gettysburg Campaign: A Study in Command* (New York: Charles Scribner's Sons, 1968; reprint, Dayton, Ohio: Morningside Bookshop, 1979), 52–53.

Lee's invasion of the North had been approved by the civilian authorities in Richmond after Lee had successfully argued that: 1) it would relieve the pressure on Richmond and, by inference, the pressure on Vicksburg, then under siege by Grant; 2) it would allow the Confederates to gather supplies in the lush Pennsylvania farmland; 3) it would encourage the northern peace movement; 4) it would reopen the possibility of European recognition; 5) it might result in the capture of Baltimore or Washington; 6) thence, it might lead to the destruction of the Union army in a decisive battle on northern soil. Faust, *Historical Times*, 307–8.

8. On June 6, the rest of Howe's division crossed and took up positions at the bridgehead. On June 7, Wright's division relieved Howe. On June 9, Newton's division relieved Wright. Welcher, *The Eastern Theater*, 1:401.
9. The mansion WB refers to was "Old Mansfield." The owner at the time was Mr. Bernard, the owner of a large number of slaves, who had objected bitterly to his house being used as a field headquarters at the beginning of the Fredericksburg campaign. His objections had earned him a free trip over the river, the loss of his slaves, and the destruction of his property. Davis, *Military Atlas*, plate 39.3; Stackpole, *Fredericksburg Campaign*, 162.
10. H. J. Gifford is actually listed as a 2nd sergeant in the article cited in chap. 1, n. 4.
11. A tapis is a tapestry used as a curtain, tablecloth, carpet, or the like; thus, to be on or upon the tapis is to be on the table, under consideration.
12. The pullout of the VI Corps from the bridgehead was accomplished by midnight on June 12. By that time most of the Army of the Potomac had left the Fredericksburg area and was stretched out along a forty-mile front from Falmouth to Beverly Ford. Coddington, *Gettysburg Campaign*, 68–70; Welcher, *The Eastern Theater*, 1: 401.
13. There is no mention of this incident in *OR*. It is certainly possible that the supposed raid was simply a rumor.
14. Capt. E. Vern Jewell, 146th New York, enlisted in Rome on September 3, 1862, and was discharged September 23, 1863. The 146th lost 18 men killed, wounded, and missing at Fredericksburg, 50 men killed, wounded, and missing at Chancellorsville, 28 men killed, wounded, and missing at Gettysburg and 654 men killed, wounded, and missing total during the war. Phisterer, *New York in the War of the Rebellion*, 5: 3688, 3698.
15. General Stuart's cavalry had moved into Loudoun and Fauquier Counties in Virginia to guard the gaps in the Blue Ridge Mountains and screen Lee's movement north. There was a fight at Aldie on June 17, and on June 19 a battle raged over the Virginia countryside around Upperville, Middleburg, and Aldie between Stuart's cavalry and Union cavalry under General Pleasonton. Although this fight took place over twenty miles from the river, it was large enough and close enough to cause alarm. Burke Davis, *Jeb Stuart: The Last Cavalier* (New York: Rinehart, 1957; reprint, New York: Fairfax Press, 1988), 315–18; Wert, *Mosby's Rangers*, 89; Davis, *Military Atlas*, plate 7.1.

16. I could find no mention of Noland's Ferry being in possession of Confederate forces on June 19. There is a strong possibility that this was either deliberate misinformation spread by partisans or simply a rumor caused by all the cavalry actions in the immediate area.
17. The distance from Edward's Ferry to Washington over the roads of the period was very close to the thirty-five-mile estimate given by WB. He and Spaulding would have had to average four miles per hour on horseback, if they did indeed go nonstop, a pace that would have conserved their horses' strength. Davis, *Military Atlas*, plate 27.1.
18. On June 15 Rodes's division of Ewell's corps crossed the Potomac at Williamsport, leading the Confederate invasion of the North. Coddington, *Gettysburg Campaign*, 103.

    On June 24 the XI Corps of the Army of the Potomac crossed the Potomac River at Edward's Ferry on the ponton bridge thrown over by the engineers left there by WB and Spaulding when they were ordered back to Washington. Coddington, *Gettysburg Campaign*, 123.
19. By June 25 General Ewell's corps was well into Pennsylvania, skirmishing with Federals near McConnellsburg. On June 26, his advance elements under General Early marched through Gettysburg on their way to York. Lee's invasion of Pennsylvania caused a panic, and Governor Curtin called for sixty thousand three-month volunteers to repel the invasion. Long and Long, *Civil War Day By Day*, 370–71.

CHAPTER FIFTEEN

1. When General Lee invaded Pennsylvania he made use of every bit of subterfuge he could to confuse the Union high command. By June 27, the bulk of Lee's army was concentrated at Chambersburg with two advance wings stretching out toward York and Harrisburg. Governor Curtin was screaming in wild alarm, farmers were fleeing before the advancing Confederates with their horses in tow, and the rumors were flying fast and furious. On the twenty-seventh, Hooker was directing that cavalry be sent in advance of the Army of the Potomac to feel out Lee's positions. Even the commanding officer of the army was not sure where Lee was and what he was doing. Coddington, *Gettysburg Campaign*, 127.
2. Gen. Wade Hampton, in command of the vanguard of Stuart's cavalry, crossed the Potomac early on June 28 at Rowser's Ford, sometimes called Coon's Ford, located near Great Falls and on the road from Dranesville, Virginia, to Rockville, Maryland, Davis, *Jeb Stuart*, 325; Davis, *Military Atlas*, plate 27.1.

    There is some disagreement about the strength of Stuart's force. Coddington states that Stuart had approximately four thousand troopers with him on his raid. Coddington, *Gettysburg Campaign*, 203. Burke Davis, in his biography of Stuart, states that the number was closer to two thousand. Davis, *Jeb Stuart*, 325. Others deal with the problem by not giving a number for the size of Stuart's force, simply stating that Stuart took three brigades on his raid.
3. When Stuart's cavalry had gotten across the Potomac they had to pause to rest

their mounts. While on the north bank they destroyed as much of the C&O Canal as they could, even breaking part of the bank into the Potomac in an attempt to drain the water out of the canal. Davis, *Jeb Stuart*, 326–27.

4. The town is named Buckeystown on period maps. Davis, *Military Atlas*, plate 27.1.
5. General Hooker and General Halleck, general in chief of all Union armies, did not agree on many things; their feud went to prewar days in California. During the pursuit of Lee, Hooker demanded that the ten-thousand-man garrison at Harper's Ferry be placed under his command to bolster the manpower of the Army of the Potomac. Halleck refused this request, stating his desire to maintain a strong presence there. Hooker felt that essential troops were being withheld from him at a critical juncture and tendered his resignation. Halleck forwarded it immediately to Lincoln and Hooker was sacked. For a discussion of the motives of both Hooker and the Lincoln administration during this crisis, see Coddington, *Gettysburg Campaign*, 131–33.

   When Gen. George Gordon Meade replaced Hooker at the helm of the Army of the Potomac, he became its fourth commander in eight months. Meade was a respected corps commander and had the support of most of the high-level officers in the army, which made his job easier in the days to come. Coddington, *Gettysburg Campaign*, 209–14.
6. William McCandless began the war as a private and mustered out as a brigadier general in June 1864, after some hard fighting with the Army of the Potomac through all its campaigns. He led his brigade in a charge on the second day at Gettysburg that cleared Plum Run Valley of Confederates and captured the colors of the 15th Georgia Infantry. He was wounded in the Wilderness. Boatner, *Civil War Dictionary*, 523; Welcher, *The Eastern Theater*, 373.
7. Johnsville is twenty-one miles from Gettysburg and New Market is thirty-two miles away. It is certainly possible that WB heard the cannonading and saw smoke rising from the battlefield. Davis, *Military Atlas*, plate 27.1.
8. There seems to be a serious discrepancy in WB's narrative at this point. Colesville is a settlement eleven miles north east of Washington and not on a direct route from New Market, twenty-five miles away, hardly a march that WB's column could have made in one day. Then WB states that after Colesville they stopped at a settlement nineteen miles from Washington. I do not think that they countermarched to get to this settlement. Davis, *Military Atlas*, plate 27.1.

   There are a good many towns that might be the unnamed Quaker settlement, including Goshen, Unity, or Urbana, but none seems to quite fit.

   There are no records of Confederate cavalry being in this vicinity on July 2. General Imboden commanded a detachment of cavalry guarding Lee's rear, many miles from where WB faced mysterious enemies. There were stragglers and irregulars in the area in small numbers, and a town might have been "held" by half a dozen armed men, which could explain WB's report of the warning from a Negro.
9. Again, these Confederate troopers could only have been a small group of stragglers, as General Stuart and his entire command were at Gettysburg on July 2. Coddington, *Gettysburg Campaign*, 206.

10. Meade ordered the Harper's Ferry Garrison to Frederick, Maryland, on June 28 to guard the Army of the Potomac's lines of communication with Washington and to remove them from harm's way. Maryland Heights was reoccupied on July 7. Coddington, *Gettysburg Campaign*, 222–24, 482, 554, 556.
11. Lt. John R. Meigs was the son of Union army Quartermaster General Montgomery Meigs. Twenty-two-year-old Lieutenant Meigs graduated at the top of the United States Military Academy class of 1863 on June 11 and went into the corps of engineers. Although assigned to the Army of the Tennessee around Vicksburg, Meigs was diverted to Baltimore and the Middle Department. He was in the party that destroyed the Confederate ponton bridge at Falling Waters, delaying Lee's retreat. It was shortly after this that he barely escaped capture after an ambush. Russell F. Weigley, *Quartermaster General of the Union Army: A Biography of* M. C. *Meigs* (New York: Columbia Univ. Press, 1959), 303.

    Meigs went on to serve on General Sheridan's staff in 1864 and was killed by Confederate irregulars on October 3, 1864, in the Shenandoah Valley. Sheridan ordered all houses within a five-mile radius of Dayton, Virginia, burned in retaliation. Boatner, *Civil War Dictionary*, 542.
12. General Meade did order his cavalry to harass and slow down Lee's retreating columns and he did order General French at Frederick to destroy Lee's ponton bridge at Williamsport. Meade also ordered his own ponton trains up from Washington. All these orders were made before July 5 and point to an aggressive commander anticipating cutting off his adversary to bring on a major battle. Coddington, *Gettysburg Campaign*, 482, 542.

    However, Meade also directed his corps commanders not to bring on a major engagement during Lee's withdrawal, and, even though the ponton trains were at Harper's Ferry by July 9, Meade did not order a cavalry division to cross to the right bank of the Potomac to move against Lee's expected point of crossing from the Virginia side, a natural tactic to try to prevent Lee from getting away. This is evidence that Meade did not want another battle with Lee just yet. Edward J. Stackpole, *They Met at Gettysburg: A Stirring Account of the Battles at Gettysburg* (New York: Bonanza Books, 1956), 291–314.
13. Lee retreated from Gettysburg late on July 4, and, after some skillful marching, his army reassembled at the Potomac River near Hagerstown, Maryland. The ambulance train leading the retreat found the ponton bridge at Falling Waters destroyed by French's troopers. There was little for Lee to do but simply wait for the unseasonable high water to recede so he could cross while his engineers worked to repair the pontons. Lee got his army into entrenchments around Williamsport and invited the Union army to attack his works. Meade had ordered an immediate pursuit, but after a false start and some lower-level bungling, he had fallen behind the Confederates. Meade was hampered somewhat by Lincoln's directive that his movements always cover Washington by staying between Lee and the city. Meade was also suffering from the loss of twenty-three thousand veteran troops and two of his best corps commanders, Reynolds (killed) and Hancock (wounded). Meade then held a council of corps commanders that voted five to two against an attack. This held Meade back for two days, but he finally ordered the attack, which came too late, as Lee and the

Army of Northern Virginia had crossed over into Virginia and escaped on July 13–14. Meade was highly censured for allowing Lee to escape, but an examination of his actions from July 4 to July 14 place him in a more favorable light than McClellan after Antietam. Coddington, *Gettysburg Campaign*, 547–74; Stackpole, *Gettysburg*, 295.

The Army of the Potomac crossed the river as follows: I Corps, July 18 at Berlin; II Corps, July 18 at Harper's Ferry; III Corps, July 17 at Harper's Ferry; V Corps, July 17 at Berlin; VI Corps, July 19 at Berlin; XI Corps, July 19 at Berlin; XII Corps, July 19 at Harper's Ferry. Welcher, *The Eastern Theater*, 300–472.

CHAPTER SIXTEEN

1. Lee had brought a ponton train with his army and this train was at Williamsport on the Potomac when Meade ordered its destruction by Major General French's command at Frederick on July 4. French reported to Halleck on July 5 that the Confederate pontoon bridge was destroyed the night of July 3–4 by a cavalry expedition that also captured the guard, a lieutenant and thirteen men, and destroyed an ammunition train. When Lee's ambulance wagons arrived on July 6, the pontons were unusable and engineers immediately began to piece them back together, although they had few tools and materials to work with. This caused a delay of seven days. When Lee's army finally got across on the rebuilt ponton bridge, the rear guard cut the bridge loose to prevent close following by Union cavalry. Stackpole, *Gettysburg*, 291, 295, 316; James L. Nichols, *Confederate Engineers*, Confederate Centennial Studies, no. 5 (Tuscaloosa, Ala.: Confederate Publishing Company, 1957; reprint, Gaithersburg, Md.: Olde Soldier Books, 1987), 99; *The New York Times*, July 6, 1863; Coddington, *Gettysburg Campaign*, 542.
2. When the war began Peter A. Porter was serving in the New York House of Representatives from the 2nd District of Niagara. He helped raise the regiment and was its first colonel. The 8th was on duty at Harper's Ferry, West Virginia, from July 10, 1863, to August 3, 1863. The regiment did not see action until the Wilderness Campaign of 1864. Phisterer, *New York in the War of the Rebellion*, 1: 29, 2: 1397.
3. There are no records of Lincoln observing ponton drills in July, August, or September 1863. Miers, *Lincoln Day By Day*, 3:194–210.
4. The statue, *Armed Freedom*, was raised by steam hoist to the newly finished dome of the Capitol building on December 2, 1863. There was no ceremony, and President Lincoln was sick in bed with "varioloid" (a mild form of smallpox), although a large crowd gathered to witness the event and a battery of artillery fired a thirty-five-gun salute, which was answered by the forts surrounding the Capitol. Leech, *Reveille*, 345; French, *Witness*, 439.

   The statue was designed by Thomas Crawford (1813–1857), who also designed *Past and Present of America* on the Senate wing pediment, the bronze doors of the Senate portico, and the Washington Monument at Richmond. *Armed Freedom* had been approved by Jefferson Davis, who at the time was secretary of war and in charge of Capitol construction. Davis objected to Crawford's original design, which called for a headdress fashioned after those worn by Rome's emancipated slaves. Crawford then changed the design to a feathered headdress. The statue was modeled in Crawford's Rome, Italy, studio and shipped to America for casting. Gorton Carruth, *The Encyclopedia of American Facts and Dates*, 9th ed. (New York: HarperCollins Publishers, 1993), 256; Lonnelle Aikman, *We, the People: The Story of the United States Capitol, Its Past and Its Promise* (Washington, D.C.: The United States Capitol Historical Society, 1965), 54–57.

The nineteen-and-one-half-foot bronze form of *Armed Freedom* was cast in Clark Mills's foundry in Bladensburg, Maryland, in five parts, transported to the Capitol grounds, assembled, and put on display before the pieces were disassembled again and raised to the finished dome to be reassembled. Mills (1810–1883) was a resourceful and innovative South Carolinian who had designed and cast the equestrian bronze of Andrew Jackson for Lafayette Park. It is ironic that *Armed Freedom* was cast with the aid of Mills's slave laborers. Carruth, *American Facts*, 244; Aikman, *We, the People*, 57; French, *Witness*, 229.

5. There was no Major Arden in either the 50th or the 15th. It is possible that the Major Arden that WB refers to was an acquaintance from another regiment but the name does not appear anywhere else in the memoir and there are no other clues to his identity.

Secretary of State William Seward had three sons.

Augustus Henry Seward was born in 1826, showed an early interest in military life, and was educated at the United States Military Academy. He graduated in the class of 1847, finishing thirty-fourth out of thirty-eight. He served in Mexico and during the 1850s was with the Coast Survey. "Gus" was a paymaster in Washington during the Civil War and died in New York in 1876. *USMA*, 240; John M. Taylor, *William Henry Seward: Lincoln's Right Hand* (New York: HarperCollins, 1991), 21, 71.

Frederick William Seward was born in 1830, went into politics, and served as his father's legislative assistant while William Seward was in the U.S. Senate. "Fred" became an assistant secretary of state when his father went to the State Department. Both Fred and Gus were injured during the assassination attempt on their father by Lewis Payne. Taylor, *Seward*, 28, 103, 142, 242–45.

William Henry Seward Jr. (b. 1838) was learning the banking trade when war broke out and went into the military. By the time of the Gettysburg campaign, "Will" was the colonel of the 9th New York Heavy Artillery and assigned to the Washington defenses. In 1864 he was cited for gallantry at Cold Harbor and participated in the Battle of the Monocacy, where he was slightly wounded and then more seriously injured when his horse fell on him. Toward the end of the war, Will led a brigade in the Department of West Virginia. Taylor, *Seward*, 58, 80, 121, 164, 222, 233; Welcher, *The Eastern Theater*, 194–95, 689.

It was Will Seward who commanded at Fort Foote, located on Rozier's Bluff on the east bank of the Potomac just across from Hunting Creek and Alexandria. This area was easily reached by roads and bridges from the Navy Yard. Will survived dysentery and the war and died in 1920. Davis, *Military Atlas*, plate 89.1; Benjamin Franklin Cooling III, *Symbol, Sword and Shield: Defending Washington During the Civil War*, 2nd rev. ed. (Pennsylvania: White Mane Publishing, 1991), 161; Sifakis, *Who Was Who in the Union*, 361.

6. On September 23, 1863, after the Battle of Chickamauga, Lincoln, Stanton, Halleck, and the Cabinet made the decision to reinforce Rosecrans with Gen. O. O. Howard's XI Corps and Gen. John Slocum's XII Corps, Army of the Potomac. Stanton took command of the transfer and moved twenty thousand men and their equipment with ten batteries of artillery from Virginia to Bridgeport, Alabama (1,159 miles) in an astounding seven to nine days. Gen. Joe Hooker, the loser at Chancellorsville, was put in command of the reinforcements. Long and Long, *Civil War Day By Day*, 413–17; Boatner, *Civil War Dictionary*, 142.

CHAPTER SEVENTEEN

1. After the Army of the Potomac sent General Hooker and 2 Corps to reinforce General Rosecrans in Chattanooga, General Lee sensed that by weakening their force, the Union army might be vulnerable. Lee began the Bristoe Campaign on October 9, 1863, in an attempt to turn General Meade's right flank and fall on his lines of communications to Washington. A five-day race ensued, punctuated by continuous skirmishing between flank and lead elements of Lee's columns and flank and rear-guard elements of Meade's army. Faust, *Historical Times*, 80.
2. General Lee sent A. P. Hill's Corps in a wide, looping march, hoping to catch part of Meade's army in the flank. General Hill found the Union III Corps in retreat at Bristoe Station on the Orange & Alexandria Railroad and attacked without adequate reconnaissance. As two Confederate brigades hurled themselves at the Federals, elements of Union General Warren's II Corps, who had been concealed behind the railroad embankment, fired on the charging southerners, inflicting heavy casualties. Hill was defeated with losses of thirteen hundred men. Faust, *Historical Times*, 80–81.
3. The Battle of Rappahannock Station was part of General Meade's plan to regain ground lost during the Bristoe Campaign. On November 7, 1863, Meade's army fought at Rappahannock Station and Kelly's Ford, both major crossing points, and was successful at forcing the Confederates guarding those places to retreat. With the loss of these two places, General Lee was forced to withdraw behind the Rapidan River on November 10. The engineers were placed on alert to provide an alternate crossing opportunity in case elements of the Army of the Potomac got into trouble and needed support that would have to cross on the pontons. Faust, *Historical Times*, 615.
4. After General Meade had forced General Lee below the Rapidan River, he then made plans to turn Lee's right flank. On November 26 elements of the Army of the Potomac maneuvered into position, but General French, commanding the leading corps, was slow in developing his attack, giving Lee's troops time to

entrench on a ridge behind Mine Run. By November 30, Lee's position was too formidable to attack and Meade withdrew and went into winter quarters. The engineers were put on alert during these movements in case a river crossing needed to be performed. Faust, *Historical Times*, 497.

5. The New Year's Day reception at the White House was a tradition going back to George Washington, who held the first levee in New York City in 1790. The tradition ended in 1933 when lame duck President Hoover went on a vacation instead of staying in town for the traditional reception. Betty Boyd Caroli, *Inside the White House: America's Most Famous Home, the First 200 Years* (New York: Canopy Books, 1992), 168–79.

   Guests were expected to arrive in accordance with a strict schedule. Members of the diplomatic corps at 10 A.M., official families at 11, officers of the army and navy at 11:30, and the general public at noon. The reception was over at 2 P.M. Miers, *Lincoln Day By Day*, 3:231.

6. Below-zero temperatures were recorded as far south as Memphis, Tennessee, and Cairo, Illinois. Long and Long, *Civil War Day By Day*, 452; Bowman, *Civil War Almanac*, 180.

7. WB is referring to *Military Law and Courts Martial*, by Capt. S. V. Benet, and "Observations on Military Law, and the Constitution and Practice of Courts Martial," by Capt. William Chetwood De Hart, the two leading authorities on the conduct of military courts at the time.

8. It is hard to determine just what WB is referring to here. President Davis delivered his annual message to the Confederate Congress on December 7, 1863, and he mentioned many of the reverses of the year but praised the patriotism of the southern people. He also condemned the "savage ferocity" of the Federals. This speech has been described as being full of wishful thinking and rationalizations, flat and uninspiring. Davis was ill at the time and depressed over the situation. Long and Long, *Civil War Day By Day*, 443–44; George C. Rable, *The Confederate Republic: A Revolution Against Politics* (Chapel Hill: Univ. of North Carolina Press, 1994), 240. Davis's December 7 speech was reported verbatim in the December 12 supplement to the *New York Times*.

   On February 3, 1864, Davis called the attention of the Confederate Congress to the morale problems rampant in the South, pointed out the "discontent, disaffection and disloyalty" among noncombatants, and recommended the suspension of the writ of habeas corpus to combat spying, desertion, associating with the enemy, and disloyal meetings and activities. The Confederate Congress responded by authorizing a further six-month suspension of the writ of habeas corpus and extended and expanded conscription. Long and Long, *Civil War Day By Day*, 464; Rable, *Confederate Republic*, 240; Current, *Encyclopedia of the Confederacy*, 4: 1540. The resulting acts by the Confederate Congress were reported in the *New York Times* on February 11, 1864.

   On March 12, 1864, Davis issued a proclamation calling for a day of "humiliation, fasting and prayer" on April 8. In the proclamation Davis cited the deliverance of the government from the "nefarious scheme" to burn Richmond and murder the government leaders (Kilpatrick-Dahlgren Raid, February 28–March 4). Davis then said, "Our armies have been strengthened, our finances promise rapid progress to a satisfactory condition, and our whole country is animated with a hopeful spirit and a fixed determination to achieve independence," an unrealistic claim, given the situation then existing and given his own former entreaties to the Confederate Congress to pass laws to combat the morale problems in the South. James D. Richardson, ed. and comp., *The Messages and Papers of Jefferson Davis and the Confederacy: Including Diplomatic Correspondence 1861–1865* (1905; reprint, with a comprehensive introduction by Allan Nevins, New York: Chelsea House-Robert Hector, 1966), 412–14.

It seems that it is this latter incident that WB is citing. It is certainly possible that WB read a report of this proclamation in northern newspapers, although none could be found.

CHAPTER EIGHTEEN

1. The vice president of the Confederacy, Alexander Stephens, was home in Georgia visiting his family during the month of December 1862. Thomas E. Schott, *Alexander H. Stephens of Georgia: A Biography* (Baton Rouge: Louisiana State Univ. Press, 1988), 365.
2. Irving Gage Brainerd, to whom this memoir was written, was born on December 4, 1864.
3. Col. Ulric Dahlgren (1842–1864), son of Rear Admiral John Dahlgren, was a born adventurer who often volunteered for hazardous duty. He lost a leg after being badly wounded during the pursuit of Lee following Gettysburg. Barely recuperated, he joined Gen. Judson Kilpatrick and his reinforced 3rd Cavalry Division on a raid on lightly held Richmond on February 28, 1864. He was separated from the main force with a group of five hundred men, surrounded, and killed on March 2. Papers found on Dahlgren's body led Confederate authorities to conclude that the raid was for the purpose of torching Richmond and killing Davis and his Cabinet. Dahlgren was branded a criminal, castigated in the southern press, and buried in an unmarked grave. The controversy quieted down only after General Kilpatrick denied any knowledge of the supposed plot, leaving Dahlgren to bear all the blame. Boatner, *Civil War Dictionary*, 218–19; Faust, *Historical Times*, 202–3. Welcher, *The Eastern Theater*, 277.
4. There were a number of prominent citizens in Rome named Stevens during this time. Edward L. Stevens was the last mayor of the village of Rome before it was incorporated as a city in 1870. He also served as mayor from 1877 to 1878. Samuel B. Stevens was mayor from 1875 to 1877. There are no records that might indicate when the vote that WB refers to was taken, by whom, and what the outcome was. In 1860 Rome had a population of 6,246, and in the presidential election of 1864, 1,239 voted for McClellan (Democrat) and 769 for Lincoln (Republican). Copperheads were northern peace Democrats who opposed the Union's war policy and favored a negotiated peace. It may be inferred that Mr. Stevens was a Democrat, as he voted against the soldier vote issue. Soldiers were thought to be pro-Lincoln, as a vote against Lincoln was a vote against the administration's war policies. Virginia V. Herrmann and Fritz Updike, letter to

editor, February 12, 1994; David E. Long, *The Jewel of Liberty: Abraham Lincoln's Re-election and the End of Slavery* (Mechanicsburg, Pa.: Stackpole Books, 1994), 219–24, 285; Boatner, *Civil War Dictionary*, 175.

5. "By the pricking of my thumbs,/Something wicked this way comes./Open, locks,/ Whoever knocks!" (*Macbeth*, IV.i.ll. 45–48).
6. Francis Amasa Walker (1840–1897) began the war as a private in the 15th Massachusetts and ended it a brevet brigadier general. He was wounded and captured at Chancellorsville but was later paroled. The twenty-five-year-old Walker was described as the ideal staff officer—pragmatic, methodical, and efficient in the handling of a bewildering mass of detail. He served most of his time as a staff officer with General Hancock and after the war wrote *History of the Second Army Corps* (1887) and *General Hancock* (1895). He was the son of a famous economist and himself became a foremost economic theorist, Yale professor, MIT president, and an internationally renowned statistician. Sifakis, *Who Was Who in the Union*, 434; Boatner, *Civil War Dictionary*, 884; George M. Fredrickson, *The Inner Civil War: Northern Intellectuals and the Crisis of the Union* (New York: Harper and Row, 1965), 203.
7. Canvas pontons were lighter and more easily handled than the 1,300-pound wooden ponton boats. A 640-pound canvas ponton consisted of two side frames, cross frames, and a waterproofed canvas covering that was stretched over the assembled frame and lashed tight. A train of wagons carrying canvas pontons was faster because it carried a lighter load, although the chess and balks were the same. A well-drilled company of engineers could assemble a canvas ponton bridge as fast as a regular ponton bridge. Duane, *Manual for Engineer Troops*, 7–48.
8. Maj. Gen. Winfield Scott Hancock (1824–1886) was the most conspicuous general in the Union army who never exercised independent command. He graduated from the United States Military Academy in 1844 and was renowned for his military bearing, professionalism, and attention to detail. Hancock was a hero at Gettysburg, where he was severely wounded. He was nominated for the presidency on the Democratic ticket in 1880 but narrowly lost to Garfield. He was known throughout the army and the country as "Hancock the Superb." Faust, *Historical Times*, 337–38.

Chapter Nineteen

1. On March 4, 1864, General Meade recommended to the secretary of war that the Army of the Potomac be reorganized from five corps to three. Meade argued that the reduced strength of all the infantry regiments of the army made consolidation a good option. This reorganization was effected by an order dated March 23, 1864, and was carried out as follows: 1) two divisions of III Corps were transferred into the II Corps as the 3rd and 4th divisions and commanded by General Hancock; 2) the three divisions of the I Corps were consolidated

into two divisions and transferred into the V Corps as the 2nd and 4th divisions, commanded by General Warren; 3) the 3rd division, III Corps, was transferred into the VI Corps as 3rd division, commanded by General Sedgwick; 4) I and III Corps were discontinued and their artillery batteries were distributed among the other corps so that each corps had eight batteries.

Generals Sykes, French, Newton, Meredith, Kenly, Spinola, and Caldwell were relieved of their duties with the Army of the Potomac and reported to the adjutant general of the army for reassignment. Welcher, *The Eastern Theater*, 278.

During the Peninsula Campaign, General Kearny had the men in his 3rd Division, III Corps, wear a red diamond device to identify them easily on the battlefield and to inspire esprit de corps. When General Hooker took over the Army of the Potomac he had his chief of staff, Gen. Dan Butterfield, design corps badges for all the corps of that army. The following year, western armies followed suit. Boatner, *Civil War Dictionary*, 37, 105.

2. The losses in killed, wounded, and missing suffered by the II Corps during active operations in 1864 were as follows: Wilderness, 5,092; Spotsylvania, 6,642; North Anna, Totopotomoy, Cold Harbor, 3,510; Petersburg, June 15–30, 6,624; Petersburg, July 1–31, 206; Petersburg, August 1–31, 3,677 (Reams's Station was fought August 25); Petersburg, September 1–30, 390; Petersburg, October, November, and December, 1,625; the total was 27,766.

   These attrition figures do not include troops mustered out at the end of their terms of service or losses due to disease or desertion. Fox, *Regimental Losses*, 69; *OR*, ser. 1, vol. 40, pt. 1: 222, 254; vol. 42, pt. 1: 132, 144, 161.
3. The oratorio *Saul* was written in 1739 by George Friederic Handel (1685–1759). The "Dead March in Saul" was the popular name given to the funeral march section of the oratorio. It was used on state occasions such as the funeral of a sovereign. Denis Arnold, ed., *The Oxford Companion to Music* (New York: Oxford Univ. Press, 1983), 1: 802–7.
4. "Maddens" appears on two maps in Davis's *Military Atlas* and seems to refer to a general area near Paoli Mills on one map, and on the other it refers to the name of a resident, W. Madden, who lived near Sheppard's Grove. In the winter of 1863–64, there was a cavalry brigade stationed at Sheppard's Grove with pickets out to the north shore of the Rapidan River as far as Richardsville, Virginia. Davis, *Military Atlas*, plates 23.4, 44.3, 87.2.

   Willis Madden was a prosperous free Negro who ran a tavern on the main road from Fredericksburg to Culpeper. Gordon C. Rhea, *The Battle of the Wilderness May 5–6, 1864* (Baton Rouge: Louisiana State Univ. Press, 1994), 60.

   Col. John R. Brooke commanded the 4th Brigade, 1st Division, II Corps. William D. Matter, *If It Takes All Summer: The Battle of Spotsylvania* (Chapel Hill: Univ. of North Carolina Press, 1988), 352.
5. Charles Hale Morgan (1834–1875) graduated from the United States Military Academy in 1857 (twelfth in a class of thirty-eight) and was assigned to the artillery. He served as chief of artillery, inspector general, and chief of staff

for the II Corps and commanded the corps artillery for most of 1863 and 1864. He rose to brevet brigadier general rank during the war. Boatner, *Civil War Dictionary*, 565.

6. A thorough search of *OR* and numerous other histories of the movement of the II Corps on May 3–4 failed to turn up any mention of this incident. That it was embarrassing to the artillerymen responsible probably was one reason that it was not mentioned in official reports. Also, the fact that the engineers were able to force their way through the impasse and arrive on time at the rendezvous relegated the incident to a non-event. Had the delay caused serious problems, there certainly would have been official mention and someone would have been brought up on charges.
7. Lt. Col. Ira Spaulding, commanding the 50th New York Engineers, reported that WB arrived at Ely's Ford shortly after 6 A.M. on May 4 and completed his 190-foot bridge by 9:15, at which time the marching column was transferred to that bridge and the canvas ponton bridge under command of Captain Folwell was taken up. WB's bridge was dismantled at 2 P.M. on May 5, and the command moved to Chancellorsville. *OR*, ser. 1, vol. 36, pt. 1: 304–6.
8. The V, VI, and IX Corps of the Army of the Potomac crossed at Germanna Ford, farther to the west. The army supply trains crossed at Culpeper Ford. *OR*, ser. 1, vol. 36, pt. 1: 305–6.
9. In 1519, Hernan Cortes landed on the gulf shore of Mexico to begin his conquest of the Aztecs. After subjugating the native population in the coastal regions, Cortes began his overland campaign to conquer all of Mexico. He believed that to prevent any of his conquistadors from deserting, he had to cut off all retreat and, after salvaging useful hardware, rigging, and sails, burned his ships. F. A. Kirkpatrick, *The Spanish Conquistadores*, 2nd ed. (1934; New York: Barnes and Noble, 1967), 74–75.
10. The 50th New York Engineers went into battle with 32 officers and 1,010 enlisted men. *OR*, ser. 1, vol. 36, pt. 1: 307.
11. General Crawford was a division commander in the V Corps, and his brigades were in front of the engineers. Rhea, *Wilderness*, 406, Appendix.
12. The 146th New York Infantry was in the 1st Brigade (Brig. Gen. Romelyn Ayres), 1st Division (Brig. Gen. Charles Griffin). On the afternoon of May 5, the 146th, a flashy Zouave regiment, had been horribly chewed up in the confusion of the battle, and Col. David T. Jenkins had disappeared with most of his subordinate officers in the fighting. Jenkins's fate was never established. Rhea, *Wilderness*, 150–51n. 157.
13. Late on the afternoon of the May 6, Confederate General Gordon led an attack against the weakened right flank of the Union army. The attack was carried out with three brigades, Pegram's, Gordon's, and Johnston's, and routed the Union VI Corps brigades of Generals Shaler and Seymour, both of whom were taken prisoner along with five hundred of their men. Neill's brigade, the next Union unit in line, refused its right and held against the onslaught. In the darkness and confusion of the Wilderness, the Confederate attack ran out of steam and the tired troops withdrew back to their former lines. General Gordon led a late evening attack but was repulsed. Rhea, *Wilderness*, 404–25.

CHAPTER TWENTY

1. As news of the impending disaster flooded into Federal headquarters, Grant was calm and imperturbable as he gave the orders to meet the crisis. He then went and sat down on a stool, lit a cigar, and continued to take reports. When a general officer approached him and suggested that Confederate General Lee was a serious threat, Grant rose and said, "Oh, I am heartily tired of hearing about what Lee is going to do. Some of you always seem to think he is suddenly going to turn a double somersault and land in our rear and on both of our flanks at the same time. Go back to your command, and try to think what we are going to do ourselves, instead of what Lee is going to do." Horace Porter, *Campaigning With Grant* (New Jersey: The Blue and Grey Press, 1984), 69–70.
2. The engineers had been relieved from the firing lines and ordered by General Warren to build communications across Wilderness Run. By noon they had completed twelve corduroy bridges and two fords with approaches. *OR*, ser. 1, vol. 36, pt. 1: 307.
3. There was a short fight between some of Warren's artillery and advancing Confederate skirmishers at 6:30 A.M. on May 7 that Warren dismissed as a trifle. Skirmishing took place along Warren's front until 11 A.M. Porter, *Campaigning With Grant*, 74–75.
4. After General Burnside was relieved following the Mud March and his ill-advised ultimatum to Lincoln, he was assigned to command the Department of Ohio. While there, Burnside arrested the Copperhead Clement Vallandigham and put him on trial for sedition and captured Confederate raider Gen. John Morgan. He went on to defend Knoxville successfully against Gen. James Longstreet and returned to the East in the Spring to command the IX Corps. Faust, *Historical Times*, 97.
5. After Jackson died on May 10, 1863, of pneumonia, his body was removed to Richmond, where it lay in state. It was then taken to Lexington, Virginia, where Jackson's body was buried on May 15. Jackson's amputated arm was buried in the Lacy family plot at Elwood, the family's plantation near Wilderness Tavern, and a marker was placed on the grave. At the time WB observed the site, the marker had evidently been removed. Because of the density of fighting in the area over the previous two years it might be surmised that, because the grave was unmarked, it was the final resting place of another fallen soldier. However, WB does say "little mound of earth," so we may suppose that it was indeed the gravesite of Jackson's arm. Davis, *Military Atlas*, plate 55.1; Byron Farwell, *Stonewall: A Biography of General Thomas J. Jackson* (New York: W. W. Norton, 1992), 513n.
6. The road WB marched over that night was the Orange and Fredericksburg Turnpike, usually called the Orange Turnpike or simply the Turnpike. During the two days of the fighting in the Wilderness, the Union army suffered 2,246 killed, 12,037 wounded, and 3,383 missing. Livermore, *Number and Losses*, 110; Davis, *Military Atlas*, plate 55.1.
7. The Ny River comes close by the Plank Road, two miles south of the

Chancellorsville clearing; Davis, *Military Atlas*, plate 41.1. Lieutenant Colonel Spaulding states that the rendezvous spot was at a place called Fallen Mill Crossing, but this site does not appear on any of the maps in the *Military Atlas*. *OR*, ser. 1, vol. 36, pt. 1: 308.

8. The Aldrich house is located at the intersection of the Plank Road and the Catharpin Road, four miles from Chancellorsville. Davis, *Military Atlas*, plate 41.1.

   Maj. Gen. "Uncle John" Sedgwick was killed May 9 while making a reconnaissance and directing the placement of artillery at Spotsylvania. Confederate sharpshooters were firing at the group around Sedgwick, and, to reassure his men, he said in jest, "They couldn't hit an elephant at this distance," just before a sniper's bullet found its mark. Boatner, *Civil War Dictionary*, 730; Gregory Jaynes, *The Killing Ground: Wilderness to Cold Harbor*. The Civil War Series (Alexandria, Va.: Time-Life Books, 1986), 88.

   Brig. Gen. James Wadsworth was leading the 4th Division, V Corps, when, on May 6, as he tried to rally the 20th Massachusetts, his horse bolted and carried him into Confederate lines, where he was shot in the head. He died two days later in a Confederate hospital. Boatner, *Civil War Dictionary*, 882–83; Jaynes, *The Killing Ground*, 77–78.

   Brig. Gen. Alexander Hays commanded the 2nd Brigade, 3rd Division, II Corps, on the Union left at the Wilderness when he was mortally wounded on May 5. Boatner, *Civil War Dictionary*, 389; Jaynes, *The Killing Ground*, 71.

9. In his memoirs Grant states two reasons for this tactical movement: "First, I did not want Lee to get back to Richmond in time to attempt to crush Butler before I could get there; second, I wanted to get between his army and Richmond if possible; and, if not, to draw him into the open field." Mary Drake McFeely and William S. McFeely, eds., *Grant: Personal Memoirs of U. S. Grant Selected Letters 1839–1865* (The Library of America, 1990), 540. The reference to Butler is to Gen. Benjamin Butler's movement up the James River to attack Richmond from the south in concert with Grant's own movement against Richmond from the north.

10. *OR*, ser. 1, vol. 36, pt. 1: 308.

11. Casualty reports from the two battles at Wilderness, May 5–7, and Spotsylvania, May 10, list 15,384 wounded. These figures do not take into account the fighting on May 8 and 9 that certainly added to these numbers and the fighting after May 10. Livermore, *Number and Losses*, 110–12.

12. During the time WB was in Fredericksburg, the Union army was constantly shifting to the southeast. This uncovered the roads into Fredericksburg from the west, and Confederate cavalry patrols, sometimes at brigade strength (Rosser's brigade, Chambliss's brigade, and Cobb's legion), pushed toward the town with the object of ascertaining the position of the Federal right flank. The Union base of supplies had been shifted to Belle Plain when the ponton bridges across the Rapidan were taken up and supplies would necessarily have to come through Fredericksburg and wounded would have to be evacuated through there also, all of them over the ponton bridge WB constructed on May 10. The Union alert about the impending assault by Confederate cavalry was probably a reaction to a

Confederate scouting probe enhanced by natural command sensitivity to the Union army's main line of communication through Fredericksburg. Grant was anxious to remove the base of supplies to Port Royal because of constant Confederate pressure at Fredericksburg and Belle Plain. Matter, *If It Takes All Summer*, 283, 287–88. 313; *OR*, ser. 1, vol. 36, pt. 2: 231–32, 308–9, 752, 881.

13. Dorothea Dix (1802–1887), Union superintendent of women nurses, had authority over the selection and management of all women nurses employed by the army. Slightly built and soft-spoken, Dix managed to succeed in a male world by using high-handed, arbitrary methods with her enemies and solid management skills. She enforced rigid standards and strict rules on her corps of nurses, refusing to hire any woman who might be considered "young and pretty." Tireless and practical, Dix conducted incessant tours of hospitals and interceded for her nurses with uncooperative doctors. Her efforts during the war were voluntary, as she received no pay. Faust, *Historical Times*, 222.
14. Confederate records for Spotsylvania are incomplete, but Hancock's attack on the "Mule Shoe" salient on May 12 netted between 2,000 and 4,000 prisoners, depending on different estimates. Boatner, *Civil War Dictionary*, 788; Livermore, *Number and Losses*, 113; Current, *Encyclopedia of the Confederacy*, 4: 1520.

    Confederate General Ewell's Corps lost 4,600 men captured from May 10 to May 20. Matter, *If It Takes All Summer*, 348.

    Federal records indicate that between May 13 and May 18 approximately 7,500 Confederate prisoners passed through the Punch Bowl at Belle Plain on their way to Point Lookout prison. William A. Frassanito, *Grant and Lee: The Virginia Campaigns 1864–1865* (New York: Charles Scribner's Sons, 1983), 57.
15. Maj. Gen. Edward "Old Allegheny" Johnson and Brig. Gen. George "Maryland" Steuart were both captured in the II Corps attack on the "Mule Shoe" salient on the morning of May 12. Matter, *If It Takes All Summer*, 197.
16. The formal exchange of prisoners between North and South had been going on since February 1862. Exchange differed from parole in that it was a man-for-man transfer, while parole was the release of a captive on his word not to return to duty until a captive was released from the other side. Both forms of prisoner release were confusing and caused estrangement between the commissioners from each side. Halleck suspended the exchange of prisoners in May 1863, but the practice continued under special circumstances until Grant took overall command, when all exchanges ended. By this time, exchange had become the Confederacy's principal means of maintaining troop strength. The prisoners taken at Spotsylvania would not be returning to the South until the end of the war. Boatner, *Civil War Dictionary*, 270, 620; Faust, *Historical Times*, 558.
17. The Confederates who caused the panic were soldiers in Brig. Gen. Tom Rosser's cavalry brigade looking for the right flank of the Federal army. Rosser's probe got him into a spirited little fight at Aldrich's, after which he followed the Ny River almost to the Fredericksburg Road. His presence so close to the Union supply wagons at Salem Church caused their relocation to Fredericksburg. Matter, *If It Takes All Summer*, 291–93; *OR*, ser. 1, vol. 36, pt. 1: 309.
18. At Cold Harbor, June 3, 1864, the following heavy artillery regiments suffered serious losses: 8th New York, 505; 7th New York, 418; 2nd Connecticut, 325;

2nd New York, 215. Of course, it must be remembered that the "Heavies" were large units—their enrollments often exceeding 1,500 men—but their inexperience as infantry added to their vulnerability in battle. Fox, *Regimental Losses*, 450.

19. The final Union assault at Spotsylvania on May 18 was carried out by Gibbon's and Barlow's divisions of II Corps and two divisions of VI Corps. The II Corps divisions got to within 220 yards of the Confederate position, but heavy Confederate artillery fire and a thick layer of abatis kept them from going any farther. The attack was called off by 9 A.M. with heavy losses. Total Union losses in the Wilderness from May 5 to May 7 were over 17,666. At Spotsylvania, May 8 through May 21, Union losses were 18,399. Only Gettysburg had more—23,000. Matter, *If It Takes All Summer*, 308–12; Fox, *Regimental Losses*, 543–49.

Chapter Twenty-One

1. General Grant's strategy in the East for 1864 involved three major elements: Gen. Franz Sigel's movement up the Shenandoah Valley against the breadbasket of the Confederacy; Gen. Benjamin Butler's movement up the James River toward Richmond and Petersburg; and Grant's own movement toward Richmond overland. These simultaneous movements would prevent all of them from being countered and increase the chances of one or more of them being successful. General Sigel, a political general of dubious military talent, was defeated at the Battle of New Market on May 15, 1864, by Confederate Gen. John C. Breckinridge. General Butler, another political general of even less talent than Sigel, was defeated at the Battle of Drewry's Bluff on May 16, 1864, by Confederate General Beauregard and was then "bottled up" at Bermuda Hundred. With the loss of these two elements of his strategy, General Grant was forced to fight it out against General Lee alone during the summer of 1864. Faust, *Historical Times*, 57–58, 527–28.
2. Richard Napoleon Batchelder (1832–1901) began the war as a first lieutenant in the 1st New Hampshire Infantry. Quickly working his way up through the ranks, he became a colonel in the quartermaster corps by the end of the war. During the summer of 1864, he was lieutenant colonel, chief quartermaster of the II Corps. After the war he was accepted into the regular army as a captain and rose to become quartermaster general of the army, retiring in 1896 as a brigadier general. He was a Medal of Honor recipient for action against the Confederates at Catlett's and Fairfax Stations, October 13–15, 1863. Boatner, *Civil War Dictionary*, 49; Sifakis, *Who Was Who in the Union*, 24.
3. The still-smoking, burned railroad bridge can be clearly seen in a photograph taken by Timothy O'Sullivan on May 25, 1864. On that date Union troops finished the destruction of the bridge begun by the Confederates during the fight for the river crossings the day before. Frassanito, *Grant and Lee*, 133–39.
4. General Grant was after the railroad intersection at Hanover Junction, south of the North Anna River. General Lee devised a very strong defensive line that would force the attacking Federals to divide their forces to cross the

river, isolating their two wings and making support impossible. Faust, *Historical Times*, 535.

5. General Grant had crossed his forces over the North Anna without knowing exactly the disposition of Confederate forces. Once over, with his army split into two wings, Grant could see that he would have to cross the river twice to support one wing with the other. General Lee had interior lines and could support any part of his army quickly, thus bringing superior numbers to bear against an isolated portion of the Union army. This was untenable and Grant pulled out. McFeely and McFeely, *Grant*, 564–69.
6. WB is placing too much importance on this incident. A bridge or two carried away certainly could have been recovered, repaired, rebuilt, or replaced in short order. The engineers were too resourceful to suffer such a small loss and not recover quickly. However, it is unlikely that the structure that WB describes as being built by Captain McGrath could have seriously damaged a well-built wooden pontoon bridge with anchored, heavy pontoon boats lashed together with heavy chess and balk. These bridges were designed to carry loads of artillery, infantry, and supply trains. Even if Captain McGrath's flimsy foot bridge did carry away and strike the first of the pontoon bridges and damage it beyond repair, there was a second bridge that probably would have escaped damage. But, even if both bridges had been destroyed, Captain Personius was within two miles of the site with another operationally ready bridge train awaiting orders. And, if all this were not enough, the Chesterfield bridge was captured intact by Birney's division of the II Corps and stood one-half mile away ready for use in an emergency. *OR*, ser. 1, vol. 36, pt. 1: 305–16; Davis, *Military Atlas*, plates 91.2, 96.1; Frassanito, *Grant and Lee*, 132–35.
7. General Hancock had been a captain in the regular army in California before the war, and he had served as chief quartermaster, Southern District, so he had extensive experience in logistics matters. Boatner, *Civil War Dictionary*, 372.
8. During the early part of the summer campaign, the Union army supply base was at Belle Plain on the Potomac and then through Fredericksburg. On May 19, General Grant alerted General Halleck that he would be moving the supply base to Port Royal on the Rappahannock soon, and on May 22 Grant ordered Halleck to effect the change of base to Port Royal. On May 26, General Grant ordered Halleck to change the supply base to White House Landing on the Pamunkey River. All this certainly interrupted the flow of supplies to the Army of the Potomac, which was constantly on the move and in combat. *OR*, ser. 1, vol. 36, pt. 1: 6–9.
9. This conference was held at Bethesda Church, Virginia, on June 2, 1864. Frassanito, *Grant and Lee*, 121.
10. The XVIII Corps was commanded by Maj. Gen. William F. Smith and was part of the Army of the James under General Butler that remained bottled up on Bermuda Hundred. On May 28 Grant ordered the XVIII Corps to move via the James and York Rivers to White House Landing and march to join the Army of the Potomac at Cold Harbor. The corps arrived on June 1 and participated in the attacks on June 2 and 3. Welcher, *The Eastern Theater*, 478–79.
11. Brig. Gen. Robert Tyler commanded the 4th Brigade, 2nd Division, II Corps, at

Cold Harbor, where he was severely wounded in the ankle, an injury he never recovered from, although he continued on active duty until his death in 1874. Noah Andre Trudeau, *Bloody Roads South: The Wilderness to Cold Harbor, May–June 1864* (Boston: Little, Brown, 1989), 285.

12. Col. Elisha S. Kellogg commanded the eighteen-hundred-man 2nd Connecticut Heavy Artillery. The "Heavies" went in with the first wave and were shattered in a matter of minutes; Colonel Kellogg was shot in the head and killed at the abatis in front of Brigadier General Clingman's North Carolina brigade of Hoke's division. Trudeau, *Bloody Roads South*, 268–71.
13. Barlow's division had the good fortune to hit a portion of the Confederate line that was thinly held because a recent rain had turned the ground into a mudhole. His lead brigades burst through this first line of defense, captured three cannon, two hundred prisoners, and a stand of colors. Barlow's attack bogged down under a punishing crossfire from the flanks and had to withdraw after taking horrendous casualties. Lee counterattacked Barlow's Division with Finegan's Florida Brigade. Jaynes, *The Killing Ground*, 158–59; Trudeau, *Bloody Roads South*, 284–92.
14. The Cold Harbor fighting lasted from June 1 to June 12. During that time Union casualties were 1,905 killed, 10,570 wounded, and 2,456 missing for a total of 14,931. Dyer, *A Compendium*, 592. There were few accurate accountings of the casualties resulting from the ill-fated charge on June 3, but General Humphreys, Meade's chief of staff, estimated that, based on hospital records, the carnage for that day was 4,517 total with at least 1,100 killed. Confederate casualties for June 1–12 were 3,765 killed and wounded and 1,082 captured, for a total of 4,847. Trudeau, *Bloody Roads South*, 298, 341.
15. The troops in the ranks were tired of charging earthworks and said so in loud voices after the failure at Cold Harbor. Jaynes, *The Killing Ground*, 167.

    Grant toured the lines at midday and interviewed the corps commanders. Shortly thereafter, he wrote the following to General Meade: "The opinion of corps commanders not being sanguine of success in case an assault is ordered, you may direct a suspension of farther advance for the present." In his memoirs, Grant commented, "I have always regretted that the last assault at Cold Harbor was ever made." McFeely and McFeely, *Grant*, 585, 588.

CHAPTER TWENTY-TWO

1. The II Corps lost 7 colonels in the short, bloody attack at Cold Harbor: McKeen, 81st Pennsylvania; Haskell, 36th Wisconsin; McMahon, 164th New York.; Morris, 66th New York; Byrnes, 28th Massachusetts; Morris, 7th New York Heavy Artillery; Porter, 8th New York Heavy Artillery. *OR*, ser. 1, vol. 36, pt. 1: 367.

    The 8th New York Heavy Artillery was in the 4th Brigade (Tyler), 2nd Division (Gibbon), II Corps, and lost 7 officers, 73 enlisted men killed; 16 officers, 323 enlisted men wounded; 1 officer, 85 enlisted men missing during the attack at Cold Harbor, a total of 505 casualties in one day. Phisterer, *New York in the War of the Rebellion*, 1: 235, 2: 1397.

2. The only person that I could find that fits the description WB gives is Ranald Mackenzie, who was promoted to brigadier general on October 19, 1864. Mackenzie graduated first in his class from the United States Military Academy in 1862, was assigned to the engineers, and rose to command a corps late in the war. He went on to become a famous Indian fighter in the West, spent some time in an asylum, and died in 1889. Mackenzie came from an illustrious family and was highly praised by U. S. Grant in his memoirs. Boatner, *Civil War Dictionary*, 499–500; McFeely and McFeely, *Grant*, 772.

   Mackenzie was on duty as a lieutenant of engineers (a regular army rank, although Boatner states that Mackenzie was promoted to captain on November 6, 1863) with the II Corps from June 3 to June 10, at which time he was given volunteer rank of colonel and placed in command of the 2nd Connecticut Volunteer Artillery. *OR*, ser. 1, vol. 36, pt. 1: 302; Boatner, *Civil War Dictionary*, 499.

   I could not find any reference to the incident that WB describes.
3. Hi-bob-a-ree is a term that puzzled every researcher whom I asked. WB's penmanship, while often hard to make out, is very clear on the spelling of this term, so we cannot blame it. Finally, a librarian brought to my attention that it might be a corruption of High Barbaree, the name of a British sailor's song popular in America in the 1840s. The last verse is: "'Oh, quarter, oh, quarter!' these pirates did cry/Blow high, blow low, and so sailed we;/But the quarters that we gave them—we sunk them in the sea,/Cruising down along the coast of the High Barbaree." As WB was in school and probably playing games during the 1840s, perhaps we can assume that he and his schoolmates enjoyed singing this popular song, especially since it refers to the defeat of the Barbary pirates in 1815 by the upstart United States Navy. Albert B. Friedman, ed., *The Viking Book of Folk Ballads of the English-Speaking World* (New York: Viking Press, 1956), 407–9.
4. On June 7 a truce was arranged from 6 to 8 P.M., when the Union army was able to remove its wounded and bury the dead from between the lines at Cold Harbor. *OR*, ser. 1, vol. 36, pt. 1: 368.
5. These coehorn mortars were served by Battery D, 4th New York Heavy Artillery. *OR*, ser. 1, vol. 36, pt. 1: 512–13, 527–28.
6. Confederate attacks were made on II Corps entrenchments on June 3 at 7:30 P.M., June 4 at 8 P.M., and June 5 at 7 P.M. *OR*, ser. 1, vol. 36, pt. 1: 367–68.
7. British Army Major (later General) Shrapnel (1761–1842) invented a segmented artillery shell in 1803 that contained numerous small metal balls and a charge of explosive set to detonate over enemy troops. Small, segmented, hand-thrown explosives called grenades were first used in 1594, and the first regular troops armed with them were called grenadiers and appeared in 1667. Lawrence L. Gordon, *Military Origins*, ed. J. B. R. Nicholson (New York: A. S. Barnes, 1971), 211, 240–41.

   The rough sketch in the memoir depicts a shell that closely resembles a modern-day hand grenade.
8. The 2nd New York Heavy Artillery lost 3 officers and 64 enlisted men killed and 6 officers and 228 enlisted men wounded or missing during the assault on Petersburg, June 15–19. The 126th New York Volunteer Infantry lost 4 officers

and 4 enlisted men killed and 3 officers and 17 enlisted men wounded or missing during the same period. Phisterer, *New York in the War of the Rebellion*, 2: 1236, 4: 3498.

9. The Chickahominy at this point, the site of the old "Long Bridge" of the Peninsula Campaign, was 100 feet wide, with a 250-foot-wide island, then another channel of 60 feet with swamps bordering the approaches and numerous sunken piles and timber that had to be removed. The bridging took two and one-half hours and was delayed by enemy fire. *OR*, ser. 1, vol. 40, pt. 1: 296.
10. The bridge over the James River was one of the largest engineering feats achieved during the war. It consisted of 101 wooden pontons and was 2,200 feet long, the longest in the history of warfare. Three schooners were anchored in the channel to stabilize the bridge, and a 100-foot removable raft was left in the center so boats could pass up and down the river. It was constructed in 8 hours by 450 engineers under Major Duane and later under General Benham. Boatner, *Civil War Dictionary*, 434–35.
11. General Benham sent WB six ponton boats with balks and chess and a detachment from the 15th New York Volunteer Engineers. *OR*, ser. 1, vol. 40, pt. 1: 297.
12. The article by Swinton was answered in an official report by Lt. Col. Charles H. Morgan, assistant inspector general, U.S. Army, chief of staff, Second Army Corps, of operations June 15–16.

> HEADQUARTERS SECOND ARMY CORPS,
> June 25, 1864
>
> "GENERAL: I have the honor to submit the following statement of occurrences preceding and attending the march of this corps to Petersburg, June 14 and 15:
>
> About 11 o'clock on the night of the 14th a telegraphic message was received from Major-General Meade, stating that 60,000 rations had been ordered from City Point; that as soon as they were issued the corps would take the nearest and most direct route to Petersburg, taking position with its left on the City Point Railroad, where the road from Windmill Point crosses, and extending along Harrison's Creek toward its mouth. On the receipt of this order I sent the chief commissary, Colonel Smith, to the south bank of the river to make all necessary arrangements for the receipt and delivery of the rations, and directed the quartermaster, Captain McEntee, to send the transport to the upper wharf, then in process of repair, as soon as it arrived. At 2 o'clock I went down to the wharf to expedite the crossing. About 8 a.m. Major Brainerd, engineer detachment, who had been repairing the wharf where the rations were to be received, returned to the north bank and reported to me that Colonel Smith was at the wharf with his details, and that the transport containing the rations had just arrived. I saw a transport then lying at the

> wharf, and after watching it for a length of time sufficient to allow of it being unloaded it disappeared. I reported, therefore, to Major-General Hancock that the rations had come and were being issued. (It is proper to state that Major Brainerd now says he stated only to me his impression that the rations had arrived. It was conveyed to me in so positive a manner, indeed as a message from the commissary, that I had no doubt of the fact.) At ——— a.m., when the order came for the corps to march without its rations, an answer was returned that they had arrived. The mistake was discovered at ——, and the order was at once given for the corps to move. I understood that General Hancock had sent it by signal telegraph, but my recollection is that when I arrived at General Birney's headquarters he had not received it. The column was put in motion about 11:30, as I learn from the memorandum I made at the time."

Morgan went on to report on the confusion about marching routes, erroneous maps, lost communications while on the march to Petersburg, frustration with being out of touch with army headquarters, and not finding anyone who could answer his questions about where the corps should go. It all added up to a snarled operation. *OR*, ser. 1, vol. 51, pt. 1: 269–70.

General Hancock, in his official report, mentioned only that he had delayed his march three hours in anticipation of the arrival of badly needed rations, but when the rations had not been forthcoming, he had ordered the march toward Petersburg to commence without them. This order miscarried in reaching the head of the column, causing a further delay. Then General Hancock realized that the map he had been given was faulty, his objective was behind enemy lines and not where he had been led to believe it was, and *he had not even been told by army headquarters that there was to be an attack on Petersburg that day*. He stated that, had he been told to support Smith sooner, he certainly could have done so. *OR*, ser. 1, vol. 40, pt. 1: 303–4.

The first discrepancy between WB's account and Colonel Morgan's seems to be the transport that WB said did not stop and the transport that Colonel Morgan saw stopped at the wharf. There very possibly might have been two transports. Once WB delivered his message, he might have simply gone on to his tent for some much-needed rest. After WB and the first transport both moved on and the signal was given as WB described, a second transport could have tied up at the wharf, leading Morgan to believe that this transport was the one delivering rations. Given the sleep-deprived condition that everyone was suffering from, memories might have been fuzzy all around about the details of the incident.

The second discrepancy concerns the delay caused by the mixup over the rations. WB explained that the delay was no more than a few moments. Colonel Morgan clearly stated that a delay of three and one-half hours was caused by the erroneous assumption that the rations had arrived, and General Hancock seemed to support this. Swinton claimed that the delay was five and one-half hours. However, it is obvious from the reports that the real delay occurred after the corps left the crossing area, not at the wharf.

WB's role in all of this is shown to have been quite small and his culpability for any delay in the march to Petersburg is problematic to determine and was accidental.

13. William Swinton (1833–1892) was a correspondent for the *New York Times*, while his brother, John, was managing editor (of military coverage). It was to Swinton that General Hooker had confided his belief that "the country needs a dictator." Swinton was caught surreptitiously listening in on a high-level meeting between Grant and General Meade at General Grant's headquarters during the Wilderness campaign, but he was let off with a reprimand. Weeks later, General Burnside caught Swinton, who had written some uncomplimentary things about Burnside and the IX Corps, in his camps and threatened to shoot him for spying. Grant, after realizing that Burnside was serious, stepped in, and Swinton was ordered out of the Army of the Potomac by Circular Order on July 6, 1864. It can be presumed that his article about WB and the II Corps simply added to the ledger against the newsman but certainly was not the impetus for his loss of credentials with the Army of the Potomac. Swinton went on to write *Campaigns of the Army of the Potomac*, 1866, and in it handles Burnside with vicious sarcasm and gives Grant less than his due. Louis M. Starr, *Bohemian Brigade: Civil War Newsmen in Action* (New York: Alfred A. Knopf, 1954; reprint, Madison: Univ. of Wisconsin Press, 1987), 194, 279, 357–58; McFeely and McFeely, *Grant*, 486–88; Marvel, *Burnside*, 381, 475n. II:11.

Epilogue

1. *OR*, ser. 1, vol. 40, pt. 1: 299.
2. Floyd, *Dear Friends at Home*, 46.
3. *OR*, ser. 1, vol. 40, pt. 1: 300.
4. Ibid., vol. 42, pt. 1: 164.
5. Ibid., 167–68.
6. Phisterer, *New York in the War of the Rebellion*, 373. Brainerd Papers, leave of absence authorization.
7. Phisterer, *New York in the War of the Rebellion*, 1650–57.
8. *OR*, ser. 1, vol. 46, pt. 1: 72.
9. Ibid.
10. Ibid., pt. 1: 72, pt. 2:395, 436–39, 461–63, 490–93, 521, 537.
11. Ibid., pt. 1: 73.
12. Ibid., 641–42.
13. Ibid.
14. Ibid., 74. Brainerd Papers, muster-out documents.
15. *Portrait and Biographical Record*, 306.

16. *Evanston Index*, May 25, 1901.
17. *Portrait and Biographical Record*, 306.
18. *Portrait and Biographical Record*, 306–08.
19. Johnson and Buel, *The Tide Shifts*, 121.
20. Eleanor M. Gehres, Western History Department of Denver Public Library, letter to editor, December 6, 1990. *Portrait and Biographical Record*, 308.
21. Brainard, *Genealogy of the Brainerd Family*, 130.
22. Ibid.; Historical Society of Douglas County, letter to editor, February 6, 1991.
23. Kathryn M. Harris, Reference and Technical Services of Illinois State Historical Library, letter to editor, April 22, 1991. Obituary, Chicago *Tribune*, January 25, 1927. *The Biographical Dictionary and Portrait Gallery of Representative Men of Chicago, Minnesota Cities and the World's Columbian Exposition* (Chicago: American Biographical Publishing Company, 1892), 8–11.
24. *Evanston Index*, November 14, 1908.
25. *Portrait and Biographical Record*, 308.

Bibliography

Aikman, Lonnelle. *We, the People: The Story of the United States Capitol, Its Past and Its Promise*. Washington, D.C.: The United States Capitol Historical Society, 1965.

Annual Report of the Adjutant General of the State of New York for the Year 1898: Registers of the First, Fifteenth, and Fiftieth Engineers and the First Battalion of Sharpshooters. Serial No. 16. New York: Wynkoop Hallenbeck Crawford Co., 1899.

Arnold, Denis, editor. *The New Oxford Companion to Music*, 2 vols. New York: Oxford Univ. Press, 1983.

Barnard, J. G., and W. F. Barry. *Report of the Engineer and Artillery Operations of the Army of the Potomac, From Its Organization to the Close of the Peninsular Campaign*. New York: D. Van Nostrand, 1863.

Basler, Roy P., editor. *The Collected Works of Abraham Lincoln*. 9 vols. New Brunswick, N.J.: Rutgers Univ. Press, 1953.

The Biographical Dictionary and Portrait Gallery of Representative Men of Chicago, Minnesota Cities and the World's Columbian Exposition. Chicago: American Biographical Publishing Company, 1892.

Boatner, Mark M., III. *The Civil War Dictionary*. New York: David McKay Company, 1959.

Boritt, Gabor S., editor. *Lincoln's Generals*. New York: Oxford Univ. Press, 1994.

Bowman, John S., executive editor. *The Civil War Almanac*. New York: Gallery Books, 1983.

Brainard, Lucy Abigail. *The Genealogy of the Brainerd-Brainard Family in America 1649–1908*. Vol. 1. Hartford Press: The Case, Lockwood and Brainard Company, 1908.

Caroli, Betty Boyd. *Inside the White House: America's Most Famous Home, the First 200 Years*. New York: Canopy Books, 1992.

Carruth, Gorton. *The Encyclopedia of American Facts and Dates*. 9th ed. New York: HarperCollins Publishers, 1993.

Catton, Bruce. *The Army of the Potomac: Glory Road*. New York: Doubleday, 1952.

———. *The Army of the Potomac: Mr. Lincoln's Army*. New York: Doubleday, 1962.

Chapple, Joe Mitchell, compiler. *Heart Songs*. Boston, Mass.: Chapple Publishing, 1909.

Coddington, Edwin B. *The Gettysburg Campaign: A Study in Command*. New York: Charles Scribner's Sons, 1968. Reprint, Dayton, Ohio: Morningside Bookshop, 1979.

Commager, Henry Steele, and Robert B. Morris, editors. *The Spirit of 'Seventy Six*. New York: Bonanza Books, 1983.

Cooling, Benjamin Franklin, III, *Symbol, Sword and Shield: Defending Washington During the Civil War*. 2d rev. ed. Pennsylvania: White Mane Publishing, 1991.

Coski, John M. *The Army of the Potomac at Berkeley Plantation: The Harrison's Landing Occupation of 1862*. N.p., 1989.

Craighill, William P. *The Army Officer's Pocket Companion and Manual for Staff Officers in the Field*. New York: D. Van Nostrand, 1863.

Current, Richard N., editor in chief. *Encyclopedia of the Confederacy*. 4 vols. New York: Simon and Schuster, 1993.

Davis, Burke. *Jeb Stuart: The Last Cavalier*. New York: Rinehart, 1957. Reprint, New York: Fairfax Press, 1988.

Davis, Maj. George B., Leslie J. Perry, and Joseph W. Kirkley. *The Official Military Atlas of the Civil War*. Compiled by Capt. Calvin D. Cowles. New York: The Fairfax Press, 1983.

Davis, William C. *First Blood: Fort Sumter to Bull Run*. The Civil War Series. Alexandria, Va.: Time-Life Books, 1983.

Duane, J. C. *Manual for Engineer Troops*. New York: D. Van Nostrand, 1862.

Dyer, Frederick H. *A Compendium of the War of the Rebellion*. 2 vols. 1908. Reprint, Dayton, Ohio: Morningside Press, 1994.

Farwell, Byron. *Stonewall: A Biography of General Thomas J. Jackson*. New York: W. W. Norton, 1992.

Faust, Patricia L., editor. *Historical Times Illustrated Encyclopedia of the Civil War*. New York: Harper and Row, 1986.

Finfrock, Bradley. *Across the Rappahannock: From Fredericksburg to the Mud March*. Maryland: Heritage Books, 1994.

Floyd, Dale E., editor. Introduction to *"Dear Friends at Home . . .": The Letters and Diary of Thomas James Owen, Fiftieth New York Volunteer Engineer Regiment, During the Civil War*. Corps of Engineers Historical Studies, no. 4. Washington, D.C., 1985.

Ford, Paul Leicester. *The True George Washington*. 1896. Reprint, Freeport, N.Y.: Books for Libraries Press, 1971.

Fox, William F. *New York at Gettysburg*. 3 vols. Albany, N.Y.: State of New York, 1900.

———. *Regimental Losses in the American Civil War 1861–1865*. 1989, 1974. Reprint, Dayton, Ohio: Morningside Bookshop, 1985.

Frassanito, William A. *Grant and Lee: The Virginia Campaigns 1864–1865*. New York: Charles Scribner's Sons, 1983.

Fredrickson, George M. *The Inner Civil War: Northern Intellectuals and the Crisis of the Union*. New York: Harper and Row, 1965.

French, Benjamin Brown. *Witness to the Young Republic: A Yankee's Journal, 1828–1870*. Edited by Donald B. Cole and John J. McDonough. Hanover: Univ. Press of New England, 1989.

Friedman, Albert B., editor. *The Viking Book of Folk Ballads of the English-Speaking World*. New York: Viking Press, 1956.

Furgurson, Ernest B. *Chancellorsville 1863: The Souls of the Brave*. New York: Alfred A. Knopf, 1992.

Glover, Michael. *The Napoleonic Wars: An Illustrated History 1792–1815*. New York: Hippocrene Books, 1978.

Gordon, Lawrence L. *Military Origins*. Edited by J. B. R. Nicholson. New York: A. S. Barnes, 1971.

Goss, Warren Lee. *Recollections of a Private*. New York: Thomas Y. Crowell, 1890.

Hard, Abner. *History of the Eighth Cavalry Regiment, Illinois Volunteers*. N.p., 1868.

Haydon, Frederick Stansbury. *Aeronautics in the Union and Confederate Armies*. Vol. 1. New York: Arno Press, 1980.

Jaynes, Gregory. *The Killing Ground: Wilderness to Cold Harbor*. The Civil War Series. Alexandria, Va.: Time-Life Books, 1986.

Johnson, Robert Underwood, and Clarence Clough Buel, editors. *The Struggle Intensifies*. Vol. 2 of *Battles and Leaders of the Civil War*. N.p.: Castle, n.d.

———. *The Tide Shifts*. Vol. 3 of *Battles and Leaders of the Civil War*. N.p.: Castle, n.d.

Katcher, Philip. *The Civil War Source Book*. New York: Facts on File, 1992.

Kirkpatrick, F. A. *The Spanish Conquistadores*. 2nd ed. 1934. Reprint, New York: Barnes and Noble, 1967.

Leech, Margaret. *Reveille in Washington: 1860–1865*. Alexandria, Va.: Time-Life Books, 1980.

Livermore, Thomas L. *Numbers and Losses in the Civil War in America: 1861–65*. 1900. Reprint, with an introduction by Edward E. Barthell Jr., Bloomington, Ind.: Indiana Univ. Press, 1957.

Long, David E. *The Jewel of Liberty: Abraham Lincoln's Re-election and the End of Slavery*. Mechanicsburg, Pa.: Stackpole Books, 1994.

Long, E. B., and Barbara Long. *The Civil War Day By Day: An Almanac 1861–1865*. New York: Doubleday, 1971.

Lonsdale, Roger, editor. *The New Oxford Book of Eighteenth Century Verse*. New York: Oxford Univ. Press, 1989.

Lossing, Benson J. *Pictorial History of the Civil War in the United States of America*. 3 vols. Philadelphia, Pa.: David McKay Publishers, 1894.

Luvaas, Jay, and Harold W. Nelson, editors. *The U.S. Army War College Guide to the Battles of Chancellorsville and Fredericksburg*. Carlisle, Pa.: South Mountain Press, 1988.

Magnus, Philip. *King Edward the Seventh*. New York: E. P. Dutton, 1964.

Malone, Dumas, editor. *Dictionary of American Biography*. 11 vols. New York: Charles Scribner's Sons, 1928.

Marvel, William. *Burnside*. Chapel Hill: Univ. of North Carolina Press, 1991.

Matter, William D. *If It Takes All Summer: The Battle of Spotsylvania*. Chapel Hill: Univ. of North Carolina Press, 1988.

McFeely, Mary Drake, and William S. McFeely, editors. *Grant: Personal Memoirs of U. S. Grant Selected Letters 1839–1865*. The Library of America, 1990.

McPherson, James M., editor. *The Atlas of the Civil War*. New York: Macmillan, 1994.

Miers, Earl Schenck, editor. *Lincoln Day By Day: A Chronology 1809–1865*. 3 vols. Dayton, Ohio: Morningside, 1991.

Moore, Frank, editor. *The Rebellion Record: A Diary of American Events*. 12 vols. 1861–68. Reprint, New York: Arno Press, 1977.

Naisawald, L. Van Loan. *Grape and Canister: The Story of the Field Artillery of the Army of the Potomac*. 1960. Reprint, Washington, D.C.: Zenger Publishing, 1983.

Nesbitt, Mark. *Rebel Rivers: A Guide to Civil War Sites on the Potomac, Rappahannock, York, and James*. Pennsylvania: Stackpole Books, 1993.

Nevins, Allan. *A House Dividing, 1852–1856*. Vol. 2 of *The Ordeal of the Union*. New York: Charles Scribner's Sons, 1947.

Nevins, Allan, James I. Robertson Jr., and Bell Irvin Wiley, editors. *Civil War Books: A Critical Bibliography*. 2 vols. Baton Rouge, La.: Louisiana State Univ. Press, 1967.

Newton, John, and Gerald Simons, editors. *Lee Takes Command: From Seven Days to Second Bull Run*. The Civil War Series. Alexandria, Va.: Time-Life Books, 1984.

Nichols, James L. *Confederate Engineers*. Confederate Centennial Studies. No. 5. Tuscaloosa, Ala.: Confederate Publishing Company, 1957. Reprint, Gaithersburg, Md.: Olde Soldier Books, 1987.

Nicholson, John P., editor and compiler. *Pennsylvania at Gettysburg: Ceremonies at the Dedication of the Monuments*. 3 vols. Harrisburg, Pa.: Wm. Stanley Ray, State Printer, 1914.

Ordway, Frederick I., Jr. *Register of the Society of Colonial Wars in the District of Columbia*. Washington, D.C., 1967.

Phisterer, Frederick. *New York in the War of the Rebellion: 1861 to 1865*. 3rd ed. 5 vols. Albany, N.Y.: State of New York, 1912.

Porter, Horace. *Campaigning with Grant*. New Jersey: The Blue and Grey Press, 1984.

Portrait and Biographical Record of the State of Colorado. Chicago: Chapman Publishing Company, 1899.

Rable, George C. *The Confederate Republic: A Revolution Against Politics*. Chapel Hill: Univ. of North Carolina Press, 1994.

Reece, J. N. *Report of the Adjutant General of the State of Illinois*. Vol. 8. N.p., 1901.

Reed, P. Bradley, editor. *Coin World Almanac*. 6th ed. Sidney, Ohio: Amos Press, 1990.

Register of Graduates and Former Cadets of the United States Military Academy. West Point, N.Y.: The West Point Alumni Foundation, 1970.

Register of Officers and Agents, Civil, Military and Naval in the Service of the United States on the Thirtieth of September, 1861. Washington, D.C., 1861.

Rhea, Gordon C. *The Battle of the Wilderness May 5–6, 1864*. Baton Rouge, La.: Louisiana State Univ. Press, 1994.

Richardson, James D., editor and compiler. *The Messages and Papers of Jefferson Davis and the Confederacy: Including Diplomatic Correspondence 1861–1865*. 1905. Reprint, with a comprehensive introduction by Allan Nevins, New York: Chelsea House-Robert Hector, 1966.

Schott, Thomas E. *Alexander H. Stephens of Georgia: A Biography*. Baton Rouge, La.: Louisiana State Univ. Press, 1988.

Sears, Stephen W. *George B. McClellan: The Young Napoleon*. New York: Ticknor and Fields, 1988.

———. *Landscape Turned Red: The Battle of Antietam*. New Haven: Ticknor and Fields, 1983.

———. *To the Gates of Richmond: The Peninsula Campaign*. New York: Ticknor and Fields, 1992.

Sifakis, Stewart. *Who Was Who in the Union*. Vol. 1 of *Who Was Who in The Civil War*. New York: Facts on File, 1988.

———. *Who Was Who in the Confederacy*. Vol. 2 of *Who Was Who in the Civil War*. New York: Facts on File, 1988.

Silber, Irwin, compiler and editor. *Songs of Independence*. Harrisburg, Pa.: Stackpole, 1973.

Smith, Rhea Marsh. *Spain: A Modern History*. Ann Arbor, Mich.: Univ. of Michigan Press, 1965.

Stackpole, Edward J. *The Fredericksburg Campaign: Drama on the Rappahannock*. New York: Bonanza Books, 1957.

———. *They Met at Gettysburg: A Stirring Account of the Battles at Gettysburg*. New York: Bonanza Books, 1956.

Starr, Louis M. *Bohemian Brigade: Civil War Newsmen in Action*. New York: Alfred A. Knopf, 1954. Reprint, Madison, Wisc.: Univ. of Wisconsin Press, 1987.

Taylor, John M. *William Henry Seward: Lincoln's Right Hand*. New York: HarperCollins, 1991.

Trudeau, Noah Andre. *Bloody Roads South: The Wilderness to Cold Harbor, May–June 1864*. Boston: Little, Brown, 1989.

Walls, Matthew S. "Northern Hell on Earth." *America's Civil War* 3 (6) (March 1991).

The War of the Rebellion: A Compilation of the Official Records of the Union and Confederate Armies. 128 vols. Washington, D.C., 1887.

The Washington Directory Showing the Name, Occupation, and Residence of Each Head of a Family and Person in Business Together With Other Useful Information. Washington, D.C.: S. A. Elliot, 1827.

Webb, Alexander S., LL.D. *The Peninsula: McClellan's Campaign of 1862* of *Campaigns of the Civil War*. Introduction by William F. Howard. Charles Scribner's Sons, 1881. Reprint, Wilmington, N.C.: Broadfoot Publishing, 1989.

Weigley, Russell F. *Quartermaster General of the Union Army: A Biography of M. C. Meigs*. New York: Columbia Univ. Press, 1959.

Welcher, Frank J. *The Eastern Theater*. Vol. 1 of *The Union Army 1861–1865: Organization and Operations*. Bloomington, Ind.: Indiana Univ. Press, 1989.

Wert, Jeffry D. *Mosby's Rangers*. New York: Simon and Schuster, 1990.

Wheeler, Richard. *A Rising Thunder: From Lincoln's Election to the Battle of Bull Run: An Eyewitness Account*. New York: HarperCollins Publishers, 1994.

Wiley, Bell Irvin. *The Life of Billy Yank: The Common Soldier of the Union*. Baton Rouge, La.: Louisiana State Univ. Press, 1986.

Wimsatt, William K., editor. *Alexander Pope: Selected Poetry and Prose*. New York: Holt, Rinehart and Winston, 1972.

Winstock, Lewis. *Songs & Music of the Redcoats: A History of the War Music of the British Army 1642–1902*. Harrisburg, Pa.: Stackpole, 1970.

Index

Photographs are depicted in *italics*.